T0174896

GENETICS, GENOMICS AND BREEDING OF SUGARCANE

Genetics, Genomics and Breeding of Crop Plants

Series Editor
Chittaranjan Kole
Department of Genetics and Biochemistry
Clemson University
Clemson, SC
USA

Books in this Series:

Published or in Press:
- Jinguo Hu, Gerald Seiler & Chittaranjan Kole: *Sunflower*
- Kristin D. Bilyeu, Milind B. Ratnaparkhe & Chittaranjan Kole: *Soybean*
- Robert Henry & Chittaranjan Kole: *Sugarcane*

Books under preparation:
- Jan Sadowsky & Chittaranjan Kole: *Vegetable Brassicas*
- Kevin Folta & Chittaranjan Kole: *Berries*
- C.P. Joshi, Stephen DiFazio & Chittaranjan Kole: *Poplar*
- James M. Bradeen & Chittaranjan Kole: *Potato*
- Jose Miguel Martinez Zapater, Anne-Françoise Adam Blondon & Chittaranjan Kole: *Grapes*

GENETICS, GENOMICS AND BREEDING OF SUGARCANE

Editors

Robert J. Henry

Centre for Plant Conservation Genetics
Southern Cross University
Lismore, NSW
Australia

Chittaranjan Kole

Department of Genetics and Biochemistry
Clemson University
Clemson, SC
USA

CRC Press
Taylor & Francis Group
Boca Raton London New York

CRC Press is an imprint of the
Taylor & Francis Group, an **informa** business

A SCIENCE PUBLISHERS BOOK

CRC Press
Taylor & Francis Group
6000 Broken Sound Parkway NW, Suite 300
Boca Raton, FL 33487-2742

First issued in paperback 2019

Copyright reserved © 2010 by Taylor & Francis Group, LLC
CRC Press is an imprint of Taylor & Francis Group, an Informa business

No claim to original U.S. Government works

ISBN-13: 978-1-57808-684-9 (hbk)
ISBN-13: 978-0-367-38370-1 (pbk)

This book contains information obtained from authentic and highly regarded sources. Reasonable efforts have been made to publish reliable data and information, but the author and publisher cannot assume responsibility for the validity of all materials or the consequences of their use. The authors and publishers have attempted to trace the copyright holders of all material reproduced in this publication and apologize to copyright holders if permission to publish in this form has not been obtained. If any copyright material has not been acknowledged please write and let us know so we may rectify in any future reprint.

Except as permitted under U.S. Copyright Law, no part of this book may be reprinted, reproduced, transmit- ted, or utilized in any form by any electronic, mechanical, or other means, now known or hereafter invented, including photocopying, microfilming, and recording, or in any information storage or retrieval system, without written permission from the publishers.

For permission to photocopy or use material electronically from this work, please access www. copyright. com (http://www.copyright.com/) or contact the Copyright Clearance Center, Inc. (CCC), 222 Rosewood Drive, Danvers, MA 01923, 978-750-8400. CCC is a not-for-profit organization that provides licenses and registration for a variety of users. For organizations that have been granted a photocopy license by the CCC, a separate system of payment has been arranged.

Trademark Notice: Product or corporate names may be trademarks or registered trademarks, and are used only for identification and explanation without intent to infringe.

Library of Congress Cataloging-in-Publication Data

Genetics, genomics and breeding of sugarcane / editors: Robert J. Henry, Chittaranjan Kole. -- 1st ed.

p. cm. -- (Genetics, genomics and breeding of crop plants)

ISBN 978-1-57808-684-9 (hardcover)

1. Sugarcane--Genetics. 2. Sugarcane--Genome mapping. 3. Sugarcane--Breeding. I. Henry, Robert J. II. Kole, Chittaranjan. III. Series: Genetics, genomics and breeding of crop plants.

SB231.G46 2010

633.6'1233--dc22

2010011559

Visit the Taylor & Francis Web site at
http://www.taylorandfrancis.com

and the CRC Press Web site at
http://www.crcpress.com

Preface to the Series

Genetics, genomics and breeding has emerged as three overlapping and complimentary disciplines for comprehensive and fine-scale analysis of plant genomes and their precise and rapid improvement. While genetics and plant breeding have contributed enormously towards several new concepts and strategies for elucidation of plant genes and genomes as well as development of a huge number of crop varieties with desirable traits, genomics has depicted the chemical nature of genes, gene products and genomes and also provided additional resources for crop improvement.

In today's world, teaching, research, funding, regulation and utilization of plant genetics, genomics and breeding essentially require thorough understanding of their components including classical, biochemical, cytological and molecular genetics; and traditional, molecular, transgenic and genomics-assisted breeding. There are several book volumes and reviews available that cover individually or in combination of a few of these components for the major plants or plant groups; and also on the concepts and strategies for these individual components with examples drawn mainly from the major plants. Therefore, we planned to fill an existing gap with individual book volumes dedicated to the leading crop and model plants with comprehensive deliberations on all the classical, advanced and modern concepts of depiction and improvement of genomes. The success stories and limitations in the different plant species, crop or model, must vary; however, we have tried to include a more or less general outline of the contents of the chapters of the volumes to maintain uniformity as far as possible.

Often genetics, genomics and plant breeding and particularly their complimentary and supplementary disciplines are studied and practiced by people who do not have, and reasonably so, the basic understanding of biology of the plants for which they are contributing. A general description of the plants and their botany would surely instill more interest among them on the plant species they are working for and therefore we presented lucid details on the economic and/or academic importance of the plant(s); historical information on geographical origin and distribution; botanical origin and evolution; available germplasms and gene pools, and genetic and cytogenetic stocks as genetic, genomic and breeding resources; and basic information on taxonomy, habit, habitat, morphology, karyotype, ploidy level and genome size, etc.

Classical genetics and traditional breeding have contributed enormously even by employing the phenotype-to-genotype approach. We included detailed descriptions on these classical efforts such as genetic mapping using morphological, cytological and isozyme markers; and achievements of conventional breeding for desirable and against undesirable traits. Employment of the in vitro culture techniques such as micro- and megaspore culture, and somatic mutation and hybridization, has also been enumerated. In addition, an assessment of the achievements and limitations of the basic genetics and conventional breeding efforts has been presented.

It is a hard truth that in many instances we depend too much on a few advanced technologies, we are trained in, for creating and using novel or alien genes but forget the infinite wealth of desirable genes in the indigenous cultivars and wild allied species besides the available germplasms in national and international institutes or centers. Exploring as broad as possible natural genetic diversity not only provides information on availability of target donor genes but also on genetically divergent genotypes, botanical varieties, subspecies, species and even genera to be used as potential parents in crosses to realize optimum genetic polymorphism required for mapping and breeding. Genetic divergence has been evaluated using the available tools at a particular point of time. We included discussions on phenotype-based strategies employing morphological markers, genotype-based strategies employing molecular markers; the statistical procedures utilized; their utilities for evaluation of genetic divergence among genotypes, local landraces, species and genera; and also on the effects of breeding pedigrees and geographical locations on the degree of genetic diversity.

Association mapping using molecular markers is a recent strategy to utilize the natural genetic variability to detect marker-trait association and to validate the genomic locations of genes, particularly those controlling the quantitative traits. Association mapping has been employed effectively in genetic studies in human and other animal models and those have inspired the plant scientists to take advantage of this tool. We included examples of its use and implication in some of the volumes that devote to the plants for which this technique has been successfully employed for assessment of the degree of linkage disequilibrium related to a particular gene or genome, and for germplasm enhancement.

Genetic linkage mapping using molecular markers have been discussed in many books, reviews and book series. However, in this series, genetic mapping has been discussed at length with more elaborations and examples on diverse markers including the anonymous type 2 markers such as RFLPs, RAPDs, AFLPs, etc. and the gene-specific type 1 markers such as EST-SSRs,

SNPs, etc.; various mapping populations including F_2, backcross, recombinant inbred, doubled haploid, near-isogenic and pseudotestcross; computer software including MapMaker, JoinMap, etc. used; and different types of genetic maps including preliminary, high-resolution, high-density, saturated, reference, consensus and integrated developed so far.

Mapping of simply inherited traits and quantitative traits controlled by oligogenes and polygenes, respectively has been deliberated in the earlier literature crop-wise or crop group-wise. However, more detailed information on mapping or tagging oligogenes by linkage mapping or bulked segregant analysis, mapping polygenes by QTL analysis, and different computer software employed such as MapMaker, JoinMap, QTL Cartographer, Map Manager, etc. for these purposes have been discussed at more depth in the present volumes.

The strategies and achievements of marker-assisted or molecular breeding have been discussed in a few books and reviews earlier. However, those mostly deliberated on the general aspects with examples drawn mainly from major plants. In this series, we included comprehensive descriptions on the use of molecular markers for germplasm characterization, detection and maintenance of distinctiveness, uniformity and stability of genotypes, introgression and pyramiding of genes. We have also included elucidations on the strategies and achievements of transgenic breeding for developing genotypes particularly with resistance to herbicide, biotic and abiotic stresses; for biofuel production, biopharming, phytoremediation; and also for producing resources for functional genomics.

A number of desirable genes and QTLs have been cloned in plants since 1992 and 2000, respectively using different strategies, mainly positional cloning and transposon tagging. We included enumeration of these and other strategies for isolation of genes and QTLs, testing of their expression and their effective utilization in the relevant volumes.

Physical maps and integrated physical-genetic maps are now available in most of the leading crop and model plants owing mainly to the BAC, YAC, EST and cDNA libraries. Similar libraries and other required genomic resources have also been developed for the remaining crops. We have devoted a section on the library development and sequencing of these resources; detection, validation and utilization of gene-based molecular markers; and impact of new generation sequencing technologies on structural genomics.

As mentioned earlier, whole genome sequencing has been completed in one model plant (Arabidopsis) and seven economic plants (rice, poplar, peach, papaya, grapes, soybean and sorghum) and is progressing in an array of model and economic plants. Advent of massively parallel DNA sequencing using 454-pyrosequencing, Solexa Genome Analyzer, SOLiD system, Heliscope and SMRT have facilitated whole genome sequencing in many other plants more rapidly, cheaply and precisely. We have included extensive coverage on the level (national or international) of collaboration and the strategies and status of whole genome sequencing in plants for which sequencing efforts have been completed or are progressing currently. We have also included critical assessment of the impact of these genome initiatives in the respective volumes.

Comparative genome mapping based on molecular markers and map positions of genes and QTLs practiced during the last two decades of the last century provided answers to many basic questions related to evolution, origin and phylogenetic relationship of close plant taxa. Enrichment of genomic resources has reinforced the study of genome homology and synteny of genes among plants not only in the same family but also of taxonomically distant families. Comparative genomics is not only delivering answers to the questions of academic interest but also providing many candidate genes for plant genetic improvement.

The 'central dogma' enunciated in 1958 provided a simple picture of gene function—gene to mRNA to transcripts to proteins (enzymes) to metabolites. The enormous amount of information generated on characterization of transcripts, proteins and metabolites now have led to the emergence of individual disciplines including functional genomics, transcriptomics, proteomics and metabolomics. Although all of them ultimately strengthen the analysis and improvement of a genome, they deserve individual deliberations for each plant species. For example, microarrays, SAGE, MPSS for transcriptome analysis; and 2D gel electrophoresis, MALDI, NMR, MS for proteomics and metabolomics studies require elaboration. Besides transcriptome, proteome or metabolome QTL mapping and application of transcriptomics, proteomics and metabolomics in genomics-assisted breeding are frontier fields now. We included discussions on them in the relevant volumes.

The databases for storage, search and utilization on the genomes, genes, gene products and their sequences are growing enormously in each second and they require robust bioinformatics tools plant-wise and purpose-wise. We included a section on databases on the gene and genomes, gene expression, comparative genomes, molecular marker and genetic maps, protein and metabolomes, and their integration.

Notwithstanding the progress made so far, each crop or model plant species requires more pragmatic retrospect. For the model plants we need to answer how much they have been utilized to answer the basic questions of genetics and genomics as compared to other wild and domesticated species. For the economic plants we need to answer as to whether they have been genetically tailored perfectly for expanded geographical regions and current requirements for green fuel, plant-based bioproducts and for improvements of ecology and environment. These futuristic explanations have been addressed finally in the volumes.

We are aware of exclusions of some plants for which we have comprehensive compilations on genetics, genomics and breeding in hard copy or digital format and also some other plants which will have enough achievements to claim for individual book volume only in distant future. However, we feel satisfied that we could present comprehensive deliberations on genetics, genomics and breeding of 30 model and economic plants, and their groups in a few cases, in this series. I personally feel also happy that I could work with many internationally celebrated scientists who edited the book volumes on the leading plants and plant groups and included chapters authored by many scientists reputed globally for their contributions on the concerned plant or plant group.

We paid serious attention to reviewing, revising and updating of the manuscripts of all the chapters of this book series, but some technical and formatting mistakes will remain for sure. As the series editor, I take complete responsibility for all these mistakes and will look forward to the readers for corrections of these mistakes and also for their suggestions for further improvement of the volumes and the series so that future editions can serve better the purposes of the students, scientists, industries, and the society of this and future generations.

Science publishers, Inc. has been serving the requirements of science and society for a long time with publications of books devoted to advanced concepts, strategies, tools, methodologies and achievements of various science disciplines. Myself as the editor and also on behalf of the volume editors, chapter authors and the ultimate beneficiaries of the volumes take this opportunity to acknowledge the publisher for presenting these books that could be useful for teaching, research and extension of genetics, genomics and breeding.

Chittaranjan Kole

Preface to the Volume

Sugarcane is a major crop of tropical and sub-tropical regions. This fast growing plant is a major source of sugar (sucrose). The high productivity of the sugarcane plant makes it a key target for use as an energy crop. The fibre of the plant is often used to generate electricity. The sugarcane plant is also used to produce ethanol as a fuel.

Sugarcane is a hybrid of two species each of which is genetically complex. The high level of genetic complexity in sugarcane creates challenges in the application of both conventional and molecular breeding to the genetic improvement of sugarcane as a sugar and energy crop. Recent technology developments indicate the potential to greatly increase our understanding of the sugarcane plant by application of emerging genomic technologies. This should result in an increased rate of improvement of sugarcane for human uses.

The chapters in this volume have been written by experts with extensive experience in the genetics and genomics of this complex and important crop plant. Chapter 1 provides a general introduction. The highly polyploidy nature of sugarcane is explained in chapter 2 covering cytogenetics. Diversity in sugarcane is reviewed in chapter 3, association genetics in chapter 4 and linkage analysis in chapter 5. The use of markers for simple and complex traits is covered in chapters 6 and 7 respectively. Structural genomics and sequencing is described in chapter 8. Transcriptomics is outlined in chapter 9 and proteomics and metabolomics in chapter 10. The important role of bioinformatics in such a complex system is described in chapter 11. Future prospects for sugarcane as crop are discussed in the final chapter (chapter 12).

Sugarcane is arguably the leading industrial crop with very large scale global production. The technologies described in this book support the continued use and improvement of sugarcane as source of food and energy.

We thank Linda Hammond for assistance in coordination with chapter authors in the preparation of this manuscript and in compilation of the contents and index of the book.

Robert Henry
Chittaranjan Kole

Contents

List of Contributors

Karen Aitken
CSIRO Plant Industry, Queensland Bioscience Precinct, 306 Carmody Road, St. Lucia, Queensland, 4067, Australia.
Email: *Karen.Aitken@csiro.au*

Sreedhar Alwala
Dow Agro Sciences, York, USA.
Email: *SAlwala@dow.com*

Frederik C Botha
BSES Limited, PO Box 86, Indooroopilly Qld 4068, Australia.
Email: *fbotha@bses.org.au*

Mike Butterfield
SA Sugarcane Research Institute, Private Bag X02, Mount Edgecombe, 4300, South Africa.
Email: *m.butterfield@cgiar.org*

Rosanne E. Casu
CSIRO Plant Industry, Queensland Bioscience Precinct, 306 Carmody Road, St. Lucia, QLD, 4067, Australia.
Email: *Rosanne.Casu@csiro.au*

Angélique D'Hont
CIRAD, Centre de Coopération Internationale en Recherche Agronomique pour le Développement, UMR1098/DAP, TAA96/03 avenue Agropolis, 34398 Montpellier cedex 5, France.
Email: *dhont@cirad.fr*

Antonio Augusto Franco Garcia
Departamento de Genética, Escola Superior de Agricultura Luiz de Queiroz (ESALQ), Universidade de São Paulo (USP), CP 83, CEP 13400-970, Piracicaba-SP, Brazil.
Email: *aafgarci@esalq.usp.br*

Olivier Garsmeur
CIRAD, UMR 1096, TA40/03, Avenue Agropolis, 34 398 Montpellier, cedex 5, France.
Email: *garsmeur@cirad.fr*

Andrew George
CSIRO Mathematical and Information Sciences, 306 Carmody Rd, St. Lucia, Queensland, 4070, Australia.
Email: *Andrew.George@csiro.au*

Robert J. Henry
Current Address: Queensland Alliance for Agriculture and Food Innovation (QAAFI), University of Queensland, Brisbane QLD 4072 Australia.
Email: *robert.henry@uq.edu.au*

Previous Address: Centre for Plant Conservation Genetics, Southern Cross University, PO Box 157 Lismore NSW, 2480 Australia.

Carlos Takeshi Hotta
Instituto de Química, Departamento de Bioquímica, Universidade de São Paulo. Av. Prof. Lineu Prestes 748, B9S, sala 954, São Paulo, SP, 05508-900, Brazil.
Email: *carlos.hotta@cantab.net*

Emma Huang
CSIRO Mathematical and Information Sciences, 306 Carmody Rd, St. Lucia, Queensland, 4070, Australia.
Email: *Emma.Huang@csiro.au*

Barbara Huckett
SA Sugarcane Research Institute, Private Bag X02, Mount Edgecombe, 4300, South Africa.
Email: *huckett@vodamail.co.za*

Collins A. Kimbeng
School of Plant, Environmental and Soil Sciences, M. B. Sturgis Hall, Louisiana State University Agricultural Center, Baton Rouge, LA 70803, USA.
Email: *CKimbeng@agcenter.lsu.edu*

Lynne McIntyre
CSIRO Plant Industry, 306 Carmody Rd, St Lucia, QLD 4067, Australia.
Email: *Lynne.McIntyre@csiro.au*

Meredith McNeil
CSIRO Plant Industry, Queensland Bioscience Precinct, 306 Carmody Road, St. Lucia, Queensland, 4067, Australia.
Email: *Meredith.McNeil@csiro.au*

Ray Ming
Department of Plant Biology, University of Illinois at Urbana-Champaign, Urbana, IL 61801, USA.
Email: *rming@life.uiuc.edu*

Karine Miranda Oliveira
Centro de Tecnologia Canavieira—CTC, Caixa Postal 162, 13400-970, Piracicaba, São Paulo, Brazil.
Email: *karine@ctc.com.br*

Maria Marta Pastina
Departamento de Genética, Escola Superior de Agricultura Luiz de Queiroz (ESALQ), Universidade de São Paulo (USP), CP 83, CEP 13400-970, Piracicaba-SP, Brazil.
Email: *mmpastin@carpa.ciagri.usp.br*

Andrew H Paterson
Plant Genome Mapping Laboratory, University of Georgia, 111 Riverbend Road, Athens, GA 30602, USA.
Email: *paterson@uga.edu*

Luciana Rossini Pinto
Centro Avançado da Pesquisa Tecnológica do Agronegócio de Cana—IAC/Apta, Anel Viário Contorno Sul, Km 321, CP 206, CEP 14.001-970, Ribeirão Preto-SP, Brazil.
Email: *lurossini@iac.sp.gov.br*

George Piperidis
BSES Limited, PMB 57, Mackay Mail Centre, Mackay, Queensland, 4741, Australia.
Email: *GPiperidis@bses.org.au*

Nathalie Piperidis
BSES Limited, PMB 57, Mackay Mail Centre, Mackay, Queensland, 4741, Australia.
Email: *NPiperidis@bses.org.au*

Anete Pereira de Souza
Centro de Biologia Molecular e Engenharia Genética (CBMEG), Instituto de Biologia, Departamento. de Biologia Vegetal - Universidade Estadual de Campinas (UNICAMP), Cidade Universitária Zeferino Vaz, CP 6010, CEP 13083-875, Campinas-SP, Brasil.
Email: *anete@unicamp.br*

Glaucia Mendes Souza
Instituto de Química—Departamento de Bioquímica, Universidade de São Paulo. Av. Prof. Lineu Prestes, 748, B9S, sala 954. São Paulo, SP. Brazil 05508-900.
Email: *glmsouza@iq.usp.br*

Marie-Anne Van Sluys
GaTE lab, Departamento de Botânica-IB, USP, rua do Matão 277, 05508-090, São Paulo, SP-Brazil.
Email: *mavsluys@gmail.com*

Derek Watt
SA Sugarcane Research Institute, Private Bag X02, Mount Edgecombe, 4300, South Africa.
Email: *derek.watt@sugar.org.za*

Basic Information on the Sugarcane Plant

Robert J. Henry

ABSTRACT

Sugarcane is a major world crop supplying sugar and energy. This highly efficient crop from the grass family is grown in tropical and subtropical environments globally. The genome of modern cultivated sugarcane is large and complex originating from hybrids between two wild polyploid relatives, *Saccharum officinarum* and *Saccharum spontaneum*. This chapter introduces a volume that describes the genetics and genomics of sugarcane and consequences for sugarcane breeding.

Keywords: Sugarcane production, importance

1.1 Economic Importance

Sugarcane is probably the crop produced in the greatest quantities globally. More than 1,000 million tons of sugarcane are harvested each year. This exceeds the level of production of the major grain food crops, wheat, rice and maize, each of which has production of around 600 million tons per annum. Sugarcane is the source of most of the sugar produced in the world greatly exceeding sugar beet as a source of sugar (Cordeiro et al. 2007).

Sugarcane is arguably the most important industrial crop being used for energy production (ethanol and electricity; Tew and Cobill 2008) in addition to sugar and fiber products (e.g., paper, cardboard and fiber board).

Current Address: Queensland Alliance for Agriculture and Food Innovation (QAAFI), University of Queensland, Brisbane QLD 4072 Australia; *e-mail: robert.henry@uq.edu.au*
Previous Address: Centre for Plant Conservation Genetics, Southern Cross University, PO Box 157 Lismore NSW, 2480 Australia.

Sugarcane is a highly efficient plant utilizing the C4 pathway of photosynthesis that supports efficient photosynthesis especially at high temperatures.

Sugarcane can be produced in high yield with the production in farmers' fields in some countries averaging around 40 tons of dry matter production per hectare with the best farmers achieving as much as 70 tons per hectare and 100 tons per hectare being possible under ideal experimental conditions. This probably makes sugarcane the highest yielding of crop species.

Sugarcane production has expanded and may continue to expand in subtropical and tropical environments if the use of sugarcane as an energy crop continues to increase. Thirty years ago India was the largest producer of sugarcane. Brazil is now the country with the largest production of sugarcane and export of sugar but production is widespread in tropical areas of the world. China for a long time ranked third in sugarcane production. Sugarcane production is continuing to expand into areas that are more marginal for sugarcane growth. The production of varieties of sugarcane adapted to drought and possibly cold will be a key determinant of the extent of success in expansion of sugarcane production in many areas. Production of sugarcane in different countries is given in Table 1-1.

The stalk of the sugarcane plant contains around 13% sugar (90% sucrose), 12% fiber and 75% water with a very low ash content (Tew and Cobill 2008). The sugar content is around 50% on a dry weight basis making sugarcane an attractive source of sugar.

1.2 Academic Importance

Sugarcane is unusual in having such a high level of polyploidy (8-14x). This makes sugarcane a difficult species to work with at the genetic and molecular genetic levels. Sugarcane may be considered an extreme model for understanding the impact of polyploidy. The application of cytogenetic and genomic techniques in such a complex genome has been more challenging than in simpler (e.g., diploid) genomes. For example, analysis of gene families for agronomic traits of importance has required cloning and sequencing (McIntyre et al. 2006) rather than simple amplification and sequencing that is possible in other species. Simple sequence repeat (SSR) and single nucleotide polymorphism (SNP) markers have been analyzed using appropriately adapted techniques (Cordeiro et al. 2000, 2003, 2006a). The approaches taken in these analyses provide tools for use in other polyploidy or complex genomes (e.g., wheat). Sequencing of the sugarcane genome (Chapter 8) requires complex assembly of data and will be aided by sequence data from the progenitor species genomes. Techniques for discovery of genetic variation at the SNP level by ecotilling have been refined using the challenges of the sugarcane genome (Cordeiro et al. 2006b). The power

Table 1-1 Sugarcane production (in tons) (after FAO 2008).

	1965	1970	1975	1980	1985	1990	1995	2000	2005
Brazil	75,852,860	79,752,940	91,524,560	148,650,600	247,199,500	262,674,100	303,699,500	327,705,000	420,121,000
India	122,077,000	135,024,000	144,288,900	128,833,400	170,319,200	225,569,200	275,540,000	299,230,000	232,320,000
China	23,118,700	19,702,960	24,565,420	31,977,590	58,711,260	63,451,070	70,278,810	69,298,730	88,730,000
Thailand			14,592,300	12,826,660	25,055,020	33,561,000	50,597,000	54,052,120	49,572,000
Pakistan	18,667,010	26,369,500	21,242,000	27,497,700	32,139,600	35,493,600	47,168,400	46,332,600	47,244,100
Mexico	30,955,680	34,651,420	35,840,580	35,278,620	34,430,840	39,919,370	44,452,950	44,100,000	45,126,500
Colombia	12,720,450	12,700,000	20,500,000	26,100,000	25,364,400	27,790,740	32,000,000	33,400,000	39,849,240
Australia	14,382,230	17,644,820	21,959,010	23,975,650	24,401,600	24,369,940	34,943,000	38,164,690	38,246,000
The Philippines	21,150,000	26,140,000	35,868,000	30,900,000	22,842,000	25,482,000	24,590,000	24,491,000	31,000,000
United States of America	21,466,690	21,768,690	25,713,100	24,460,290	25,594,300	25,524,000	27,938,000	32,762,070	25,803,960
Indonesia	11,224,800	9,747,900	13,079,200	17,133,300	22,621,170	27,979,630	28,998,800	23,900,000	25,500,000
South Africa	8,406,000	12,143,900	16,813,540	14,062,200	18,803,010	18,083,490	16,713,650	23,876,160	21,725,100
Argentina	13,950,000	9,700,000	15,600,000	17,200,000	14,105,000	15,700,000	17,700,000	18,400,000	19,300,000
Guatemala				5,700,000	6,580,000	9,603,100	15,443,780	16,552,400	18,000,000
Egypt		6,934,000	7,902,000	8,618,000	9,684,000	11,095,260	14,104,770	15,705,800	16,335,000
Vietnam							10,711,100	15,044,300	15,000,000
Cuba	50,695,300	82,900,000	52,389,010	63,977,410	67,400,000	81,800,000	33,600,000	36,400,000	12,500,000
Venezuela,Bolivar Rep of			5,486,162		5,673,165	6,618,905		8,831,520	8,800,000
Peru	8,000,000	8,067,975	9,439,616	6,054,521	8,063,000	6,700,000	7,000,000	7,750,000	7,100,000
Iran, Islamic Rep of									6,500,000
Bangladesh	6,330,696	7,518,800	6,741,283	6,676,010	6,878,040	7,423,345	7,445,650	6,910,000	
Dominican Republic	5,544,081	8,654,779	9,337,018	9,055,700	8,419,497	6,511,584			
Ecuador	8,086,730	6,500,000	7,723,420	6,615,197			6,750,000		
Jamaica	4,775,421								
Mauritius	5,984,489	5,119,995							
Puerto Rico	7,989,509	5,343,975							

of deep sequencing to identify SNP in genes in a complex highly polypolid genome has also been demonstrated for the sugarcane genome (Bundock et al. 2009).

Cytogenetics has an important role to play in a species with a complex hybrid genome. The application of cytogenetic tools to sugarcane is described in Chapter 2 (Piperidis et al. 2009).

Sugarcane breeding is widespread but is largely based upon working within the domesticated sugarcane germplasm pool. Accessing more diverse germplasm may be aided by the development of efficient molecular tools for selection and gene transfer. Sugarcane transformation has been possible for some time. However the complexity of the sugarcane genome makes the routine production of sugarcane plants with stable levels of transgene expression very difficult. Advances in knowledge of the sugarcane genome and the control of gene expression in this complex polyploidy may aid the development of more robust transformation systems. Sequencing of the genome is currently in progress and this should form a new basis for sugarcane genomics. Sugarcane is vegetatively propagated and most commercial sugarcane varieties are probably only a few sexual generations from wild plants suggesting significant potential for further genetic improvement.

1.3 Brief History of Sugarcane as a Crop

The centre of origin of sugarcane is South East Asia and New Guinea. The history of domesticated sugarcane is not well known. Sugarcane has been grown in India and later China more than 2,000 years ago. Sugarcane has been produced worldwide for more than 500 years and was introduced to the New World by Columbus. The origin and domestication of sugarcane are outlined in Chapter 3 (Aitken and McNeil 2009) with a detailed analysis of the molecular and other evidence. *Saccharum officinarum* was probably domesticated in New Guinea from wild *S. robustum*. Humans distributed *S. officinarum* widely. Hybrids between *S. officinarum* and *S. spontaneum* became *S. barberi* in India and *S. sinense* in China. Modern sugarcane varieties with this hybrid origin have been distributed worldwide to the tropical and subtropical areas. Breeding is conducted in many countries.

1.4 Germplasm and Gene Pools

The *Saccharum* complex includes the *Saccharum* genus and several related genera.

Sugarcane is a member of the Andropogoneae tribe, which also includes sorghum. Sorghum and sugarcane are each other's closest relatives amongst cultivated plants (Dillon et al. 2007). *Erianthus* has been used as a source of novel genetic variation for sugarcane improvement programs. *Miscanthus* is now being developed as a bioenergy crop. Hybrids between *Miscanthus* and cultivated sugarcane, and *Sorghum* and cultivated sugarcane have recently been reported. Sugarcane with *Saccharum spontaneum* as a maternal genome has been produced and verified using molecular markers (Pan et al. 2004).

1.4.1 Sorghum Sequence

The genome sequence of Sorghum (*Sorghum bicolor*) has been analyzed (Paterson et al. 2009a). This immediately became an important tool for analysis of the related sugarcane genome. The close relationship between sorghum and sugarcane genomes is found when the sequences of sorghum and sugarcane genes are compared.

1.4.2 Genetic Resource Collections

Living collections of vegetatively propagated sugarcane clones are maintained in the field. This contrasts to genetic resources for most crop species that are maintained as seed. A world collection of sugarcane germplasm has been established in the field in the US and India. These collections and their diversity are described in Chapter 3 (Aitken and McNeil 2009).

1.5 Basic Botany of Sugarcane

Sugarcane is a member of the grass family, Poaceae. It is a member of the Andropogoneae tribe along with maize and sorghum. The Saccharine subtribe includes the *Saccharum* genus and related genera such as *Erianthus* and *Miscanthus*. Six species are often recognized within *Saccharum* (Table 1-2).

Modern sugarcane varieties have arisen from introgression of *S spontaneum* germplasm into the domesticated *S. officinarum*.

Table 1-2 The genus *Saccharum* (after Cordeiro et al. 2007).

Species	Chromosome number (2*n*)	
S. barbari	111–120	Indian cane
S. edule	60–80	Domesticated form with aborted flowers
S. officinarum	80	Domesticated sweet cane of New Guinea and South Pacific
S. robustum	60–80	Probable ancestor of *S. officinarum*
S. sinense	80–140	Chinese cane
S. spontaneum	40–128	Wild cane

Sugarcane is a very large perennial grass growing 2–6 meters high and being harvested for up to five years before requiring replanting. The stems especially the lower parts have a high sugar (sucrose) content. Sugarcane is a tropical plant and is not tolerant of frost.

Modern sugarcane varieties are highly polyploid and vary in the number of chromosomes (around 100). Chromosomes can be identified (D'Hont et al. 1995) as deriving from *S. officinarum* (around 80%), *S. spontaneum* (about 10%) or being a result of recombinations between chromosomes derived from the two progenitor species (about 10%). The progenitor species have high polyploidy (Table 1-2).

The genome of sugarcane is large because of this high level of polyploidy. The genome size is likely to vary between genotypes especially those with different chromosome numbers. The genome is of the order of 10,000 Mbp in size. The sequencing of sugarcane (Paterson et al. 2009b) and related genomes will improve the accuracy of our knowledge regarding the genome size.

References

Aitken K, McNeil M (2009) Diversity analysis. In: RJ Henry, C Kole (eds) Genetics, Genomics and Breeding of Sugarcane. Science Publ, Enfield, New Hampshire, USA, pp 19–42.

Bundock PC, Eliott F, Ablett G, Benson AD, Casu R, Aitken K, Henry RJ (2009) Targeted SNP discovery in sugarcane using 454 sequencing. Plant Biotechnol J 7(4): 347–354.

Cordeiro GM, Taylor GO, Henry RJ (2000) Characterisation of microsatellite markers from Sugarcane (*Saccharum* sp.) a highly polyploid species. Plant Sci 155: 161–168.

Cordeiro GM, Pan YB, Henry RJ (2003) Sugarcane microsatellites for the assessment of genetic diversity in sugarcane germplasm. Plant Sci 165: 181–189.

Cordeiro GM, Eliott F, McIntyre L, Casu RE, Henry RJ (2006a) Characterisation of single nucleotide polymorphisms in sugarcane EST's. Theor Appl Genet 113: 331–343.

Cordeiro GM, Eliott F, Henry RJ (2006b) An optimised ecotilling protocol for polyploids or pooled samples using a capillary electrophoresis system. Anal Biochem 355: 145–147.

Cordeiro GM, Amouyal O, Eloitt F, Henry RJ (2007) Sugarcane. In: C Kole (ed) Genome Mapping and Molecular Breeding in Plants, vol 3: Pulses, Sugar and Tuber Crops. Springer, Berlin, Heidelberg, New York, pp 175–204.

D'Hont A, Rao P, Feldmann P, Grivet L, Islam-Faridi N, Taylor P, Glaszmann JC (1995) Identification and characterisation of sugarcane intergeneric hybrids, *Saccharum officinarum* × *Erianthus arundinaceus* with molecular markers and DNA in situ hybridization. Theor Appl Genet 91: 320–326.

Dillon SL, Shapter FM, Henry RJ, Cordeiro G, Izquierdo L, Lee LS (2007) Domestication to crop improvement: Genetic resources for *Sorghum* and *Saccharum* (Andropogoneae). Ann Bot 100: 975–989.

FAO (2008) Food and agricultural commodities production: *www.fao.org/es/ess/top/commodity.html* (accessed 2 Oct 2008).

McIntyre CL, Jackson M, Cordeiro GM, Amouyal O, Elliott F, Henry RJ, Casu RE, Hermann S, Aitken KS (2006) The identification and characterization of alleles of sucrose phosphate synthase gene family III in sugarcane. Mol Breed 18: 39–50.

Pan YB, Burner DM, Wei Q, Cordeiro GM, Legendre BL, Henry RJ (2004) New *Saccharum* hybrids in *S. spontaneum* cytoplasm developed through a combination of conventional and molecular breeding approaches. Plant Genet Resour 2: 131–139.

Paterson AH, Bowers JE, Bruggmann R, Dubchak I, et al. (2009a) The *Sorghum bicolor* genome and the diversification of the grasses. Nature 457: 551–556.

Paterson AH, Souza G, Van Sluys M-A, Ming R, D'Hont A (2009b) Structural genomics and genome sequencing. In: RJ Henry, C Kole (eds) Genetics, Genomics and Breeding of Sugarcane. Science Publ, Enfield, New Hampshire, USA, pp 149–166.

Piperidis N, Piperidis G, D'Hont A (2009) Molecular cytogenetics. In: RJ Henry, C Kole (eds) Genetics, Genomics and Breeding of Sugarcane. Science Publ, Enfield, New Hampshire, USA, pp 9–18.

Tew TL, Cobill RM (2008) Genetic improvement of sugarcane (*Saccharum* spp.) as an energy crop. In: W Vermerris (ed) Genetic Improvement of Bioenergy Crops. Springer, New York, USA, pp 249–272.

Molecular Cytogenetics

Nathalie Piperidis,[1*] George Piperidis[1] and Angélique D'Hont[2]

ABSTRACT

The progress that has been achieved in understanding sugarcane (*Saccharum* spp.) genome structure over the last 15 years, thanks to the development of molecular cytogenetics is reviewed here. The basic chromosome numbers of the species *S. spontaneum, S. officinarum* and *S. robustum* have been determined using fluorescent in situ hybridization (FISH) with rDNA probes. Genomic in situ hybridization allowed the differentiation of chromosomes from *S. officinarum* and *S. spontaneum*, the two species involved in the origin of modern sugarcane cultivars, assessment of the proportion of chromosomes of these species in the genome of the modern cultivars and assessment of the extent of interspecific chromosome recombination. GISH has also been used to clarify the origin of the related taxa *S. barberi* and *S. sinense*. These techniques also proved to be very useful in monitoring introgression from the related genera.

Keywords: GISH, FISH

2.1 Introduction

Many classical cytological studies have been performed in sugarcane (review by Sreenivasan et al. 1987). These studies delivered important information about the complexity of the genome of sugarcane and were also very useful to help unravel the taxonomy of the *Saccharum* genus. Chromosome numbers

[1]BSES Limited, PMB 57, Mackay Mail Centre, Mackay, Queensland, 4741, Australia.
[2]CIRAD, Centre de Coopération Internationale en Recherche Agronomique pour le Développement, UMR1098/DAP, TAA96/03 avenue Agropolis, 34398 Montpellier cedex 5, France.
*Corresponding author: *NPiperidis@bses.org.au*

were determined, uncovering highly polyploid and, frequently, aneuploid members in this genus. The vast majority of *S. officinarum* clones display $2n = 80$ chromosomes. It has been suggested that the rare clones with higher chromosome numbers are hybrids (Bremer 1923; Sreenivasan et al. 1987). *S. robustum* largely encompasses clones with $2n = 60$ or 80, but also includes many other forms that may have up to 200 chromosomes (Price 1965). Chromosome number in *S. spontaneum* varies from 40 to 128, with five major cytotypes: $2n = 64$, 80, 96, 112, and 128 (Panje and Babu 1960). Pairing behavior was studied in these species that showed mainly bivalents, but various types of meiotic irregularities are also found (Bremer 1923; Price 1963; Burner 1991). The occurrence of $2n$ gamete transmission in hybrids between *S. officinarum* and *S. spontaneum* and their first backcross with *S. officinarum* was demonstrated (Bremer 1923, 1961).

However, because the species are highly polyploid and the chromosomes within and between species are similar, information such as basic chromosome number, as well as the precise behavior of wild chromosomes during introgression processes and their exact contribution to cultivars, could not be definitely established.

In the last 20 years, genomic in situ hybridization (GISH) and fluorescence in situ hybridization (FISH) have been developed and used extensively in plants. The former technique was used to identify parental chromosomes in interspecific hybrids, to test the origin of natural amphiploids, to track down the introgression of alien chromosomes or to test the occurrence of exchange between the genomes involved (Jiang and Gill 1994, 2006). The latter technique can reveal the physical location of repeated, low-copy-number or unique DNA sequences, provide useful cytological markers and enable comparisons between physical and genetic maps.

These techniques proved to be particularly relevant to refine our understanding of the genome structure of sugarcane and its taxonomy. They have been applied to sugarcane to answer specific questions regarding:

1) Origin of *S. barberi* and *S. sinense*
2) Basic chromosome number in *S. spontaneum*, *S. officinarum* and *S. robustum*
3) Genome structure of modern sugarcane cultivars
4) Introgression with other genera

2.2 Origin of *S. barberi* and *S. Sinense*

Sugarcanes indigenous from North India and China, referred to as *S. barberi* ($2n = 81$–124) and *S. sinense* ($2n = 116$–120), respectively have been cultivated from prehistoric times. Extraction of sugar was most probably developed from these canes in India and China (Daniels and Daniels 1975). These two groups of canes contrast with *S. officinarum*; they have thinner stalks, lower sucrose concentration, higher fiber content and a greater tolerance to biotic and abiotic conditions, reflecting closer characteristics to wild species. The taxonomic status of "North Indian" and "Chinese" sugarcanes has long been controversial (Daniels and Roach 1987; Daniels 1996; Irvine 1999). Three major hypotheses for their origin have been suggested: (i) *S. barberi* and *S. sinensi* arose from introgression of *S. officinarum* with *S. spontaneum* in India and China, respectively since prehistoric times; (ii) *S. barberi* was developed from *S. spontaneum* in India; and (iii) *S. barberi* and *S. sinensi* arose through introgression between *S. officinarum*, *S. spontaneum* and other genera such as *Miscanthus* or *Erianthus*. Although "North Indian" and "Chinese" sugarcanes are no longer commercially cultivated, there has been a continuing interest in elucidating their origin, because the hypothesis that they arose independently from the true sugarcane *S. officinarum* would imply that there are two or more possible genetic pathways for the improvement of sugarcane (Daniels and Daniels 1975).

GISH was used to test these hypotheses on a representative sample of the *S. barberi* and *S. sinense* clone collections using total genomic DNA from *S. officinarum* and *S. spontaneum*. The results clearly showed two distinct populations of chromosomes or chromosome fragments demonstrating the interspecific origin of *S. barberi* and *S. sinense* (D'Hont et al. 2002), The hypothesis that other genera were involved was ruled out because GISH results show a homogenous hybridization signal on all chromosomes: if another genus was involved we would expect to see entire chromosomes or portions of chromosomes with very weak hybridization signals. These results are corroborated by Southern data obtained with genus-specific sequences (Alix et al. 1998, 1999).

These GISH results demonstrated the interspecific origin of *S. barberi* and *S. sinense*. Together with recent cytoplasmic (D'Hont et al. 1993) and nuclear molecular marker analyses (Glaszmann et al. 1990; Burnquist et al. 1992; Lu et al. 1994), they were in agreement with an origin involving hybridization between *S. officinarum* (female) and *S. spontaneum* (male).

The chromosomal compositions of *S. barberi* and *S. sinense* were variable, with 61% of *S. officinarum* and 39% of *S. spontaneum* for clones with $2n = 82$; 68% of *S. officinarum* and 32% of *S. spontaneum* for clones with $2n = 91$ and 66% of *S. officinarum* and 33% of *S. spontaneum* with $2n = 116$. None or very few chromosomes appeared to be derived from interspecific recombination. The absence or scarceness of intra-chromosomal exchanges between the two species suggests that no or only a few meioses have occurred since the interspecific hybridization. Indeed, in the well-known modern cultivars, which derive from 3 to 6 generations of intercrossing after the interspecific hybridization, we observed between 10 and 20% of interspecific recombinant chromosomes (see below, D'Hont et al. 1996 and unpublished results).

Further restriction fragment length polymorphism (RFLP) analyses indicated that the S. *barberi* and S. *sinense* clones are clustered into a few groups, each derived from a single interspecific hybrid that has subsequently undergone somatic mutations (D'Hont et al. 2002). These groups correspond quite well with those already defined by morphological characters and chromosome numbers (reviewed by Daniels et al. 1991). However, the calculated genetic similarities do not support the existence of two distinct taxa. Therefore "North Indian" and "Chinese" sugarcanes represent a set of horticultural groups rather than established species.

2.3 Basic Chromosome Numbers in S. *spontaneum*, S. *officinarum* and S. *robustum*

Basic chromosome numbers of $x = 5, 6, 8, 10$ and 12 have been suggested for the *Saccharum* species (Sreenivasan et al. 1987). For S. *officinarum* and S. *robustum*, for which the major cytotype are $2n = 80$, and $2n = 80$ and 60, respectively, a basic number of 10 was proposed; $x = 10$ is also the most common basic chromosome number in the Andropogoneae (Bremer 1961). For S. *spontaneum*, a basic chromosome number of $x = 8$ seems to be more likely according to the major cytotypes of this species $2n = 64, 80, 96, 112$ and 128 (Panje and Babu 1960), which are all multiples of eight.

FISH with probes corresponding to the 45S and 5S rRNA (ribosomal RNA) genes was used to investigate the basic chromosome number of S. *spontaneum*, S. *officinarum* and S. *robustum* (D'Hont et al. 1998). The ribosomal genes are generally organized in separate loci and arranged in long tandem arrays of repeat units incorporating coding sequences and intergenic spacers. The basis unit for the 45S represents several kilobases while the 5S genes are much smaller and less repeated. The 45S locus is often associated with secondary restrictions and nucleolus organizer regions. Various cytotypes of S. *officinarum*, S. *robustum* and S. *spontaneum* were analyzed. The 5S rRNA genes were located at an interstitial position on the chromosomes for the three species, while the 45S rRNA genes were interstitial for S. *spontaneum* and terminal for S. *officinarum* and S. *robustum*. The intensity of the signal varies among chromosomes suggesting the presence of major and minor sites.

For S. *officinarum* and S. *robustum*, both probes (45S and 5S) revealed eight hybridization sites for the cytotypes with $2n = 80$ and six sites for S. *robustum* cytotype with $2n = 60$ (Table 2-1; D'Hont et al. 1996, 1998). The number of sites and identical position on the chromosomes were interpreted as the presence of one locus per gene category (45S and 5S), carried by 8 and 6 homologous chromosomes, respectively. A deduced basic chromosome number of $x = 10$ for S. *officinarum* and S. *robustum* was obtained by dividing the number of chromosomes per cytotype by the number of sites detected.

Table 2-1 Number of rDNA sites revealed by FISH in *S. officinarum*, *S. robustum* and *S. spontaneum* cytotypes.

		$2n$ cell	45S sites	5S sites	Source
S. officinarum	BNS 3066	80	8	/	D'Hont et al. 1996
	Black Cheribon	80	8	8	D'Hont et al. 1998
S. robustum					
	NG 77230	80	8	8	D'Hont et al. 1998
	Mol 4503	60	6	6	D'Hont et al. 1998
	IM 76234	60	6	6	D'Hont et al. 1998
S. spontaneum					
	Haploid of SES208	32	4	4	Ha et al. 1999
	SES14	64	8	8	D'Hont et al. 1996
	Mol 5801	80	10	10	D'Hont et al. 1998
	Mandalay	96	12	12	D'Hont et al. 1998
	Glagah	112	10	14	D'Hont et al. 1998

/ indicates data not known

For *S. spontaneum*, cytotypes with $2n$ = 64, 80, 96 and 112 were analyzed (D'Hont et al. 1996, 1998 and Ha et al. 1999). Both probes (45S and 5S) revealed a number of hybridization sites directly proportional to the total number of chromosomes except in one case. The number of sites and the identical position on the chromosome were interpreted as the presence of one locus for each category of genes (45S and 5S). A deduced basic chromosome number of $x = 8$ for *S. spontaneum* was obtained by dividing the number of chromosomes of the cytotype by the number of sites detected. The only anomaly was obtained with probe 45S on the cytotype with the highest number of chromosomes, Glagah ($2n = 112$). In the various clones studied, minor as well as major rDNA sites were detected, probably reflecting a reduction in the number of repeats; in the clones with very high chromosome numbers, this phenomenon may have been accentuated to the point where sites are very difficult to detect or have even been completely deleted.

These studies have shown that the two major species involved in modern cultivars have different basic chromosome number; $x = 10$ for *S. officinarum* and $x = 8$ for *S. spontaneum* and thus some structural differences are to be expected. At the very least, some rearrangements, such as fusion or fission, are expected between chromosomes (linkage groups in genetic maps) of these species. If the differences in linkage group constitution between species are complex, the interspecific intra-chromosomal recombination that was observed by GISH (D'Hont et al. 1996) will have major implications, such as bridging linkage groups that are distinct within each species. Progressively, each cultivar could appear to have its specific basic map. The genetic map of a modern cultivar, R570 (Grivet et al. 1996; Rossi et al. 2003; A. D'Hont, pers. com.; *http://tropgenedb.cirad.fr/index_fr.html*), supports a general colinearity between *S. officinarum* and *S. spontaneum* genomes, with a coherent single organization of co-segregation groups into putative basic linkage groups. This suggests that the structural differences between the two founder species are not too complex. R570 mapping data strongly suggest that, in the case of linkage group VI and VIII, one chromosome of *S. spontaneum* corresponds to two chromosomes of *S. officinarum* (Grivet et al. 1996; Glaszmann et al. 1997; Rossi et al. 2003; D'Hont et al. 2008).

2.4 Genome Structure of Modern Sugarcane Cultivars

The interspecific origin of modern sugarcane cultivars is well known and has been documented for quite some time (Arceneaux 1965; Price 1965). Modern sugarcane cultivars derived from hybridizations, performed a century ago, between domesticated species *S. officinarum*, and the wild species *S. spontaneum*. A series of backcrosses were then performed with *S. officinarum* to recover the agronomically favorable traits inherited from this species and dilute the unfavorable traits inherited from the wild canes. However, the proportion of chromosomes contributed to modern cultivars by the wild species, *S. spontaneum*, remained unknown until molecular cytogenetic techniques were applied by sugarcane researchers.

D'Hont et al. (1996) applied GISH with genomic DNA from *S. officinarum* and *S. spontaneum* to the French cultivar R570 and were able to differentiate the chromosomes from the two species. They showed that R570 had about $2n = 115$ chromosomes of which 80% were *S. officinarum* chromosomes, 10% were derived from *S. spontaneum* while 10% were coming from interspecific exchange between the two species. These results demonstrated for the first time the occurrence of chromosomal exchange between the *S. officinarum* and *S. spontaneum* genomes and therefore contradict the previous assumption that no exchange between these two species occurs during meiosis (Price 1963, 1965; Berding and Roach 1987).

These results together with the basic chromosome number in *S. officinarum* and *S. spontaneum* allowed D'Hont (2005) to propose a schematic representation of the genome of a typical modern cultivar (Fig. 2-1). Since then, other researchers have used GISH to analyze the chromosome composition of modern cultivars or modern breeding clones of importance for their breeding programs. Piperidis and D'Hont (2001) analyzed seven modern cultivars and the proportion of *S. spontaneum* chromosomes comprised between 10 to 20% while the proportion of the recombined chromosomes comprised between 5 to 17%. Three modern breeding clones were analyzed by Cuadrado et al. (2004) and about 16% of the chromosomes were from *S. spontaneum* chromosomes while less than 5% chromosomes appeared to be recombined.

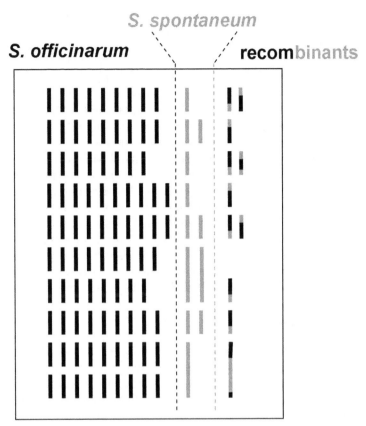

Figure 2-1 Schematic representation of the genome of sugarcane cultivars as deduced from FISH and GISH (from D'Hont et al. 2005).

2.5 Introgression with Other Genera

Encouraged by the breakthrough obtained in sugarcane breeding a century ago by crossing the wild species *S. spontaneum* with *S. officinarum*, breeders have tried to exploit favorable traits from related genera, in particular *Miscanthus* and *Erianthus*. The genus *Erianthus* has been proposed to be part of the *Saccharum* complex based on their intercrossability (Mukherjee 1957). The genus *Miscanthus* was added to the complex by (Daniels and Daniels 1975). The basic chromosome number for *Erianthus* is $x = 10$ and was revealed by FISH experiments using the 45S and the 5S rRNA genes (D'Hont et al. 1995).

Although many attempts to cross *Erianthus* with sugarcane have been undertaken, it has proven very difficult to identify genuine hybrids based only on their morphological characters. The development of *Erianthus*-specific molecular markers (D'Hont et al. 1995; Besse et al. 1996; Alix et al.

1998), molecular fingerprinting (Nair et al. 1999) and GISH (D'Hont et al. 1995; Alix et al. 1998; Piperidis et al. 2000) have overcome this difficulty and since then, several validated *Saccharum-Erianthus* hybrids have been reported (D'Hont et al. 1995; Besse et al. 1996, Piperidis et al. 2000; Cai et al. 2005).

GISH allowed the visualization of the chromosome complement of F_1 hybrids between *Saccharum* and *Erianthus* (D'Hont et al. 1995; Piperidis et al. 2000) and also between *Saccharum* and *Miscanthus* (A. D'Hont, pers. com.). In the GISH experiment of the *Saccharum-Erianthus* hybrids, very little cross hybridization between DNA of the two genera was observed, which suggests that these two species are not closely related. This is in contrast to the result obtained between *S. officinarum* and *S. spontaneum*, where cross hybridization does occur because of their relatedness. These results are in agreement with phylogenetic studies based on molecular markers showing that *Saccharum* and *Erianthus* species are more distant (Besse et al. 1997). This may also explain the difficulty to obtain viable and fertile progenies from the crosses between *Saccharum* and *Erianthus*. GISH also revealed, in some cases but not all, the occurrence of chromosome losses in *Saccharum–Erianthus* F_1 hybrids (Piperidis and D'Hont 2001; N. Piperidis, pers. com.).

2.6 Conclusion

In this chapter, we presented the progress achieved using molecular cytogenetics in understanding sugarcane taxonomy and genome structure.

GISH is now well established in sugarcane and could be used to monitor new programs aimed at exploiting wild species. It also could be used to help resolve remaining taxonomic questions such as the origin of *S. officinarum* clones with more than 80 chromosomes or the origin of *S. edule*. Another possibility offered by FISH is the location of bacterial artificial chromosome (BAC) clones on chromosomes. BAC-FISH with BAC anchored to genetic maps could help refine our understanding of sugarcane genome structure. For example, it could be used to determine the number of chromosomes in the various homoeologous groups and verify the hypothesis regarding the different chromosome structures between *S. officinarum* and *S. spontaneum*.

References

Alix K, Baurens FC, Paulet F, Glaszmann JC, D'Hont A (1998) Isolation and characterization of a satellite DNA family in the *Saccharum* complex. Genome 41: 854–864.

Alix K, Paulet F, Glaszmann JC, D'Hont A (1999) Inter-Alu like species-specific sequences in the *Saccharum* complex. Theor Appl Genet 6: 962–968.

Arceneaux G (1965) Cultivated sugarcanes of the World and their botanical derivation. Proc Int Soc Sugar Cane Technol 12: 844–854.

Berding N, Roach BT (1987) Germplasm collection, maintenance, and use. In: DJ Heinz (ed) Sugarcane Improvement through Breeding. Elsevier, Amsterdam, The Netherlands, pp 143–210.

Besse P, Mc Intyre L, Berding N (1996) Ribosomal DNA variations in *Erianthus*, a wild sugar cane relative (Andropogoneae–Saccharinae). Theor Appl Genet 92: 733–743.

Besse P, Mc Intyre CL, Burner D M, De Almeida CG (1997) Characterisation of *Erianthus* sect. Ripidium and *Saccharum* germplasm (Andropogonae-Saccharinae). Theor Appl Genet 92: 733–743.

Bremer G (1923) A cytological investigation of some species and species-hybrids of the genus *Saccharum*. Genetica 5: 273–326.

Bremer G (1961) Problems in breeding and cytology of sugar cane. Euphytica 10: 59–78.

Burner DM (1991) Cytogenetic analysis of sugarcane relatives (Andropogoneae: Saccharinae. Euphytica 54: 125–133.

Burnquist WL, Sorrels ME, Tanksley S (1992) Characterization of genetic variability in *Saccharum* germplasm by means of restriction fragment length polymorphism (RFLP) analysis. Proc Int Sugar Cane Technol 21: 355–365.

Cai Q, Aitken K, Deng H, Chen X, Fu C, Jackson P, McIntyre C (2005) Verification of the introgression of *Erianthus arundinaceus* germplasm into sugarcane using molecular markers. Plant Breed 124: 322–328.

Cuadrado A, Acevedo R, Moreno Díaz de la Espina S, Jouve N, de la Torre C (2004) Genome remodeling in three modern *S. officinarum* × *S. spontaneum* sugarcane cultivars. J Exp Bot 55: 847–854.

Daniels C (1996) Biology and biological technology. In: J Needham (ed) Science and Civilization in China, vol 6. Part III Agro-industries and forestry. Cambridge Univ Press, Cambridge, UK.

Daniels J, Daniels C (1975) Geographical, historical and cultural aspect of the origin of the Indian and Chinese sugarcanes *S. barberi* and *S. sinense*. Sugarcane Breed Newsl 36: 4–23.

Daniels J, Roach BT (1987) Taxonomy and evolution. In: DJ Heinz (ed) Sugarcane Improvement through Breeding. Elsevier, New York, Amsterdam, pp 7–84.

Daniels J, Roach B, Daniels C, Panton N (1991) The taxonomic status of *Saccharum Barberi* Jeswest and *S. Sinense* Roxb. Sugarcane 3: 11–16.

D'Hont A (2005) Unravelling the genome structure of polyploids using FISH and GISH; examples of sugarcane and banana. Cytogenet Genome Res 109: 27–33.

D'Hont A, Lu YH, Feldmann P, Glaszmann JC (1993) Cytoplasmic diversity in sugarcane revealed by heterologous probes. Sugar Cane 1: 12–15.

D'Hont A, Rao P, Feldmann P, Grivet L, Islam-Faridi N, Taylor P, Glaszmann JC (1995) Identification and characterization of intergeneric hybrids, *S officinarum* × *Erianthus arundinaceus*, with molecular markers and in situ hybridization. Theor Appl Genet 91: 320–326.

D'Hont A, Grivet L, Feldmann P, Rao S, Berding N, Glaszmann JC (1996) Characterisation of the double genome structure of modern sugarcane cultivars (*Saccharum* spp.) by molecular cytogenetics. Mol Gen Genet 250: 405–413.

D'Hont A, Ison D, Alix K, Roux C, Glaszmann JC (1998) Determination of basic chromosome numbers in the genus *Saccharum* by physical mapping of ribosomal RNA genes. Genome 41: 221–225.

D'Hont A, Lu Y H, Paulet F, Glaszmann JC (2002) Oligoclonal interspecific origin of 'North Indian' and 'Chinese' sugarcanes. Chrom Res 10: 253–262.

D'Hont A, Mendes Souza G, Menossi M, Vincentz M, Van Sluys MA, Glaszmann JC, Ulian EC (2008). Sugarcane: a major source of sweetness, alcohol, and bio-energy. In: PH Moore, R Ming (eds) Genomics of Tropical Crop Plants. Springer, New York, USA, pp 483–513.

Glaszmann JC, Lu YH, Lanaud C (1990) Variation of nuclear ribosomal DNA in sugarcane. J Genet Breed 44: 191–198.

Glaszmann JC, Dufour P, Grivet L, D'Hont A, Deu M, Paulet F, Hamon P (1997) Comparative genome analysis between several tropical grasses. Euphytica 96: 13–21.

Grivet L, D'Hont A, Roques D, Feldmann P, Lanaud C, Glaszmann JC (1996) RFLP mapping in cultivated sugarcane (*Saccharum* spp.): Genome organization in a highly polyploid and aneuploid inter-specific hybrid. Genetics 142: 987–1000.

Ha S, Moore PH, Heinz D, Kato S, Ohmido N, Fukui K (1999) Quantitative chromosome map of the polyploid *Saccharum spontaneum* by multicolor fluorescence *in situ* hybridization and imaging methods. Plant Mol Biol 39: 1165–1173.

Irvine JE (1999) *Saccharum* species as horticultural classes. Theor Appl Genet 98: 186–194.

Jiang J, Gill BS (1994) Nonisotopic in situ hybridization and plant genome mapping: the first 10 years. Genome 37(5): 717–725.

Jiang J, Gill BS (2006) Current status and the future of fluorescence in situ hybridization (FISH) in plant genome research. Genome 49(9): 1057–1068.

Lu YH, D'Hont A, Walker DIT, Rao PS, Feldmann P, Glaszmann JC (1994) Relationships among ancestral species of sugarcane revealed with RFLP using single copy maize nuclear probes. Euphytica 78: 7–18.

Mukherjee SK (1957) Origin and distribution of *Saccharum*. Bot Gaz 119: 55–56.

Nair NV, Nair S, Sreenivasan TV, Mohan M (1999) Analysis of genetic diversity and phylogeny in *Saccharum* and related genera using RAPD markers. Genet Resour Crop Evol 46: 73–79.

Panje R, Babu C (1960) Studies in *Saccharum spontaneum* distribution and geographical association of chromosome numbers. Cytologia 25: 152–172.

Piperidis G, D'Hont A (2001) Chromosome composition analysis of various *Saccharum* interspecific hybrids by genomic *in situ* hybridisation (GISH). Int Soc Sugar Cane Technol Congr 11: 565.

Piperidis G, Christopher MJ, Carroll BJ, Berding N, D'Hont A (2000) Molecular contribution to selection of inter-generic hybrids between sugarcane and the wild species *Erianthus arundinaceus*. Genome 43: 1033–1037.

Price S (1963) Cytogenetics of modern sugar canes. Econ Bot 17: 97–106.

Price S (1965) Interspecific hybridization in sugarcane breeding. Proc Int Soc Sugar Cane Technol 12: 1021–1026.

Rossi M, Araujo P, Paulet F, Garsmeur O, Dias V, Hui C, Van Sluys MA, D'Hont A (2003) Genome Distribution and Characterization of EST derived sugarcane resistance gene analogs. Mol Genet Genom 269: 406–419.

Sreenivasan TV, Ahloowalia BS, Heinz DJ (1987) Cytogenetics. In: DJ Heinz (ed) Sugarcane Improvement through Breeding. Elsevier, New York, USA, pp 211–253.

Diversity Analysis

*Karen Aitken** and *Meredith McNeil*

ABSTRACT

Sugarcane species are found over a vast geographical area stretching from India to Polynesia. They belong to the genus *Saccharum*, which is currently divided into six species, two wild species *S. spontaneum* and *S. robustum*, and four cultivated species *S. officinarum*, *S. barberi*, *S. sinense*, and *S. edule*. Of these species, *S. spontaneum*, *S. robustum*, *S. officinarum*, *S. barberi*, and *S. sinense* are interfertile and have produced interspecific hybrids. All species are polyploid and outcrossing resulting in high heterozygosity and morphological diversity. Over the past few years molecular markers have been used to establish the relationships between these species and levels of diversity within species. In this chapter we review the current literature on both morphological and molecular diversity of these *Saccharum* species and how this diversity can be exploited by sugarcane breeders.

Keywords: genetic diversity, molecular markers, *Saccharum* complex, phenotypic diversity

3.1 Introduction

The assessment of genetic diversity among cultivars is potentially an important tool for plant breeding purposes, since it can provide breeders with the means for analyzing variation available in germplasm collections. This diversity gives sugarcane breeders the opportunity to select more

CSIRO Plant Industry, Queensland Bioscience Precinct, 306 Carmody Road, St. Lucia, Queensland, 4067, Australia and CRC Sugar Industry Innovation through Biotechnology, Level 5, John Hines Building, The University of Queensland, St Lucia, QLD, 4072, Australia.
*Corresponding author: *Karen.Aitken@csiro.au*

diverse germplasm to include within their breeding programs. Sugarcane has been postulated to have evolved from a primitive form of *S. spontaneum* in the foothills of the Himalaya in northern India (Stevenson 1965). Subsequent selection, movement, and introgression resulted in two centers of diversity—*S. spontaneum* in India and *S. robustum* in New Guinea (Daniels and Roach 1987). The classification of species within the *Saccharum* genus has been the subject of great debate over the years (Daniels and Roach 1987; Irvine 1999). To facilitate the reliable classification of accessions, and lead to the identification of subsets of core accessions for the utilization with specific breeding purposes, the analysis of genetic diversity in germplasm collections has been employed. Several methods have been used to investigate the genetic diversity of sugarcane and include: pedigree data (Lima et al 2002), morphological data (Brown et al. 2002), agronomic performance data (Skinner et al. 1987), biochemical data obtained by analysis of isozymes (Glaszmann et al. 1989; Pocovi et al. 2008), and, recently, DNA-based marker data that allow more reliable differentiation of genotypes (Lu et al. 1994b; D'Hont et al. 2002; Aitken et al. 2006), all leading to the generation of diverse data sets. For the analysis of genetic diversity, the measurement of genetic distance-similarity between two genotypes, populations, or individuals may be calculated by various statistical measures depending on the data set (Box 3-1).

Box 3-1 Tools for measuring genetic diversity.

Comprehensive reviews of various distance measures are available in literature (Felsenstein 1984; Nei 1987; Beaumont et al. 1998; Mohammadi and Prasanna 2003). For molecular marker data, allele frequencies are calculated and used to generate a binary matrix for statistical analysis. The most commonly used measures of genetic distance-similarity using binary data to assess sugarcane diversity are: Nei and Li's (1979) coefficient, Jaccard's (1908) coefficient, and the simple matching coefficient (Sokal and Michener 1958). Alternatively, multivariate analytical techniques, which simultaneously analyze multiple measurements on each individual under investigation, are widely used in analysis of genetic diversity irrespective of the dataset (morphological, biochemical, or molecular marker data) (Mohammadi and Prasanna 2003). Among these algorithms, cluster analysis using the agglomerative hierarchical method UPGMA (Unweighted Paired Group Method using Arithmetic averages; Sneath and Sokal 1973), principal component analysis (PCA) and principal coordinate analysis (PCoA) are, at present, most commonly used and appear particularly useful (Aitken et al. 2006). Many statistical software packages are available for analyzing genetic diversity (Labate 2000) and include menu-driven statistical packages such as NTSYS-pc (F.J. Rohlf, State University of New York, Stony Brook, USA) and PHYLIP (J. Felsenstein, University of Washington, Seattle, USA). New packages are constantly being developed and there is still a distinct need for developing comprehensive and user-friendly statistical packages that facilitate an integrated analysis of different data sets for generating reliable information about genetic diversity.

3.2 Phenotype-based Diversity Analysis

Initially, traditional methods, which combine agronomic and morphologic characteristics, were used to characterize the sugarcane germplasm (Stevenson 1965; Skinner et al. 1987). Traditional sugarcane taxonomy divides the genus *Saccharum* into six species, the four cultivated species *S. officinarum* L., *S. barberi* Jesw., *S. sinense* Roxb., and *S. edule* Hassk., and the two wild species *S. spontaneum* L. and *S. robustum* Brandes and Jeswiet ex Grassl (Naidu and Sreenivasan 1987). These six species along with *Narenga* Bor, *Sclerostachya* (Anderss. Ex Hackel) A, Camus, and *Erianthus* Michx. and *Miscanthus* Anderss. section *Ripidium* are known as the "*Saccharum* Complex" (Mukherjee 1957). However, this classification can be compounded by the influence of the environment on these vegetative characteristics, which can lead to continuous variation and a high degree of plasticity, which often does not reflect the true genetic diversity of the *Saccharum* spp. germplasm (Lima et al. 2002). Also, recent molecular data disputes this taxonomic division of the genus *Saccharum* into six species and it has been suggested that only two species are valid, *S. officinarum* and *S. spontaneum*, the other species being interspecific hybrids between these two (Irvine 1999; D'Hont et al. 2002; Fig. 3-1). Despite this disagreement, the separation of clones into the various groups or species provides a useful method of classification for the management of the World Collection of sugarcane germplasm located at Canal Point and Miami, Florida and by the Sugarcane Breeding Institute at Cannanore and Coimbatore, India (Brown et al. 2002).

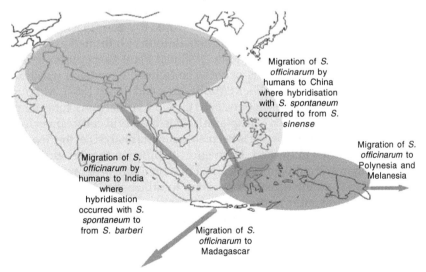

Figure 3-1 Area of origin of *S. spontaneum* (light gray), *S. sinense* and *S. barberi* (mid grey) and *S. robustum* (dark gray). Arrows indicate the dispersion of *S. officinarum* by human migration.

Within cultivated sugarcane there exists a great deal of phenotypic diversity, which has been very effectively exploited by a number of sugarcane breeding programs around the world (Skinner et al. 1987). This is despite the history of the development of modern sugarcane cultivars where only a limited number of *S. officinarum* and *S. spontaneum* clones were involved in the development of modern sugarcane cultivars compared to the large number of basic clones that exist in the *Saccharum* genus (Arceneaux 1967; Roach 1989). This limited involvement of wild clones is probably because the early hybridizations were so successful and breeders found further genetic gains from recurrent selection of progeny giving little incentive to recreate these early hybrids. This has led to a concern by sugarcane breeders of the narrow genetic base of sugarcane breeding programs. As a result of this concern, sugarcane geneticists realized that a large diverse germplasm collection was essential for sustained crop improvement and motivated a recommencement of interspecific hybridization in several countries in the 1960s (Berding and Roach 1987; Miller and Tai 1992). At least 31 organized collections of wild germplasm were made from 1892 to 1985 (Brown et al. 2002). The objectives of these expeditions were to collect genotypes that were resistant to diseases, were highly productive, or had high sugar content (Berding and Roach 1987). These clones have been organized into world collections of material of the *Saccharum* complex and related genera recognized by the International Society of Sugar Cane Technologists (ISSCT). The clones from these collections are maintained by the USDA/ARS at Canal Point and Miami, Florida and by the Sugarcane Breeding Institute at Cannanore and Coimbatore, India. There are 102 descriptors, classified into eight categories, used to define the clones within this germplasm collection and the data is available at the Germplasm Resources Information Network (GRIN) database maintained by the National Plant Germplasm System (NPGS) of the USDA. Details of these collections have been published (Balasundaram et al. 1980; Miller 1982; Berding and Roach 1987). Although there is some detailed information about the world collections, for example the catalog of *S. spontaneum* germplasm held in the world collection in India (Kandasami et al. 1983), there has been little general analysis of the diversity of these collections published. In an effort to broaden the genetic base of modern sugarcane cultivars, the Canal Point breeding program in Florida has established an introgression program to utilize the World Sugarcane Germplasm Collection to develop cultivars with increased yields and provide resistance to pests and environmental stresses. As a result of this introgression program, the cultivar CP 96-1252 was released to the Florida industry with a widened germplasm base (Miller et al. 2005). More recently, a breeding program and a research program were initiated in Australia with the aim to utilize wild *S. spontaneum* from China in sugarcane improvement for sugar or energy-from-biomass production systems (Wang et al. 2008).

Saccharum officinarum L. is the primary sugar producing species of sugarcane and is found only in cultivation. These clones are classified as "noble canes" and are characterized as having broad leaves, thick stalks with high sucrose and low fiber content (Daniels and Roach 1987). A very large collection of over 750 *S. officinarum* accessions is maintained in the *Sacchaurm* world collection at the Sugarcane Breeding Institute-Research Centre, Cannanore, India. Of the 750 accessions, 713 have been systematically characterized and documented for botanical yield and quality attributes and for disease ratings (Sreenivasan and Nair 1991). Of these, 476 accessions were collected during several expeditions from 1895 to 1977 from the primary center of diversity in Indonesia to the New Guinea region. The remaining accessions were collected from India, Fiji, Hawaii and Mauritius. A subset of the total collection, 690 accessions with complete data for 27 qualitative descriptors and 10 quantitative characters were evaluated for phenotypic diversity by means of a diversity index and principal component analysis (Balakrishnen et al. 2000). The non-hierarchical clustering of accessions resulted in 10 clusters. This clustering did not reveal any pattern or grouping of the accessions based on their geographical source. Most of the qualitative descriptors that were considered for diversity had fairly high diversity values as measured using the Shannon-Weaver Diversity Index indicating high levels of phenotypic diversity for these descriptors within this collection of *S. officinarum* accessions. Further analysis of this collection using an information measure (LEAV) formed 12 groups, which again were not related to geographic origin (Balakrishnan and Nair 2003). In another study, 184 *S. officinarum* accessions were grown in a replicated field trial for a preliminary analysis of sugar concentration, stalk weight and height. They found significant variation for all three traits (Albertson et al. 2001). A study assaying germplasm diversity among four sugarcane species for sugar composition found that *S. officinarum* clones had the widest distribution for sucrose concentration and the highest variation for brix and pol. The authors concluded that *S. officinarum* germplasm could provide sugarcane breeders with a great diversity of genetic resources for sugar traits (Tai and Miller 2002). Ram and Hemaprabha (2005) screened 53 *S. officinarum* accessions in replicated field trials at The Sugarcane Breeding Institute in Coimbatore, India. They measured 13 quantitative characters of sugar yield and its components at 12 months crop age and subjected this to diversity analysis. Analysis of variance showed significant differences among the *S. officinarum* accessions for all characters studied. The 53 clones clustered into nine groups and the authors concluded that use of parents from different clusters with maximum diversity would increase the chances of exploiting heterosis for improving sugar yield.

S. spontaneum has the widest ecogeographical distribution among *Saccharum* spp. and different clones show wide morphological variation from short bushy types with leaves reduced to midrib and practically without any cane formation to tall, erect, broad-leaved forms with long internodes (Sreenivasan et al. 1987). However, phenotypic analysis of a collection of *S. spontaneum* accessions presents some difficulties due to its rhizomatous growth habit, large plant size and the need for special growing conditions due to a noxious weed classification in some countries. If grown in field trials it is likely to become a source of a serious weed infestation. For these reasons very little phenotypic diversity analysis has been carried out on collections of *S. spontaneum*. Although a study of 342 accessions of *S. spontaneum* from the world collection was carried out by Tai and Miller (2001). Data was collected on 11 quantitative traits including flowering time, stalk diameter, leaf length and leaf width as well as quality traits (fiber content, sucrose, glucose and fructose). As expected from the extent of the geographic origin of *S. spontaneum*, a significant amount of variation was detected for all traits. Recently, an attempt to rationalize the large *S. spontaneum* collection maintained at the Sugarcane Breeding Institute, Coimbatore, India was carried out (Alphonse Amalraj et al. 2006). Out of this collection, 617 accessions were fully characterized for 21 morphological traits and 10 quantitative traits (Kandasami et al. 1983; Sreenivasan et al. 2001). Using the Shannon Diversity Index, the majority of the descriptors measured had a high level of diversity ranging from 0.26 to 1.00 with 87% (28/32) above 0.50. The authors concluded that a core size of 10% (about 60 accessions) drawn through stratified random sampling from the diversity groups was representative of the variability in the *S. spontaneum* core collection in Coimbatore and could be exploited in breeding.

S. robustum Brandes and Jeswiet ex Grassl has been described as a thick stalked wild cane forming dense cane brakes on mud flats of river banks in New Guinea (Daniels and Roach 1987). In the world collections there are 128 accessions held at Miami and 116 at Coimbatore, India. Limited information is published on the morphological diversity of this wild species although there are a number of references in the literature to its wide phenotypic variation (Berding and Roach 1987). Analysis of phenotypic variation of 10 quantitative characters of *S. robustum* showed large amounts of variation exists for all characters (Berding and Roach 1987). Analysis of the sugar composition of 27 *S. robustum* accessions resulted in them falling into five clusters, the majority (24/27) of the accessions had low sucrose and moderate levels of glucose and fructose, one had moderate sucrose and low glucose and fructose and two accessions had high glucose and fructose and moderate sucrose (Tai and Miller 2002).

S. barberi Jesw. and *S. sinense* Roxb. are indigenous sugarcanes of North India and China, respectively, and have been cultivated since prehistoric times. Bremer (1923) studied these clones from North India and China, and separated them into five morphologically different groups: Sunnabile ($2n$ = 82 and 116), Mungo ($2n$ = 82), Nargori ($2n$ = 107 and 124), Saretha ($2n$ = 90 and 92), and Pansahi group ($2n$ = ca. 118). Jeswiet (1929) refined this classification by grouping the first four groups under the species name *S. barberi* Jeswiet and the Pansahi group as *S. sinense* Roxb. Brandes (1956) postulated that they were interspecific hybrids derived from *S. officinarum* and *S. spontaneum*. He suggested that *S. officinarum* cultivars were possibly transported by humans to mainland Asia and following the natural crossing with local *S. spontaneum* gave rise to *S. barberi* and *S. sinense* in India and China, respectively. Recent chromosome in situ hybridization experiments by D'Hont et al. (2002) support this hypothesis. They are often grouped together as a single species and are distinguished from *S. officinarum* by floral characteristics, thin to medium stalks, low to moderate sucrose, higher fiber and a greater tolerance to adverse conditions (Roach 1989). Clones are no longer grown commercially and only a few have been used in sugarcane breeding due to their poor fertility and reduced flowering. In the world collections there are 53 *S. barberi* and 66 *S. sinense* at Miami and 43 and 29, respectively at Coimbatore. In a study on 30 *S. barberi* and 28 *S. sinense* accessions to determine if sugar composition could be useful in grouping clones in the germplasm collections it was found that *S. sinense* had the lowest variation in most of the four sugar traits tested. Cluster analysis revealed that *S. sinense* had the smallest dispersion into three clusters. The majority of the clones (27) had very low glucose and fructose with moderate sucrose content. The *S. barberi* accessions formed six clusters, the majority (18) of the clones were again low in glucose and fructose and moderated in sucrose. Nine accessions had low sucrose and moderately high glucose and fructose and two accessions had very high glucose and fructose and moderate sucrose (Tai and Miller 2002).

A study by Brown et al. (2002) evaluated 30 accessions, each of *S. sinense*, *S. barberi*, and *S. robustum* germplasm, for eight agronomic characters of breeding interest in two environments, and estimated the extent of genetic diversity within and between the *Saccharum* species. The authors found a significant amount of genetic variation, as measured by the percent genetic coefficient of variability (GCV) statistic (Milligan et al. 1990), both within and between species for nearly all characters. Principal component analysis of the data clustered the *S. sinense* and *S. barberi* clones together and the *S. robustum* clones formed a separate cluster. Among the *S. robustum* clones, six accessions were outliers from the rest of the population and the authors speculate that these clones are most likely hybrids between *S. officinarum* and *S. spontaneum* as they cluster principally with the *S. sinense* and *S. barberi* cluster. The authors conclude that the results of this diversity study are compatible with the hypothesis that *S. barberi* and *S. sinense* were derived from interspecific hybridization between *S. officinarum* and *S. spontaneum* (Irvine 1999; D'Hont et al. 2002).

3.3 Genotype-based Diversity Analysis

Recently, molecular markers have been widely used for the characterization of germplasm within species of the *Saccharum* complex and are said to provide a more direct measure of genetic diversity. A number of studies have used markers, such as restriction fragment length polymorphism (RFLP; Glaszmann et al. 1990; Burnquist et al. 1992), random amplified polymorphic DNA (RAPD; Nair et al. 2002), amplified fragment length polymorphism (AFLP; Aitken et al. 2006; Selvi et al. 2006), microsatellites or simple sequence repeat (SSR; Cai et al. 2005a), inter-*Alu* PCR (Alix et al. 1999), and targeted region amplified polymorphism (TRAP; Alwala et al. 2006) for the analysis of phylogeny, inter-species relationships and genetic diversity among the *Saccharum* species, related genera and their hybrids. The key driver in the use of a particular marker system to measure genetic diversity is the ability to measure variation across the entire genome. Tivang et al. (1994) stated that the use of large numbers of polymorphic markers or bands, which are uniformly distributed over the genome, will provide an increasingly more precise estimate of genetic relationships and will reduce the variance estimation of genetic relationship due to over- or under-sampling, of certain regions of the genome. For a large complex genome such as sugarcane, without prior sequence information AFLP offers the best tool for measuring genetic variability across the genome.

Since the original classification of *S. officinarum* by Linnaeus in 1753, the characterization of the *Saccharum* spp. based on morphology, chromosome number and geographic distribution has been controversial (Daniels and Roach 1987; Irvine 1999). Reviews by Grivet et al. (2004, 2006) have used the recent molecular data to help determine the domestication and early evolution of sugarcane to resolve this controversy. Glaszmann et al. (1990) presented some of the earliest molecular evidence based on restriction fragment patterns of ribosomal DNA that enabled the separation of accessions of *S. spontaneum*, which showed the widest within-species variation, from accessions of four other taxa often afforded species status: *S. robustum, S. officinarum, S. barberi* and *S. sinense*. More recently, *S. spontaneum* has been shown to be strongly differentiated and show greater diversity than among other members of the *Saccharum* spp. by nuclear RFLP (Burnquist et al. 1992; Lu et al. 1994a), cytoplasmic RFLP (D'Hont et al. 1993), chloroplast RFLP (Sobral et al. 1994), nuclear AFLP (Selvi et al. 2005), inter-*Alu* PCR (Alix et al. 1999), SSR (Cordeiro et al. 2003; Cai et al. 2005a) and RAPD analysis (Nair et al. 1999). This diversity within *S. spontaneum* is highlighted by the characterization of variable satellite DNA within *S. spontaneum* clones from different geographic regions (Alix et al. 1998). The authors found that clones from New Guinea and Molokai (Indonesia) appear to be closer to *S. officinarum* and *S. robustum* than other *S. spontaneum* forms. This observation was confirmed by a cytogenetic study that showed when these clones were crossed with *S. officinarum* they showed a large proportion of $(n + n)$ transmission, whereas other *S. spontaneum* forms usually show a large proportion of $(2n + n)$ transmission (Sreenivasan et al. 1987).

However, most of these genetic diversity studies on *S. spontaneum* have focused on inter-species relationships rather than intraspecific diversity, which have been limited (Chen et al. 2001; Fan et al. 2001). Recently, a study by Sheji et al. (2006) analyzed the genetic diversity of 40 *S. spontaneum* clones from four different geographical areas of India using 20 random, two ISSR and two telomere primers. Of the 432 bands generated, 364 bands were polymorphic, revealing a high degree of polymorphism (84.25%) among the accessions studied. Jaccard's similarity coefficient was used to calculate genetic similarity of the 40 genotypes, which was then used to cluster the genotypes using the UPGMA method. The pairwise genetic distance among the 40 accessions ranged from 29.8 to 60.0 with a mean genetic distance of 48.9%. The moderate level of genetic diversity detected in this study was comparable with a previous report of 51% (Burnquist et al. 1992) but was higher than detected for the Chinese *S. spontaneum* in studies by Chen et al. (2001) and Fan et al. (2001). However, it was lower than was reported by Lu et al. (1994a) which detected a relatively higher genetic diversity of 69% in *S. spontaneum* based on RFLP markers. This higher degree of genetic diversity was also supported by a study of Pan et al. (2004) that assessed genetic variability in a collection of 33 *S. spontaneum* clones using 17 RAPD primers. This discrepancy is probably due to the broader sampling of *S. spontaneum* clones representing five countries and a wider range of cytotypes $(2n = 48–124)$ as reported in the study by Lu et al. (1994a).

The cultivated sugarcane species *S. officinarum* is thought to have evolved from the wild sugarcane species, *S. robustum* (Brandes 1958). From its center of origin in New Guinea, clones of *S. officinarum* are believed to have been first cultivated only for chewing in SE Asia and the Pacific Islands before being dispersed by humans all over the inter-tropics between 1500 and 1000 BC (Daniels and Roach 1987). A number of molecular studies have confirmed that these two species are the most closely related among the six *Saccharum* species studied (Glaszmann et al. 1990, 2003; Besse et al. 1997). D'Hont et al. (1993) detected a single RFLP haplotype of the mitochondrial genome among 18 *S. officinarum* clones and 15 of 17 *S. robustum* clones. The conservation of this genetic haplotype between the two *Saccharum* species is in contrast to the six haplotypes revealed in a collection of *S. spontaneum* individuals sampled over a large geographic area in the same study. Furthermore, this has been supported by RFLP analysis of single copy DNA, which showed that the average similarity between *S. officinarum* and *S. robustum* is about the same as the average similarity between two *S. robustum* clones (Lu et al. 1994a), and by microsatellite (Selvi et al. 2003) and RAPD markers (Nair et al. 1999). Cytogenetic mapping of the ribosomal RNAs 45S and 5S by fluorescent in situ hybridization (FISH) established a basic chromosome number of $x = 10$ for both *S. officinarum* and *S. robustum* (D'Hont et al. 1996, 1998). In terms of genetic diversity within the *Saccharum* genus, numerous studies have shown that *S. officinarum* appears to be the least variable species (Burnquist et al. 1992; D'Hont et al. 1993; Lu et al. 1994a; Alix et al. 1999; Selvi et al. 2003; Alwala et al. 2006). Selection pressure during domestication for high sucrose types combined with the vegetative mode of propagation has been suggested as the reason for the low diversity within the *S. officinarum* species (Sobral et al. 1994). However, most of these studies used a relatively low number of *S. officinarum* accessions to determine genetic diversity. A larger study by Jannoo et al. (1999b) on a set of 53 *S. officinarum* clones suggests that a considerable amount of diversity does in fact exist in this species. Moreover, this was confirmed by a more detailed study using AFLP to assess the extent of genetic diversity of 270 *S. officinarum* clones (Aitken et al. 2006). The *S. officinarum* clones from this study were obtained from a collection that contained clones from all the major regions where *S. officinarum* is grown and included; Papua New Guinea, Indonesia, and from the Pacific Islands of Fiji, New Caledonia and Hawaii. Using five AFLP markers, 538 polymorphic bands were generated. The genetic similarity between the *S. officinarum* clones was calculated using Jaccard's coefficient and gave rise to a mean genetic similarity of 59%, significantly higher than previous diversity studies. This is in agreement with the considerable morphological diversity observed within *S. officinarum* for characters such as stalk thickness and leaf width (Artschwager and Brandes 1958) as well as sugar content (Roach 1965). The authors also found that *S. officinarum* clones from New Guinea displayed a greater diversity than *S. officinarum* clones from other regions. This is in agreement with the hypothesis that New Guinea is the center of origin of this species. Furthermore, it was found that seven clones classified as *S. officinarum* in this study were more likely to be hybrids as they had chromosome numbers greater than the expected 80 and contained a number of markers that occurred with high frequency in the cultivar collection. This demonstrates that in the interpretation of data from any diversity study, there is a need to ensure the accurate classification of clones into the appropriate *Saccharum* group.

   S. barberi and *S. sinense* cultivars are believed to have originated from an interspecific hybridization between *S. spontaneum* on one side and *S. officinarum* or *S. robustum* on the other, based on RFLP with low copy nuclear DNA (Glaszmann et al. 1990; Burnquist et al. 1992; Lu et al. 1994a). However, the results of Brown et al. (2002) from a principal component analysis of eight agronomic traits place *S. barberi* and *S. sinense* in clusters aside one another and apart from *S. robustum*. Recently, a study by Brown et al. (2007) used 15 microsatellites to generate 498 polymorphic bands to analyze the genetic diversity within and among 30 clones of *S. barberi, S. officinarum, S. robustum, S. sinense* and *S. spontaneum*. Using principal component analysis of the data, the authors found that *S. sinense* and *S. barberi* were more diverse than previously thought to be and clearly distinct from the *S. officinarum* and *S. spontaneum* groups. They suggested that *S. barberi* and *S. sinense* may have arisen through introgression between *S. officinarum, S. spontaneum,* and/or other genera such as *Erianthus* and *Miscanthus*. Although, the results of genomic in situ hybridization (GISH) study of six *S. barberi* clones clearly disputes this hypothesis and indicates that *S. officinarum* and *S. spontaneum* are the ancestors of *S. barberi* and *S. sinense* (D'Hont et al. 2002). Furthermore, mitochondrial hapotypes of *S. barberi* and *S. sinense* were shown to be the same as *S. officinarum*, indicating that this species was the maternal parent and wild *S. spontaneum* the paternal parent in the founding crosses (D'Hont et al. 1993). More recently, analyses using nuclear RAPD (Nair et al. 1999), SSR markers (Selvi et al. 2003), AFLP markers (Selvi et al. 2006) and sequence of chloroplast genomes (Takahashi et al. 2005) showed that *S. barberi* and *S. sinense* formed a single cluster when assessing genetic diversity. Although these two species are distinguishable by morphological characteristics (Daniels and Roach 1987) and chromosome number (Price 1957), the results of these studies suggest that these two species have an extremely close relationship. A review by D'Hont et al. (2008) suggests that based on molecular evidence, the *S. barberi* and *S. sinense* cultivars are most likely derived from a single founding early generation interspecific hybrid event between *S. officinarum* and *S. spontaneum*, followed by a set of somatic mutants leading to the different morpho-cytogenetic types, which occurred in different geographic regions of continental Asia.

Unlike other members of the *Saccharum* genus, there is limited data available for assessing the genetic diversity and tracing the origin of *S. edule*. This species is grown in subsistence gardens of New Guinea to Fiji for its edible, aborted inflorescence and its large, thick-stalked canes contain no sugar (D'Hont et al. 2008). Irvine (1999) suggested, based on chromosome numbers and morphological characteristics, that *S. edule* arose from intergeneric crosses between *S. officinarum* or *S. robustum* and related genera (e.g., *Miscanthus*) or was derived from *S. robustum*. Furthermore, the results of Takahashi et al. (2005) suggest that the cytoplasm of *S. edule* is similar to that of *S. officinarum* and *S. robustum*. D'Hont et al. (1993) determined the mitochondrial haplotype of a single clone and it was shown to be identical to *S. officinarum, S. barberi* and *S. sinense* cultivars and most of the *S. robustum*. The chloroplast RFLP pattern of another *S. edule* clone led to a similar conclusion (Sobral et al. 1994). D'Hont et al. (2008) speculated that these sparse data support the hypothesis that *S. edule* corresponds to a series of mutant clones, which have been derived from *S. robustum* and were preserved by humans.

Sugarcane cultivars have a relatively high level of genetic diversity probably due to their heterozygous nature and high polyploidy. It has been widely accepted that the majority of genetic variability among modern sugarcane cultivars is largely due to the introgression of the highly polymorphic *S. spontaneum* genome (D'Hont et al. 1996). Molecular analysis of cultivated sugarcane germplasm has been fairly limited and in most cases restricted to small number of accessions (Lu et al. 1994b). Initial analysis on a population of 40 sugarcane cultivars was carried out using 22 maize low copy DNA clones located on different regions of the 10 maize chromosomes. A total of 425 fragments of which 411 were polymorphic were scored and the data analyzed using the Dice formula to generate genetic similarities and factor analysis of correspondence. The cultivars had an average of 7.28 fragments per probe/enzyme combination and an average genetic similarity of 0.61. These results indicated that modern sugarcane cultivars are highly heterozygous with multiple alleles at each locus. Cultivars from the same breeding program were dispersed throughout a dendrogram. Only the two varieties from Brazil appeared to be closely associated. This lack of preferential similarity between cultivars from the same breeding program is not surprising because of the tradition of profuse exchange of parental material between sugarcane breeding stations (Lu et al. 1994b). In contrast to the Lu et al. (1994b) study, a set of 20 sugarcane varieties from a South African Sugarcane breeding program were analyzed with random amplified polymorphic DNA (RAPD) markers. From 15 RAPD primers 160 markers were generated. Genetic distances assessed following Nei and Li (1979) formula ranged from 71.82% to 89.73 % similarity (Harvey et al. 1994). The authors concluded that there was limited genetic diversity within this group of cultivars. In a larger study carried out by Jannoo et al. (1999a), 109 modern cultivars, most of which were bred in Barbados and Mauritius, were subjected to RFLP analysis using 12 sugarcane probes, which mapped to all the homologous groups of sugarcane. Using a total of 386 polymorphic fragments, the Dice index (Nei and Li 1979) was used to calculate genetic similarity between clones and generate a dendrogram of the data, multivariate analysis was also carried out. The structure revealed by the multivariate analysis confirmed the dendrogram and that the Mauritian cultivars formed one group and the Barbados cultivars another with the Reunion cultivars clustering with the Mauritian cultivars. Australian cultivars were clustered with the Barbadian cultivars and cultivars from Brazil and South Africa were in an intermediate position. Among the first interspecific hybrids that were included in the study, the four clones from

India were co-located with the Mauritian cultivars. This division between the cultivars that originated from Mauritius and Barbados were essentially due to *S. spontaneum* alleles present in Mauritian cultivars and absent in Barbadian cultivars probably due to the regular use of early generation interspecific hybrids in the breeding program employed in Mauritius.

In a study by Lima et al. (2002), amplified fragment length polymorphism (AFLP) and pedigree data were used to investigate the genetic relationship in a group of 79 cultivars from the Brazilian breeding program. The objective of this study was to both assess the level of genetic similarity (GS) among the sugarcane cultivars and to investigate the correlation between the AFLP-based GS and the coefficient of parentage (f)-estimates generated for cultivars. Twenty-one AFLP primer combinations were used to generate 1,121 polymorphic markers. GS was determined using Jaccard's similarity coefficient and a dendrogram constructed using an unweighted pair-group method using arithmetic average (UPGMA). AFLP-based GS ranged from 0.49 to 0.89 whereas f ranged from 0 to 0.503. Cluster analysis divided the cultivars into related subgroups with cultivars obtained from the same cross grouped together, suggesting that there is important genetic relationship among the cultivars. The complex structure of the dendogram with low diversity across the cultivars from different breeding programs is in agreement with the cultivar ploidy and its high heterozygosity, which allows the maintenance, in the genotypes, of a great number of alleles from ancestors incorporated in the initial interspecific crossings. AFLP-based GS and f were significantly correlated but with a moderate to low value ($r = 0.42$, P <0.001). The significance of this r value suggests that the AFLP data may help to more accurately quantify the degree of relationship among the sugarcane cultivars then the coefficient of parentage (Lima et al. 2002). Selvi et al. (2005) carried out a similar study on 28 tropical and subtropical Indian cultivars. Twelve AFLP primer combinations were used to generate 989 markers, which were used to calculate genetic dissimilarities between cultivars using the Sokal and Michener index. The resulting matrix was then used to construct a phenogram with the Neighbor Joining method. The estimated genetic distance between the cultivars ranged from 0.17–0.48. The cluster analysis separated the majority of the cultivars into groups based on tropical and subtropical classification. This major clustering was due to the subtropical cultivars having higher numbers of *S. spontaneum* specific markers. The clustering also reflected the pedigree relationships between the cultivars, with cultivars with parents in common clustering together. In a larger study of 151 Australian cultivars or important parents in the Australian breeding program, five AFLP primer pairs were used to generate 614 polymorphic markers (Aitken et al. 2006). Jaccard's similarity coefficient was used to calculate genetic similarities, which averaged 0.592 between all the cultivars. The cultivars clustered as expected based on pedigree and reflected the major contribution of QN66-2008 and NCo310 to Australian sugarcane cultivars, which divided the cultivars into two main groups. This variation appears to be due in a large part to the introgression of *S. spontaneum* in the cultivars. A similar observation was noted by other studies (Jannoo et al. 1999a; Selvi et al. 2005) that even though only a small number of *S. officinarum* clones have been used in sugarcane breeding to date, more than 80% of the markers found in the sample of clones representing this species were also found in the cultivars. A small study was carried out using nine polymorphic maize microsatellite markers, which generated a total of 95 markers across 30 Indian sugarcane cultivars (Selvi et al. 2003). These markers revealed a broad range of pair-wise similarity values from 0.32–0.83. Cluster analysis revealed a low level of correlation between the calculated genetic similarity and pedigree with progeny generated from the same cross in different clusters. This is probably due to the low level of marker coverage with only nine SSRs being scored.

3.4 Relationship with Other Cultivated Species and Wild Relatives

Mukherjee (1957) introduced the term "*Saccharum* complex" to describe the grouping of the species *Saccharum, Erianthus* sect. *Ripidium* ($2n$ = 20, 30, 40 and 60), *Miscanthus* sect. *Diandra* Keng ($2n$ = 38, 40, 76) *Sclerostachya* (Hack.) A. Camus ($2n$ = 30) and *Narenga* Bor ($2n$ = 30) together. However, Sobral et al. (1994) showed that while the genera *Miscanthus, Narenga, Sclerostachya* and *Saccharum* form a closely related group sharing a common ancestor with respect to their chloroplast genome, the *Erianthus* species, were found to have significantly different chloroplast genomes. Similarly, *Erianthus* and *Saccharum* spp. were reported to show large intergeneric distances by RFLP analyses revealed by nuclear low-copy probes (Burnquist et al. 1992; Besse et al. 1997; Coto et al. 2002). A further distinction for the *Erianthus* spp. is that as opposed to *Erianthus* sect. *Ripidium* (including seven Old World species), the rest of the genus *Erianthus* (New World species) is not considered to belong to the "*Saccharum* complex" (Grassl 1972). This classification however is still under review and variation within the ribosomal DNA of 62 *Erianthus* Michx. clones suggests that some of the New World species, such as *E. rufipilus* (Steud.) Griseb. and *E. longisetosus* Andersson, are more similar to the Old World *Erianthus* sect. *Ripidium* (Besse et al. 1996). Also, on the basis of DNA sequences from ITS nuclear ribosomal DNA the *Saccharum* spp. were found to be more closely related to *Miscanthus* spp. than other members of the "*Saccharum* complex" (Hodkinson et al. 2002).

It has been hypothesized that the genera other than *Saccharum*, particularly *Erianthus, Miscanthus, Sclerostachya* and *Narenga* have contributed to the emergence of sugarcane (reviewed in Daniels and Roach 1987). Recent molecular evidence is contrary to these claims. Analysis of isozyme, nuclear and cytoplasmic RFLP data (Glaszmann et al. 1989, 1990; Burnquist et al. 1992, Lu et al. 1994b, D'Hont et al. 1993, 1995; Besse et al. 1998), AFLP and simple sequence repeat (SSR) data (Selvi et al. 2003; Cai et al. 2005a) and sequence data (Pan et al. 2000; Hodkinson et al. 2002) have revealed that *Saccharum, Erianthus* and *Miscanthus* are distinct from each other. However, these results were established by comparison with very few accessions from the genera *Erianthus* and *Miscanthus* and recent work by Brown et al. (2007) suggest that these genera may have contributed to the genome of *S. barberi* and *S. sinense*. Recently, an extensive study by Cai et al. (2005a) investigated the extent of genetic diversity within species of the *Saccharum* complex using approximately 200 SSR and AFLP markers to assess 69 clones representing 17 species from the five genera. In all cases, principal component analysis of both the SSR and AFLP data clustered the *Saccharum* species together and away from the non-*Saccharum* species. In general, three clusters were formed, one containing most of the *Erianthus* species (group C), one the *Saccharum* species (group A) and the third group (Group B) placed between the other two groups contained two *Miscanthus* species, a *Narenga* species, *E. fulvus* and *E. rockii* (Fig. 3-2). In addition, Alix et al. (1998) hybridized clones from *Miscanthus* and *Erianthus*, containing repeated species-specific sequences with multiple dispersed loci in the genome, to the DNA of representatives of traditional cultivars and wild *Saccharum*. In all individuals tested, no trace of the *Miscanthus* or *Erianthus* specific sequences could be found (Alix et al. 1999). Grivet et al. (2006) suggested that these data support the view that the genus *Saccharum* is a well-defined lineage that has diverged over a long period of evolution from the lineages leading to the *Erianthus* and *Miscanthus* genera.

Due to the limited genetic base of modern sugarcane cultivars, sugarcane breeders have been looking to introgress genes from wild species to increase the genetic variability in sugarcane. *Erianthus* spp. have generated a lot of interest due largely to desirable phenotypic characters in this genus such as vigor, drought tolerance, waterlogging tolerance, and disease resistance. To date, only a few intergeneric hybrids have been reported and include a hybrid between *Saccharum* spp. and *Miscanthus* spp. (Alix et al. 1999) and several hybrids between *S. officinarum* and *E. arundinaceus* (D'Hont et al. 1995; Besse et al. 1997; Piperidis et al. 2000). However, most of these hybrids display unusual morphological characteristics and do not appear to be fertile. Molecular analyses have shown that *E. arundinaceus* is genetically quite distinct from the genus *Saccharum* (Alix et al. 1998, 1999; Nair et al. 1999) and it is speculated that this could explain the existence of genome incompatibility between the two species (Piperidis et al. 2000). Despite this, fertile hybrids have been developed between *S. officinarum* and *E. arundinaceus* and reported in Cai et al. (2005b). These fertile F_1 progeny were then

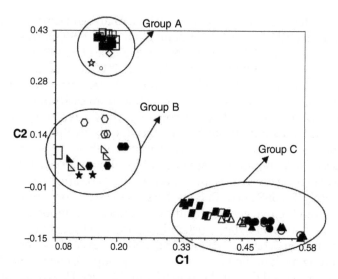

Figure 3-2 Figure taken from Cai et al. (2005a) showing principal component analysis of AFLP data generated from 17 species of 5 genera of the *Saccharum* complex. The first two components of the PCA analysis explained 20.4% and 14.3% of the total variation and three major clusters were observed (groups A, B and C).

Abbreviations

(●) = variety, (△) = *S. officinarum*, (■) = *S. spontaneum*, (◨) = *S. robustum*, (O) = *S. barberi*, (▲) = *S. sinense*, (■) = *E. arundinaceus* (Indonesia), (□) = *E. arundinaceus* (China), (▭) =*E.Procerus* (●) = *E. rockii*, (★) = *E. fulvus*, (☆) = *E. elephantinum*, (°) = *E. ravenae*, (◆) = *E. bengalense*, (◇) = *E. sarpet*, (◸) = *M. sinensis*, (◣) = *M. floridulus*, (o) = *N. porphyrocoma*, (□) = *P. schumach*.

backcrossed to a cultivar generating the first BC_1 population from introgression with *E. arundinaceus*. The crossing data indicated that fertile hybrids could only be generated between *E. arundinaceus* and *S. officinarum* and not between *E. arundinaceus* and a sugarcane cultivar, which contains *S. spontaneum* germplasm. *E. rockii* and *Saccharum* hybrids have also been reported (Aitken et al. 2007), in this case hybrids could be generated with both *S. officinarum* and a sugarcane cultivar as the female parent. Although, using *S. officinarum* the frequency of hybrid progeny was 100% as compared to using the sugarcane hybrid as the female parent, which resulted in only 10% hybrid progeny. This difference in the frequency of hybrid progeny reflects the difference in crossing two separate genomes (*S. officinarum* x *E. rockii*) to crossing three (*S. officinarum/S. spontaneum* x *E. rockii*). The relative ease in which the *S. officinarum/ E. rockii* hybrids were generated as opposed to the *S. officinarum/ E. arundinaceus* hybrids is partially explained by their relatedness as seen in Figure 3-2 where the *E. rockii* clones are more closely related to the *Saccharum* clones than the *E. arundinaceus* clones.

3.5 Relationship with Geographical Distribution

It has been suggested that the greater variability seen in *S. spontaneum* of all the *Saccharum* spp. could be related to the fact that this species has a high variability in regard to its chromosome number ($2n = 40$–128), wide geographic distribution and widely differing phenotype (Mukherjee 1957). *Saccharum spontaneum* has the widest geographic range of the *Saccharum* species, which extends across three geographic zones: 1) The East Zone, which includes South Pacific Islands, the Philippines, Taiwan, Japan, China, Vietnam, Thailand, Malaysia and Myanmar; 2) the Central Zone, which includes India, Nepal, Bangladesh, Sri Lanka, Pakistan, Turkmenistan, Afghanistan, Iran and Middle East; and 3) the West Zone, which includes Egypt, Sudan, Kenya, Uganda, Tanzania and other countries (Panje and Babu 1960; Daniels and Roach 1987). India is considered the center of origin as well as the center of diversity for this species (Mukherjee 1957). Many expeditions have collected vegetative clones of *S. spontaneum* from these zones and they are maintained in the world collection of sugarcane and related Grasses at the USDA-ARS National Germplasm Repository, Miami, Florida (Schnell et al. 1997) and the Sugarcane Breeding Institute, Coimbatore, India. Limited molecular diversity analysis has been carried out on *S. spontaneum* as a species, a recent paper by Sheji et al. (2006) analyzed 40 *S. spontaneum* clones from four regions of India using mostly random primers. Of the 40 clones, using UPGMA clustering, 39 clustered into four groups representing the four geographical regions within India from where the clones were collected . Another small study using fewer *S. spontaneum* clones and less random PCR markers demonstrated limited clustering from clones originating from different countries, although the limited number of clones and markers used could have had an effect on the results (Pan et al. 2004). In contrast to *S. spontaneum*, clustering of *S. officinarum* clones does not appear to relate to geographic distribution. A study by Jannoo et al. (1999b) showed that diversity between a collection of *S. officinarum* clones using RFLP markers did not appear to be related to geographical distribution. In a further study of a larger number of *S. officinarum* clones using AFLP markers Aitken et al. (2006) showed that there was again little clustering associated with geographic origin but clones from Hawaii and Fiji did form a separate group. The authors speculated that the absence of geographical differentiation observed is probably related to the profuse exchange of material from different subsistence gardens as well as genetic recombination. A cluster analysis revealed a very weak global structuring among sugarcane varieties (Lu at al. 1994b), again this is probably due to the exchange of basic germplasm around the world and the fact that many breeding programs have used the same ancestors in their breeding programs.

3.6 Extent of Genetic Diversity

In terms of chromosome structure, it appears that the five *Saccharum* species besides *S. spontaneum* may each contain more than 50% of the *S. officinarum* genome, due to the high chromosome number of *S. officinarum* and the ($2n + n$) transmission in *S. officinarum* x *S. spontaneum* crosses, a phenomenon known as female restitution (Bremer 1923; Price 1957). Nair et al. (1999) reported *S. sinense* to be the most diverse *Saccharum* species, followed by *S. spontaneum*, with *S. officinarum* and *S. barberi* the least diverse species based on RAPD markers. Lu et al. (1994a), on the other hand, found the genetic variability is highest in *S. spontaneum*, intermediate in *S. robustum*, and lowest in *S. officinarum*. The secondary species, *S. barberi* and *S. sinense*, displayed a combined nuclear DNA profile of *S. officinarum* and *S. spontaneum*, leading to the conclusion of an interspecific hybrid origin between these two species for *S. barberi* and *S. sinense*. Overall, the underlying genetic diversity for modern sugarcane cultivars is internationally known to be narrow. Less than 10% of the genetic variability in the "*Saccharum* Group" has been used in breeding programs. This narrow genetic base for sugarcane is likely to be one of the principal causes of the present slow rate of sugarcane breeding progress (Berding and Roach 1987). Furthermore, it has led sugarcane breeders to develop introgression programs using crosses of wild relatives of sugarcane to increase the level of genetic diversity available in sugarcane germplasm (discussed in the previous section). Although, there is molecular evidence that due to sugarcane being a polyploid, its vegetative propagation and is an outcrossing species it retains a higher level of diversity within each clone than previously thought (Aitken et al. 2006). In addition, this higher level of heterozygosity detected among modern sugarcane cultivars was revealed using RFLP markers (Lu et al. 1994b; Jannoo et al. 1999b).

On the other hand, since the first artificial crosses that gave rise to modern sugarcane cultivars there has been little opportunity to recombine founder chromosomes due to the small number of meiotic divisions that have occurred. Moreover, there were not many of these chromosomes as there were only a few founder individuals of *S. officinarum* and *S. spontaneum* used in the initial crosses. Consequently, a high level of linkage disequilibrium is still expected among modern cultivars (Grivet and Arruda 2001). This has been confirmed in a sample of Mauritian cultivars in which some chromosome haplotypes are significantly conserved over regions as long as 10 centiMorgans (Jannoo et al. 1999a). This is an important finding because it may offer original and powerful perspectives with which to identify and locate genes that are involved in traits of interest.

3.7 Conclusions

Due to the hybrid nature, heterozygosity and high polyploidy of sugarcane, there exists a large store of genetic variability within varieties. This has been effectively exploited by sugarcane breeders but there is evidence that genetic improvement of sugarcane is falling behind that of other major crops and over the last 30 years has reached a plateau (Roach 1989). The genetic base of modern sugarcanes is limited due to relatively few wild *S. spontaneum* and *S. officinarum* clones being used in the initial development of hybrid sugarcane (Arceneaux 1967). As indicated in this chapter, wild germplasm contains a large amount of genetic diversity. Incorporation of more of this diversity in existing sugarcane breeding programs will not guarantee the increase in the rate of sugarcane improvement but it will broaden the genetic base of sugarcane in breeding programs. This may provide wider and more durable disease resistance as well as increasing biomass or other traits of interest. More recently, the narrowing of the genetic base of sugarcane is being addressed in introgression programs using additional wild *S. spontaneum* clones from various germplasm collections (Miller et al. 2005; Wang et al. 2008). The larger question is how to identify the material to incorporate from the huge reservoir of genetic diversity now available. Most molecular diversity analysis has been done on only a limited number of clones and thus provides limited information. From larger studies of *S. officinarum* (Jannoo et al. 1999b; Aitken et al. 2006) we know that most of the diversity in these collections has already been exploited by sugarcane breeders. There are large collections of *S. spontaneum* clones, which have not been analyzed in any great depth and these remain to be characterized and then if they contain important traits of interest to sugarcane breeders incorporated into breeding programs. Similarly, knowledge of the extent of genetic diversity of the *Erianthus* species is limited and the recent generation of hybrids between *S. officinarum* (Cai et al. 2005b; Aitken et al. 2007) provides sugarcane breeders with additional resources to expand the genetic base for the development of improved sugarcane cultivars with traits of high interest.

While there have been a vast number of genetic diversity studies done in sugarcane as detailed in this chapter, there is little evidence that this information has been practically applied in sugarcane breeding programs. However, the accurate assessment of the levels and patterns of genetic diversity can be invaluable in crop breeding for a number of applications such as identifying diverse parental combinations to create segregating progenies with maximum genetic variability for further selection (Sheji et al. 2006), and introgressing desirable genes from diverse germplasm into the available genetic base (Wang et al. 2008). Another important use of genetic diversity is for determining the likelihood of spontaneous transfer of a transgene in a natural setting. Recently, molecular markers were used to identify the risks of transgene escape from sugarcane crops to related species (Bonnett et al. 2008). Using a combination of AFLP markers and five SSR markers, genetic diversity was determined for 79 *S. spontaneum* clones sampled in five different locations in North Queensland. The authors found that plants sampled in the majority of areas seem to be mostly establishing through vegetative spread rather than from seed. The preliminary data from this study can now be used to lead to the development of protocols for the identification of potential and actual hazards, as well as the likelihood of exposure to such hazards, for the release of genetically modified (GM) sugarcane.

References

Aitken KS, Li J-C, Jackson PJ, Piperidis G, McIntyre CL (2006) AFLP analysis of genetic diversity within *Saccharum officinarum* and comparison to sugarcane cultivars. Aust J Agri Res 57: 1167–1184.

Aitken KS, Li J, Wang L, Qing C, Fan YH, Jackson PA (2007) Introgression of *Erianthus rockii* into sugarcane and verification of intergeneric hybrids using molecular markers. Genet Resour Crop Evol 54: 1395–1405.

Albertson P, Aitken KS, Grof CPL, Jackson P, McIntyre CL (2001) Introgression of *S. officinarum*-A biochemical and molecular marker approach to improve CCS. Proc Int Sugar Cane Technol 24: 567–572.

Alix K, Baurens FC, Paulet F, Glaszmann J-C, D'Hont A (1998) Isolation and characterisation of a satellite DNA family in the *Saccharum* complex. Genome 41: 854–864.

Alix K, Paulet F, Glaszmann J-C, D'Hont A (1999) Inter-*Alu* like species-specific sequences in the *Saccharum* complex. Theor Appl Genet 99: 962–968.

Alphonse Amalraj V, Balakrishnan R, William Jebadhas A, Balasundaram N (2006) Constituting a core collection of *Saccharum spontaneum* L. and comparison of three stratified random sampling procedures. Genet Resour Crop Evol 53: 1563–1572.

Alwala S, Suman A, Arro JA, Veremis JC, Kimbeng CA (2006) Target region amplification polymorphism (TRAP) for assessing genetic diversity in sugarcane germplasm collections. Crop Sci 46: 448–455.

Arceneaux G (1967) Cultivated sugarcanes of the world and their botanical derivation. Proc Int Soc Sugar Cane Technol 12: 844–854.

Artschwager E, Brandes EW (1958) Sugarcane (*Saccharum officinarum*), Origin, Characteristics and Descriptions of Representative Clone. USDA Agriculture Handbook no 122. USDA, Washington, DC, USA.

Balakrishnan R, Nair NV (2003) Strategies for developing core collections of sugarcane (*Saccharum officinarum* L.) germplasm—comparison of sampling from diversity groups constituted by three different methods. Plant Genet Resour Newsl 134: 33–41.

Balakrishnan R, Nair NV, Sreenivasan TV (2000) A method for establishing a core collection of *Saccharum officinarum* L. germplasm based on quantitative-morphological data. Genet Resour Crop Evol 47: 1–9.

Balasundaram N, Nair NV, Somarajan KG (1980) A note on the world collection of sugarcane germplasm at Cannanore. Sugarcane Breed Newsl 43: 39–40.

Beaumont MA, Ibrahim KM, Boursot P, Bruford MW (1998) Measuring genetic distance. In: A Karp (ed) Molecular Tools for Screening Biodiversity. Chapman and Hall, London, UK, pp 315–325.

Berding N, Roach BT (1987) Germplasm collection, maintenance and use. In: DJ Heinz (ed) Sugarcane Improvement Through Breeding. Elsevier, Amsterdam, The Netherlands, pp 143–210.

Besse P, McIntyre CL, Berding N (1996) Ribosomal DNA variation in *Erianthus*, a wild sugarcane relative (Andropogoneae-Saccharinae). Theor Appl Genet 92: 733–743.

Besse P, McIntyre CL, Burner DM, de Almeida CG (1997) Using genomic slot blot hybridization to assess intergeneric *Saccharum* x *Erianthus* hybrids (*Andropogonae-Saccharinae*). Genome 40: 428–432.

Besse P, Taylor G, Carrol B, Berding N, Burner D, McIntyre CL (1998) Assessment of genetic diversity in a sugarcane germplasm collection using an automated AFLP analysis. Genetica 104: 143–153.

Bonnett GD, Nowak E, Olivares-Villegas JJ, Berding N, Morgan T, Aitken KS (2008) Identifying the risks of transgene escape from sugarcane crops to related species, with particular reference to *Saccharum spontaneum* in Australia. Trop Plant Biol 1: 58–71.

Brandes EW (1956) Origin, dispersal and use in breeding of the Melanesian garden sugarcanes and their derivatives, *Saccharum officinarum* L. Proc Int Soc Sugar Cane Technol 9: 709–750.

Brandes EW (1958) Origin, classification and characteristics. In: E Artschwager , EW Brandes (eds) Sugarcane (*Saccharum officinarum* L.). USDA Agriculture Handbook no 122. USDA, Washington DC, USA, pp 1–35.

Bremer G (1923) A cytological investigation of some species and species-hybrids of the genus *Saccharum*. Genetica 5: 273–326.

Brown JS, Schnell RJ, Tai PYP, Miller JD (2002) Phenotypic evaluation of *Saccharum barberi*, *S. robustum*, and *S. sinense* germplasm from the Miami, FL, USA world collection. Sugar Cane Int 3–16.

Brown JS, Schnell RJ, Power EJ, Douglas SL, Kuhn DN (2007) Analysis of clonal germplasm from five *Saccharum* species: *S. barberi*, *S. robustum*, *S. officinarum*, *S. sinense* and *S. spontaneum*. A study of inter- and intra species relationships using microsatellite markers. Genet Resour Crop Evol 54: 627–648.

Burnquist WL, Sorrells ME, Tanksley SD (1992) Characterisation of genetic variability in *Saccharum* germplasm by means of Restriction Fragment Length Polymorphism (RFLP) analysis. Proc Int Soc Sugarcane Technol 21: 355–365.

Cai Q, Aitken KS, Fan YH, Piperidis G, Jackson P McIntyre CL (2005a) A preliminary assessment of the genetic relationship between *Erianthus rockii* and the "*Saccharum* complex" using SSR and AFLP markers. Plant Sci 169: 976–984.

Cai Q, Aitken KS, Deng HH, Chen XW, Fu C, Jackson PA McIntyre CL (2005b) Verification of the introgression of *E. arundinaceous* germplasm into sugarcane using molecular markers. Plant Breed 124: 322–328.

Chen H, Fan YH, Shi XW, Cai Q, Zhang M, Zhang YP (2001) Research on genetic diversity and systemic evolution in *Saccharum spontaneum* L. Acta Agron Sin 27(5): 645–652.

Cordeiro GM, Pan YB, Henry RJ (2003) Sugarcane microsatellites for the assessment of genetic diversity in sugarcane germplasm. Plant Sci 165: 181–189.

Coto O, Cornide MT, Calvo D, Canales E, D'Hont A, de Prada F (2002) Genetic diversity among wild sugarcane germplasm from Laos revealed with markers. Euphytica 123: 121–130.

Daniels J, Roach BT (1987) Taxonomy and evolution. In: DJ Heinz (ed) Sugarcane Improvement through Breeding. Elsevier Press, Amsterdam, The Netherlands, pp 7–84.

D'Hont A, Paulet F, Glaszmann J-C (1993) Cytoplasmic diversity in sugarcane revealed by heterologous probes. Sugar Cane 1: 12–15.

D'Hont A, Rao PS, Feldmann F, Grivet L, Islam-Faridi N, Taylor P, Glaszmann J-C (1995) Identification and characterisation of sugarcane intergeneric hybrids, *Saccharum officinarum* x *Erianthus arundinaceus*, with molecular markers and DNA in situ hybridisation. Theor Appl Genet 91: 320–326.

D'Hont A, Grivet L, Feldmann P, Rao P, Berding N, Glaszmann J-C (1996) Characterisation of the double genome structure of modern sugarcane cultivars (*Saccharum* spp.) by molecular cytogenetics. Mol Gen Genet 250: 405–413.

D' Hont A, Ison D, Alix K, Glaszmann J-C (1998) Determination of basic chromosome numbers in the genus *Saccharum* by physical mapping of ribosomal RNA genes. Genome 41: 221–225.

D'Hont A, Paulet F, Glaszmann J-C (2002) Oligoclonal interspecific origin of 'North Indian' and 'Chinese' sugarcanes. Chrom Res 10: 253–262.

D'Hont A, Souza GM, Menossi M, Vincentz M, Van-Sluys MA, Glaszmann J-C, Ulian E (2008) Sugarcane: a major source of sweetness, alcohol and bio-energy. In: PH Moore , R Ming (eds) Genomics of Tropical Crop Plants. Springer, New York, USA, pp 483–513.

Fan YH, Chen H, Shi XW, Xai Q, Zhang M Zhang YP (2001) RAPD analysis of *Saccharum spontaneum* from different ecospecific colonies in Yunnan. Acta Bot Yunnanica 23(3): 298–308.

Felsenstein J (1984) Distance methods for inferring phylogenies: A justification. Evolution 38: 16–24.

Glaszmann J-C, Fautret A, Noyer JL, Feldmann P, Lanaud C (1989) Biochemical genetic markers in sugarcane. Theor Appl Genet 79: 537–543.

Glaszmann J-C, Lu YH, Lanaud C (1990) Variation of nuclear ribosomal DNA in sugarcane. J Genet Breed 44: 191–198.

Glaszmann J-C, Jannoo N, Grivet L, D'Hont A (2003) Sugarcane. In: P Hamon , M Sejuin , X Perrier , JC Glaszmann (eds) Genetic Diversity of Tropical Plants. CIRAD, Montpellier, France and Science Publi, Enfield, NH, USA, p 337.

Grassl CO (1972) Taxonomy of *Saccharum* relatives: *Sclerostachya, Narenga* and *Erianthus*. Proc Int Soc Sugar Cane Technol 14: 240–248.

Grivet L, Arruda P (2001) Sugarcane genomics: depicting the complex genome of an important tropical crop. Curr Opin Plant Biol 5: 122–127.

Grivet L, Daniels C, Glaszmann J-C, D'Hont A (2004) A review of recent molecular genetics evidence for sugarcane evolution and domestication. Ethnobot Res Appl 2: 9–17.

Grivet L, Glaszmann JC, D'Hont A (2006) Molecular evidences for sugarcane evolution and domestication. In: T Motley , N Zerega , H Cross (eds) Darwin's Harvest. New Approaches to the Origins, Evolution, and Conservation of Crops. Columbia Univ Press, New York, USA, pp 49–66.

Harvey M, Huckett BI, Botha FC (1994) Use of the polymerase chain reaction (PCR) and Random Amplification of polymorphic DNAs (RAPDs) for the determination of genetic distances between 21 sugarcane varieties. Proc S Afr Sugar Technol Assoc 68: 36–40.

Hodkinson TR, Chase MW, Dolores Lledo M, Salamin N, Renvoize SA (2002) Phylogenetics of *Miscanthus, Saccharum* and related genera (*Saccharinae, Andropogoneae, Poaceae*) based on DNA sequences from ITS nuclear ribosomal DNA and plastid *trnL* intron and *trnL-F* intergenic spacers. J Plant Res 115: 381–392.

Irvine JE (1999) *Saccharum* species as horticultural classes. Theor Appl Genet 98: 186–194.

Jaccard P (1908) Nouvelles researches sur la distribution florale. Bull Soc Vaudoise Sci 37: 605–613.

Jannoo N, Grivet L, Dookun A, D'Hont A, Glaszmann J-C (1999a) Linkage disequilibrium among modern sugarcane cultivars. Theor Appl Genet 99: 1053–1060.

Jannoo N, Grivet L, Seguin M, Paulet F, Domaingue R, Rao PS, Dookun A, D'Hont A, Glaszmann J-C (1999b) Molecular investigation of the genetic base of sugarcane cultivars. Theor Appl Genet 99: 171–184.

Jeswiet J (1929) The development of selection and breeding of the sugarcane in Java. Proc Int Soc Sugar Cane Technol 3: 44–57.

Kandasami PA, Sreenivasan TV, Ramana Rao TC, Palanichami K, Natarajan BV, Alexander KC, Madhusudana Rao M, Mohan Raj D (1983) Catalogue on Sugarcane Genetic Resources. 1. *Saccharum Spontaneum* L. Sugarcane Breeding Institute, Indian Council of Agricultural Research, Coimbatore, Tamil Nadu, India.

Labate JA (2000) Software for population genetic analysis of molecular marker data. Crop Sci 40: 1521–1528.

Lima MLA, Garcia AAF, Oliveira KM, Matsuoka S, Arizono H, de Souza Jr CL, de Souza AP (2002) Analysis of genetic similarity detected by AFLP and coefficient of parentage among genotypes of sugarcane (*Saccharum* spp.). Theor Appl Genet 104: 30–38.

Lu YH, D'Hont A, Walker DIT, Rao PS, Feldmann P, Glaszmann J-C (1994a) Relationships among ancestral species of sugarcane revealed with RFLP using single copy maize nuclear probes. Euphytica 78: 7–18.

Lu YH, D'Hont A, Paulet F, Grivet L, Arnaud M, Glaszmann J-C (1994b) Molecular diversity and genome structure in modern sugarcane varieties. Euphytica 78: 217–226.

Miller JD (1982) The role of the U.S. world collection of sugarcane and related grasses in the movement and quarantine of varieties. Proc Inter-Am Sugar Cane Seminar 3: 218–221.

Miller JD, Tai PYP (1992) Use of plant introductions in sugarcane cultivar development. In: HL Shands, LE Wiesner (eds) Use of Plant Introductions in Cultivar Development. Part 2. CSSA Spl Pub no 20. CSSA, Madison, WI, USA.

Miller JD, Tai PY, Edme SJ, Comstock JC, Glaz BS, Gilbert RA (2005) Basic germplasm utilization in the sugarcane development program at Canal Point, FL, USA. Int Soc Sugar Cane Technol 2: 532–536.

Milligan SB, Gravois KA, Bischoff KP, Martin FA (1990) Crop effects on broad-sense heritabilities and genetic variances of sugarcane yield components. Crop Sci 30: 344–349.

Mohammadi SA, Prasanna BM (2003) Analysis of genetic diversity in crop plants—salient statistical tools and considerations. Crop Sci 43: 1235–1248.

Mukherjee SK (1957) Origin and distribution of *Saccharum*. Bot Gaz 119: 55–61.

Naidu KM, Sreenivasan TV (1987) Conservation of sugarcane germplasm. In: Copersucar Int Sugarcane Breed Wkshp, Copersucar, Sao Paulo, Brazil, pp 33–53.

Nair NV, Nair S, Sreenivasan TV, Mohan M (1999) Analysis of genetic diversity and phylogeny in *Saccharum* and related genera using RAPD markers. Genet Resour Crop Evol 46: 73–79.

Nair NV, Selvi A, Sreenivasan TV Pushpalatha KN (2002) Molecular diversity in Indian sugarcane cultivars as revealed by randomly amplified DNA polymorphisms. Euphytica 127: 219–225.

Nei M (1987) Molecular Evolutionary Genetics. Columbia Univ Press, New York, USA.

Nei M, Li W (1979) Mathematical model for studying genetic variation in terms of restriction endonucleases. Proc Natl Acad Sci USA 76: 427–434.

Pan YB, Burner DM, Legendre BL (2000) An assessment of the phylogenetic relationship among sugarcane and related taxa based on the nucleotide sequence of 5S rRNA intergenic spacers. Genetica 108: 285–295.

Pan YB, Burner DM, Legendre BL, Grisham MP White WH (2004) An assessment of the genetic diversity within a collection of *Saccharum* L. with RAPD-PCR. Genet Resour Crop Evol 51: 895–903.

Panje RR, Babu CN (1960) Studies of *Saccharum spontaneum* distribution and geographic association of chromosome numbers. Cytologia 25: 150–152.

Piperidis G, Christopher MJ, Carroll BJ, Berding N, D'Hont A (2000) Molecular contribution to selection of intergeneric hybrids between sugar cane and the wild species *Erianthus arundinaceus*. Genome 43: 1033–1037.

Pocovi MI, Collavino NG, Locatelli FM, Pacheco MG, Diaz D, Rios RD, Mariotti JA (2008) Assessing genetic variability of subtropical hybrid sugar cane (*Saccharum* spp.) materials using isozymes and AFLP. Sugar Cane Int 26(1): 6–11.

Price S (1957) Cytological studies in *Saccharum* and allied genera II. Chromosome numbers in interspecific hybrids. Bot Gaz 118: 146–159.

Ram B, Hemaprabha G (2005) Genetic divergence of sugar yield and its components in flowering type *S. officinarum* clones. Agri Sci Digest 25 (2): 118–120.

Roach BT (1989) Origin and improvement of the genetic base of sugarcane. Proc Aust Soc Sugar Cane Technol 12: 34–47.

Roach RT (1965) Sucrose of noble cane. Sugarcane Breed Newsl 15: 2.

Schnell RJ, Tao PYP, Miller JD (1997) History and current status of the World Collection of sugarcane and related grasses maintained at National Germplasm Repository, Miami, Florida. Sugar Cane 1: 15–17.

Selvi A, Nair NV, Balasundaram N, Mohapatra T (2003) Evaluation of maize microsatellite markers for genetic diversity analysis and fingerprinting in sugarcane. Genome 46: 394–403.

Selvi A, Nair NV, Noyer JL, Singh NK, Balasundaram N, Bansal KC, Koundal KR, Mohapatra T (2005) Genomic constitution and genetic relationship among the tropical and subtropical Indian sugarcane cultivars revealed by AFLP. Crop Sci 45: 1750–1757.

Selvi A, Nair NV, Noyer JL, Singh NK, Balasundaram N, Bansal KC, Koundal KR, Mohapatra T (2006) AFLP analysis of the phenetic organisation and genetic diversity in the sugarcane complex, *Saccharum* and *Erianthus*. Genet Resour Crop Evol 53: 831–842.

Sheji M, Nair NV, Chaturvedi PK, Selvi A (2006) Analysis of genetic diversity among *Saccharum spontaneum* L. from four geographical regions in India, using molecular markers. Genet Resour Crop Evol 53: 1221–1231.

Skinner JC, Hogarth DM, Wu KK (1987) Selection methods, criteria, and indices. In: DJ Heinz (ed) Sugarcane Improvement through Breeding. Elsevier Press, Amsterdam, The Netherlands, pp 409–453.

Sneath PHA, Sokal RR (1973) Numerical taxonomy. Freeman, San Francisco, USA.

Sobral BWS, Braga DPV, LaHood ES, Keim P (1994) Phylogenetic analysis of chloroplast restriction enzyme site mutations in the *Saccharinae* Griseb. subtribe of the *Andropogoneae* Dumort. tribe. Theor Appl Genet 87: 843–853.

Sokal RR, Michener CD (1958) A statistical method for evaluating systematic relationships. Univ Kansas Sci Bull 38: 1409–1438.

Sreenivasan TV, Nair NV (1991) Catalogue of Sugarcane Genetic Resources, vol 3: *S. officinarum*. Sugarcane Breeding Institute, Coimbatore, India.

Sreenivasan TV, Ahloowalia BS, Heinz DJ (1987) Cytogenetics. In: DJ Heinz (ed) Sugarcane Improvement through Breeding. Elsevier Press, Amsterdam, The Netherlands, pp 211–253.

Sreenivasan TV, Amalraj VA, Jebadhas AW (2001) Sugarcane Genetic Resources: *Saccharum spontaneum*, vol 2. Sugarcane Breeding Institute, Coimbatore, India.

Stevenson GC (1965) Genetics and Breeding of Sugar Cane. Longmans, London, UK.

Tai PYP, Miller JD (2002) Germplasm diversity among four sugarcane species for sugar composition. Crop Sci 42: 958–964.

Takahashi S, Furukawa T, Asano T, Terajima Y, Shimada H, Sugimoto A, Kadowaki K (2005) Very close relationship of the chloroplast genomes among *Saccharum* species. Theor Appl Genet 110: 1523–1529.

Tivang JG, Nienhuis J, Smith OS (1994) Estimation of sampling variance of molecular marker data using the bootstrap procedure. Theor Appl Genet 89: 259–264.

Wang L-P, Jackson PA, Lin X, Fan Y-H, Foreman JW, Chen X-K, Deng H-H, Fu C, Ma L, Aitken KS (2008) Evaluation of sugarcane x *Saccharum spontaneum* progeny for biomass composition and yield components. Crop Sci 48: 951–961.

4

Association Studies

Emma Huang,[1] Karen Aitken[2]* and Andrew George[1]

ABSTRACT

Association analysis has recently become a popular method to detect marker-trait associations via linkage disequilibrium (LD). The development of high-throughput marker technologies to allow whole-genome analysis has contributed greatly to the increased frequency of these studies. In this chapter, we discuss the issues, which arise in planning an association study in sugarcane. We summarize the history of association studies, define linkage disequilibrium and its relevance to association analysis, and describe other issues surrounding study design and analysis. We focus on those issues, which are directly relevant to sugarcane such as correctly accounting for population structure and the suitability of available software packages. Association analysis in sugarcane has thus far been limited, but potential exists for future studies given that the structure of its unique complex polyploid genome and the development of cultivars present novel challenges for the area.

Keywords: linkage disequilibrium, association mapping, population structure, genetic marker

[1]CSIRO Mathematical and Information Sciences, 306 Carmody Rd, St. Lucia, Queensland, 4070, Australia.
[2]CSIRO Plant Industry, 306 Carmody Rd, St. Lucia, Queensland, 4070, Australia and CRC Sugar Industry Innovation through Biotechnology, Level 5, John Hines Building, The University of Queensland, St. Lucia, QLD, 4072, Australia.
*Corresponding author: *Karen.Aitken@csiro.au*

4.1 Introduction

Association studies are becoming an increasingly popular strategy for unlocking the genetic machinery controlling complex traits. Traditionally, association studies were used to fine-map genes once linkage analysis techniques had identified chromosomal regions of interest. However, with the advent of high-throughput marker platforms and the resulting dense marker maps, whole-genome association studies are being routinely performed in humans and many animal species. In the last decade, association studies have gone from being purely a refinement tool to being a new and powerful gene discovery platform.

The rising popularity of association mapping stems from its strengths over rival linkage-based approaches. First, association studies better capture the allelic variation seen within a breeding population. Second, association studies make use of many generations of historical recombination to achieve finer mapping resolution of genomic regions of interest. Third, marker-trait associations identified among elite varieties with diverse genetic backgrounds are less likely to be background-specific and hence more widely applicable to breeding. Fourth, association studies can be performed with lower cost since the material being used may already be under routine phenotypic evaluation.

In sugarcane, association studies are in their infancy. This is largely due to a lack of good quality marker data. In humans and some livestock species, sequence data and dense marker maps are available; this has led to the creation of oligonucleotide-based single nucleotide polymorphism (SNP) arrays. These arrays have revolutionized genetic discovery by enabling data to be collected on millions of loci spanning a genome. However, the complex genetics of polyploid crops such as sugarcane means that SNP arrays have yet to be developed.

Instead, simpler microarray-based technologies are emerging as viable marker collection platforms in polyploid crop plants. Data are collected not on codominant markers as in SNP arrays but instead on dominant markers. For the first time, whole-genome association studies are becoming possible in highly polyploid crops.

In this chapter, we review many of the issues facing practitioners interested in conducting an association study in sugarcane. We begin by summarizing the history of association studies, following its early use in humans, then animals, and now plants. We then define linkage disequilibrium (LD), describe how it is measured, and review important LD studies in plants. Next, we discuss issues surrounding study design and analysis of association studies, focusing on issues pertaining to sugarcane. We discuss the importance of correctly accounting for population structure in an analysis followed by a description of available association analysis software. We conclude with a discussion of the future of association studies in sugarcane.

4.2 History of Association Mapping

Association mapping has its roots in human studies aimed at identifying specific genes underlying common diseases (Lander and Schork 1994). Its development was driven by the limitation of conventional linkage analysis, which typically cannot map genes to regions smaller than 10 to 20 cM due to a limited number of recombination events. Association mapping is capable of far greater precision by capitalizing on ancestral recombination. Box 4-1 compares these two methods of detecting gene-trait relationships. The novel aspects of association mapping have been reviewed by Pritchard and Przeworski (2001), Palmer and Cardon (2005) and Slatkin (2008).

Linkage analysis and association analysis have the same ultimate goal: to detect regions of the genome associated with a trait of interest. They are also based on the same fundamental principle of the co-inheritance of adjacent DNA variants. However, the two approaches have different implementations and consequently different strengths and weaknesses.

Linkage analysis focuses on detecting recent mutations by tracing their descent through families. There have been only a few generations of recombination since the mutation event (indicated by m), so there are large blocks of the genome, which will be co-inherited. This is advantageous in that it increases the power to detect the region. Conversely, it means that the location of the mutation cannot be determined more precisely than 10–20 cM.

Association analysis focuses on much older mutations, which have spread throughout a population. The increased number of generations of recombination means that much smaller regions are co-inherited with the original mutation. Thus a causal gene can be more precisely located by examining the trait-genotype relationship in the broader population. However, much larger sample sizes may be required to detect association since there is greater variation in the population.

The area of association mapping has changed radically since its inception. Because association mapping was originally suggested as a method to refine the genome regions detected in linkage analysis, it was given the name "fine-mapping". Most studies concentrated on a small, dense set of markers in candidate genes.

It was soon realized, however, that for complex traits influenced by many loci, studies involving only a few genes would not be able to explain more than a small fraction of the total phenotypic variation. The alternative is genome-wide association studies, which achieve a comprehensive scan of variants and with sufficient sample size have greater potential in the

Box 4-1 Comparison of (a) linkage analysis and (b) association analysis (Cardon and Bell 2001).

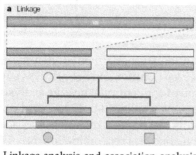

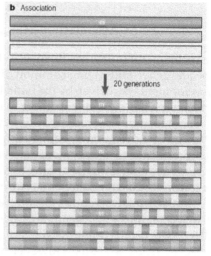

20 generations

Linkage analysis and association analysis have the same ultimate goal: to detect regions of the genome associated with a trait of interest. They are also based on the same fundamental principle of the co-inheritance of adjacent DNA variants. However, the two approaches have different implementations and consequently different strengths and weaknesses.

Linkage analysis focuses on detecting recent mutations by tracing their descent through families. There have been only a few generations of recombination since the mutation event (indicated by m), so there are large blocks of the genome, which will be co-inherited. This is advantageous in that it increases the power to detect the region. Conversely, it means that the location of the mutation cannot be determined more precisely than 10-20 cM.

Association analysis focuses on much older mutations, which have spread throughout a population. The increased number of generations of recombination means that much smaller regions are co-inherited with the original mutation. Thus a causal gene can be more precisely located by examining the trait-genotype relationship in the broader population. However, much larger sample sizes may be required to detect association since there is greater variation in the population.

Color image of this figure appears in the color plate section at the end of the book.

realm of complex traits like yield. This evolution in association studies has been encouraged by the increasing availability of genomic resources, which permit the inclusion of markers across the whole genome.

Improvements in sequencing and marker technology have had enormous impact on the scale of association studies. Ideally, markers should be densely distributed across the genome, and genotyping methods should be cheap and accurate. In humans, this is possible with single nucleotide polymorphisms (SNPs). These markers are also available for a number of highly sequenced plant genomes (Goff 2002; Clark et al. 2007). Decreases in sequencing costs across species are likely to lead to SNPs becoming the general marker of choice due to their low error rate and amenability to high-throughput analysis (Weber et al. 2007; Fusari et al. 2008). However, without genome sequence, SNP discovery can be a lengthy and costly process. As many crop plants do not yet have large amounts of sequence information available, simple sequence repeat (SSR) markers have been extensively used in mapping (Breseghello and Sorrells 2006b; Heffner et al. 2008; Jun et al. 2008). Existing marker technology for sugarcane association mapping is described in Box 4-2.

Box 4-2 Comparison of marker technology for association mapping in sugarcane.

A lack of sequence information has limited the development of marker technology in sugarcane. Thus far only SSR and amplified fragment length polymorphism (AFLP) markers have been used for published association mapping studies (Wei et al. 2006; Butterfield 2007). AFLP markers have the advantage over SSR markers for association mapping in that more markers can be generated per gel run, allowing for better coverage of the genome. This benefit is balanced by the fact that AFLP markers are dominant and hence provide less information than codominant SSRs. Both marker platforms are time-consuming and expensive compared to more automated approaches.

Diversity array technology (DArT) markers are a novel platform for genotyping in plants lacking sequence information (Jaccoud et al. 2001). DArT markers represent a high-throughput microarray hybridization-based technique that aims to simultaneously generate hundreds of polymorphic markers spread across the genome. The technique has been shown to be reproducible and has the added advantage that the same platform is used for both discovery and scoring of markers. This means that only one assay for genotyping must be developed. The markers generated by DArT are similar to AFLPs since the platforms use similar techniques to reduce the complexity of the target DNA. The advantage of the DArT technology arises from the elimination of most of the manual quality control required in marker scoring.

Sugarcane is unique in that it is the first autopolyploid species to be used to develop a DArT array. This array was recently completed and is being used to generate the data for a large scale association study in sugarcane (Jackson pers. com.). In the process of developing this array, it was found that the high level of polyploidy found in sugarcane cultivars directly affects the DArT marker scoring. The majority of markers identified using the DArT array can only be measured in terms of a continuous intensity rather than as dominant biallelic markers. Whether this can be converted to an estimation of copy number is not yet fully understood. However, even the set of dominance-scored markers is larger than would be achieved with a similar expenditure for AFLPs or SSRs. At this time, the DArT array which produces a relatively large number of markers (1,000–2,000) with good quality control measures appears to be the most appropriate marker system for association mapping in sugarcane. The first SNP markers are being developed in sugarcane but to date there are not enough to cover the genome. With the decrease in sequencing costs it is expected that many SNP markers will become available in the near future.

Progress in marker discovery is only one development, which has led to the proliferation of association mapping in recent years. Public databases of markers, insights into linkage disequilibrium, and inexpensive and accurate technologies for high-throughput genotyping have all contributed to the success of association mapping (International HapMap Consortium 2005). In humans, the foundation provided by these factors has brought genome-wide association studies to the level of sample sizes in the thousands being genotyped on 500,000 SNPs for a wide variety of diseases. Human association studies have had varying success but have generally proven their effectiveness in the identification of common SNP-based variants with modest to large effects on phenotype (McCarthy et al. 2008).

The growth of association mapping in humans has spurred similar progress in other species (Fig. 4-1). Genome sequences are now available for major domestic animal species and along with them, corresponding panels of approximately two million SNPs in size (Georges 2007). From these, SNP chips select an efficient subset for high-throughput genotyping which contains tens of thousands of SNPs. Although this resolution is inferior to that in humans, it may prove sufficient for animal populations. The level of inbreeding and selection in these populations permits a similar degree of fine-mapping with a reduced number of markers. Similar basic genetic knowledge for these organisms and humans means that many methods can be directly transferred between species.

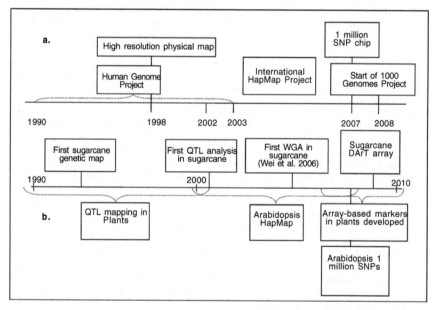

Figure 4-1 Timelines of major events in association mapping in sugarcane with reference to a) humans and b) plants.

However, there are core differences between human and animal studies which necessitate novel approaches similar to those required in plant studies. First, the structure of a population that has undergone generations of breeding does not resemble the structure arising from random mating. Second, novel genetic designs can be explored through controlled crosses to probe genetic diversity. Third, the aim of association mapping in animal populations may differ from that in humans. While there is interest in identifying specific genes, an additional motivation in quantitative trait loci (QTL) mapping is marker-assisted selection (MAS). For this, identifying a few markers with moderate to large effects may be of more interest than pinpointing the exact location of a causal gene. Hence association studies in plants and animals may not require as large sample sizes as in humans.

4.3 Association Mapping in Plants

The path of association mapping in plants, as in humans and animals, began with candidate gene studies. In one of the first association mapping studies in plants a candidate gene *dwarf8* (*d8*) was tested for association with flowering time in 92 inbred lines of maize (Thornsberry et al. 2001). Nine polymorphisms, which were identified, indicated that selection had occurred at these loci to produce earlier flowering maize. In another association study in rice the amylase content and alleles at the waxy locus were found to be associated (Olsen et al. 2006). More recently Harjes et al. (2008) used 288 maize lines to evaluate select members of gene families encoding biosynthetic enzymes of the carotenoid pathway. They showed that variation at the *lycopene epsilon cyclase* (*lcyE*) locus altered flux down α-carotene versus β-carotene branches of the carotenoid pathway.

Interest in whole-genome scans has increased as next-generation technologies have decreased genome sequencing costs. For some species, we are now at the point where whole-genome association studies can be carried out with sample sizes that enable robust inference. While this approach has enjoyed some success, it is important to learn from the lessons of human studies, where many initial discoveries have not stood up to validation (Palmer and Cardon 2005). Hence, associations should be treated as tentative until validated in target studies.

The first genome-wide association study in plants examined 95 accessions of *Arabidopsis*. The study genotyped SNPs across the genome and searched for associations with flowering time and pathogen resistance (Aranzana et al. 2005). Major genes for all phenotypes tested were identified, demonstrating the potential of genome-wide association mapping for this inbred species.

Similar studies have been performed in crop plants. In a more recent study 150 SSR markers were used to identify associations with seed protein content in soybean (Jun et al. 2008). A total of 96 soybean accessions were divided into two subpopulations and then tested for QTL. Of the 11 QTLs identified, nine mapped to regions in which linkage had previously been demonstrated; the other two were neither previously reported nor close to known maturity QTL. Larger studies have been capable of even greater precision. A study testing 8,590 SNPs for association with oleic acid content in 553 inbred maize lines detected a locus with major effects (Beló et al.

2008). The most likely candidate gene near this locus on chromosome 4 was identified as *fad2*. This represents the first genome-wide association study to report a putative gene for a quantitative trait in plants.

A newer type of design has been implemented in maize, the nested association mapping population (NAM; Yu et al. 2008). This design produces 200 inbred lines from each of 25 families with a single common reference parent. The size and structure of this population allows for the integration of both linkage and association analysis to dissect complex traits. However, it is unlikely to be appropriate or feasible in all species of plants.

4.3.1 Association Mapping in Polyploid Crops

The successes of association mapping in *Arabidopsis* and maize have provoked interest across the spectrum of plant species, however other species face obstacles to association mapping—many agricultural crop plants have polyploid genomes with complex genetics and large genome sizes. Autopolyploids, such as potato and sugarcane, are among the most challenging of these. Despite this, several association studies have so far been carried out in these crops.

In potato, studies of disease resistance have used targeted-gene association methods. Five markers, linked to known QTLs for resistance to late blight and plant maturity were used to screen a genebank collection of 600 potato cultivars and 114 accessions of 30 wild *Solanum* species (Gebhardt et al. 2004). Two markers specific to the major late blight resistance gene *R1* showed highly significant association to resistance and later plant maturity. These marker alleles were traced to an introgression from the wild species *S. demissum*. In a similar study, Simko et al. (2004) screened 139 North American cultivars with a microsatellite marker allele linked to a QTL for *Verticillium* wilt disease. The clone USDA 41956 was identified as possessing three copies of the resistance allele. This produces a high frequency of resistant offspring and contributes greatly to the resistance found in present commercial cultivars.

4.3.2 Association Mapping in Sugarcane

In sugarcane, there has been little targeted association mapping. McIntyre et al. (2005) used an association mapping approach to validate QTLs identified for pachymetra and brown rust resistance in a conventional cross. Markers were screened across a set of 154 elite clones to determine whether they would be useful in a broader genetic background. Six of the 13 markers linked to pachymetra resistance and seven of the 15 markers linked to rust resistance remained associated in the more diverse germplasm. These results held promise for the use of markers for selection among the broader sugarcane population.

Due to the limited information on candidate genes, association studies in sugarcane have focused on genome-wide approaches. Studies have attempted to select markers associated with disease resistance from a larger set of anonymous markers. Wei et al. (2006) searched for association between sugarcane diseases and 1,068 AFLP and 141 SSR markers. These markers were genotyped on a collection of 154 sugarcane clones consisting of important ancestors and cultivars. The analysis revealed that 59% of the phenotypic variation in smut ratings could be accounted for by 11 markers, 32% of the variation for leaf scald and pachymetra root rot rating by four markers and 26% of the variation for Fiji leaf gall by five markers.

A similar study by Butterfield (2007) also scanned the genome for markers associated with sugarcane disease resistance. Butterfield measured 77 sugarcane genotypes on 275 RFLP and 1,056 AFLP markers and looked for associations with resistance to sugarcane smut and to eldana stalk borer. Sixty-four markers were found to be significantly associated with smut rating and 115 markers with eldana rating. Of these, six markers explained 54% of the variation for smut resistance and 62% of the variation for eldana resistance. The number of associations detected is promising given the relatively small sample size. Larger genome-wide association studies are currently in progress, such as one by Jackson and colleagues (P.A. Jackson, pers. comm.) in a collection of 473 clones. Multiple quantitative traits are under study in this sample, including yield components, sugar content and disease resistance. The polyploid nature of sugarcane means that effects for these quantitative traits are likely to be small requiring larger sample sizes to detect them. The larger sample size has potential to result in more and replicable associations, which will prove useful in breeding programs.

4.4 LD Structure in the Genome

Linkage disequilibrium (LD) is defined as the nonrandom association of alleles at different loci (Flint-Garcia et al. 2003). In a large, randomly-mated population without selection, mutation or migration, polymorphic loci segregating independently will be in linkage equilibrium (Falconer and Mackey 1996). Considering a mutation occurring at a single locus in a population at linkage equilibrium at adjacent positions, the genome will undergo the same pressures of selection. Within a small region, recombination will be rare. Thus even after many generations, two adjacent loci will be highly correlated and the tight linkage will result in high LD. In contrast, two loci on separate chromosomes will experience different selection pressures and independent segregation and thus have lower LD (Flint-Garcia et al. 2003).

A variety of factors contribute to LD within the genome. The occurrence and pattern of LD is affected by population characteristics such as mating systems, population structure, genetic drift, admixture, directional selection, and population history (Gaut and Long 2003). LD patterns within species have been found to be variable among chromosomes and over distance (Breseghello and Sorrells 2006a; Kim et al. 2006). The most important factor for association mapping is disequilibrium due to physical linkage (Flint-Garcia et al. 2003).

The level of LD observed within a population is the key to finding an association between functional loci and markers that are physically linked. High levels of LD near a functional locus translates to correlation with nearby loci, and hence the genotype-phenotype association at the functional locus will be detectable indirectly through other loci. A lower level of LD implies a reduction in the strength of this indirect relationship between marker and phenotype, which means that association is much more difficult to detect. This is typical of diverse populations where there have been many generations of historical recombination. The structure of LD as described by the level in different regions of the genome can thus greatly affect the precision and power of association mapping. Box 4-3 describes measures of linkage disequilibrium.

Box 4-3 Measurement of linkage disequilibrium.

Several methods exist to assess the level of linkage disequilibrium (LD) in a population. LD can be estimated between two or more loci, but the multilocus measures are not in common usage (Weir 1996). Pairwise LD measures the tendency for alleles at two linked loci to be associated in the population. Define a haplotype as a combination of alleles at multiple loci that are transmitted together on the same chromosome. Then we can describe LD in terms of the deviation in haplotype frequency from that expected if the alleles were inherited independently. If p_A and p_B are the frequencies of the mutant alleles at the two loci, and p_{AB} is the frequency of the haplotype with both these alleles, then a measure of LD is

$$\delta = p_A p_B - p_{AB}.$$

The quantity δ will vary depending on the allele frequencies at the two loci, but it is constrained by the fact that all haplotypes must have non-negative probability. Consequently,

$$\delta_{max} = \begin{cases} \min[p_B(1-p_A), p_A(1-p_B)] & \text{if } \delta > 0 \\ \min[(1-p_B)(1-p_A), p_A p_B] & \text{if } \delta < 0 \end{cases}$$

This relationship can be used to standardize δ as $\delta' = |\delta|/\delta_{max}$ so that it takes a range of values between 0 and 1. Another widely used measure is r^2, which measures the correlation between the two loci. It also takes values between 0 and 1 and can be computed as

$$r^2 = \frac{\delta^2}{p_A p_B (1-p_A)(1-p_B)}$$

δ' is a common measure of historical recombination and provides the upper limit of r^2. r^2 for two biallelic loci is inversely related to the required sample size for association mapping to detect a given effect (Wang et al. 2005).

4.4.1 LD in Plants

Because the identification of markers associated with traits of interest depends on the presence of linkage disequilibrium (LD), attention has recently focused on determining the extent of LD in plant populations (Flint-Garcia et al. 2003; Gupta et al. 2005). Works by Tenaillon et al. (2001), Remington et al. (2001) and Rafalski (2002) have shown that LD can vary considerably between different populations within a species. For example, in diverse germplasm, Tenaillon et al. (2001) found that LD did not extend beyond 200 bp for 21 loci on chromosome 1 of maize. For diverse maize inbred lines, Remington et al. (2001) found LD extended up to 2,000 bp. In contrast, LD was as high as 100 kb for commercial elite inbred lines for the *adh1* and *y1* loci, and decay in LD was not detectable over a 300–500 bp range for 18 other genes (Ching et al. 2002). These results for LD patterns in maize probably reflect the fact that maize is descended from an extremely variable outcrossing wild relative. Most of the observed haplotypes were probably generated before domestication occurred and the different rates of LD reflect differing levels of population bottleneck (Flint-Garcia et al. 2003). In a diverse population of *Sorghum bicolor*, another outcrossing species, LD was examined (Hamblin et al. 2004) using 95 loci derived from mapped RFLPs. This study also found large stretches of LD, with significant decay occurring at 15 kb.

LD studies have also been carried out on inbreeding species. Studies of LD patterns have shown that LD extends much further in *Arabidopsis* (a highly inbred species) than in maize or sorghum. A study genotyping 163 SNPs across 76 *Arabidopsis* accessions found extended LD in a region of 250 kb equivalent to about one cM of recombination (Nordborg et al. 2002). Similar levels of LD were found in 14 sequenced fragments in the region of the *FRIGIDA* flowering locus (Hagenblad and Nordborg 2002). The difference in LD patterns between maize and *Arabidopsis* is consistent with the difference in population history (Nordborg 2000). Similar LD patterns have been observed in other inbred species, such as barley (Caldwell et al. 2006), rice (Garris et al. 2003), and both durum and bread wheat (Maccaferri et al. 2005; Breseghello and Sorrells 2006b).

4.4.2 LD in Sugarcane

While sugarcane is an outcrossing species like maize, its breeding history produces LD more similar to that of *Arabidopsis*. Commercial sugarcane varieties descended from a limited number of original ancestral *S. officinarum* clones contributing the majority of the germplasm in concert with an even more limited number of *S. spontaneum* clones (Arceneaux 1967). The small effective population size is demonstrated by a number of cultivars being found in the genealogy of most of the commercial cultivars worldwide. The limited number of generations (~10 or less) since the hybridizations means that there has been little opportunity for recombination between chromosomes during meiosis. Combined with the strong founder effect, this leads us to expect LD within sugarcane germplasm to be fairly extensive, in spite of its large genome.

Multiple studies in sugarcane have confirmed high levels of LD. LD was first investigated by Jannoo et al. (1999) using 38 RFLP probes on 59 cultivars in comparison with an RFLP map of a commercial variety (see Chapter 5 for details of the genetic map). Forty-two cases of bilocus association among 33 loci were observed. Most of these pairs of loci were separated by less than 10 cM. This global disequilibrium was interpreted as the result of the foundation bottleneck created by the first interpecific crosses. In a recent sugarcane LD study, 72 modern cultivars were screened with 1,537 AFLP markers (Raboin et al. 2008). To discriminate association due to linkage from other causes of association, a statistical threshold was derived from available maps. LD was common among closely linked markers and decreased within 0–30 cM. A study on LD in a large population of sugarcane clones is being carried out using DArT markers (P.A. Jackson, pers. comm.). Initial data analysis shows similar results to Raboin et al. (2008) (Fig. 4-2). These studies in sugarcane lay the foundation for association mapping, as LD structure in the genome greatly affects the number and density of markers required, sample size, and many other aspects of study design.

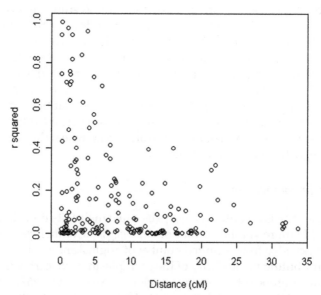

Figure 4-2 Extent of linkage disequilibrium in sugarcane as measured by DArT markers.

4.5 Association Study Design and Analysis

The scale of association studies influences data quality and the power to map novel associations. Decisions made early in the planning process have serious impact on later stages, and complicated analyses cannot compensate for poor design and data collection. General design and analysis principles have been discussed for plants by Zhu et al. (2008). Box 4-4 describes four important areas to consider when planning an association study, which we elaborate below.

Box 4-4 Checklist of considerations when planning an association study.

When planning an association study, four sets of factors must be balanced to optimize the potential to find useful results. Planning starts with the bottom line: what is available in terms of money, knowledge of the genome, marker technology and software. Characteristics of the trait will determine whether association studies are appropriate, and what type of association study has the best chance of success. Once the general type of study has been selected, consideration can be given to the details of specific markers and populations.

Selection of genotype and phenotype samples is highly intertwined. In determining the type, number and quality of marker genotyped, one must account for the cost of genotyping. Some compromise between quantity and quality may be required in populations without genome sequence and SNP arrays. Selection of markers will also be affected by the structure of population and sample, since highly inbred species may not require as many markers for the same level of resolution. While greater sample sizes are always desirable, this will be highly dependent on the cost of both genotyping and phenotyping. For a given sample size, different study designs to alter the sample structure may provide greater opportunities to detect association.

Resources
- ☐ Marker technology/databases
- ☐ Genome—linkage maps/physical maps/sequence
- ☐ Money
- ☐ Software

Markers
- ☐ Position
- ☐ Number
- ☐ Type
- ☐ Quality
- ☐ Cost

Population
- ☐ Species—outbred/inbred
- ☐ Extent of LD
- ☐ Sample structure—unrelated/families/loose interrelationship
- ☐ Sample size

Trait
- ☐ Mendelian or complex
- ☐ Candidate gene or genome-wide association
- ☐ Use of existing phenotypic data from plant breeding field trials
- ☐ Generation of new robust phenotypic data
  - o Efficient field design and replication
  - o Multiple years
  - o Multiple locations
  - o Collection of high quality data

The first consideration when contemplating association mapping is to assess existing resources and constraints. These may include publicly available databases of markers (e.g., HapMap; International HapMap Consortium 2005), prior knowledge of the trait, linkage maps, software, and of course, funding for the study—each of these affects data collection and analysis. Existing databases of markers are a valuable resource in the selection of loci for genotyping, since, if few markers are available, an additional marker discovery stage may be required before proper association mapping. Prior knowledge of the trait and potentially associated regions are required for candidate gene studies, but may be unnecessary for many complex traits, where a candidate gene study is likely to examine only a small portion of the total variation and whole-genome studies will be the method of choice. Linkage maps, physical maps or genetic sequence provide information about genome structure which can be used both in selecting markers for genotyping and analysis. All of these resources factor into the potential scope of the study. Their impact will be constrained by the costs of genotyping and phenotyping. Software resources will be discussed further below.

The second consideration is the selection of which markers will be genotyped in the study. This step follows directly from the first, since the availability of marker resources and maps will influence what can be studied. A dense and evenly spaced set of markers in the region of interest is desirable, whether that region encompasses a gene or the whole genome. If the spacing between markers is too small, many markers may provide redundant information about the trait; if the spacing is too large, the markers may not be close enough to the causal locus to detect linkage with the trait. Unfortunately, this information is not often available in sugarcane due to the lack of complete linkage maps. Markers may instead be chosen based on the type (microsatellite, AFLP, SNP, DArT) and respective quality. The number of each type of marker will be balanced by its cost.

The third consideration is the selection of a sample from the population. This is highly influenced by the second, since the number of markers and number of samples are both constrained by cost. The type of population plays an important role in this step, since outbred and inbred species will have different LD structures. Lower levels of LD will generally require larger samples to detect the same effect sizes. However, careful study design may help maximize the yield from even small samples. Selective genotyping, which preferentially examines cultivars with extreme phenotypes, has been extensively investigated in human and animal studies (e.g., Huang and Lin 2007; Ansari-Mahyari and Berg 2008). For a given sample size, it can provide more power than a random sample of individuals. Family designs have the alternate advantage of providing some level of control for population structure (McCarthy et al. 2008). While both types of designs have been implemented in other species, their efficacy and practicality have not been assessed in sugarcane. At this time, no research has been performed to investigate the power of different sample sizes and designs in sugarcane. Simulation studies have the potential to contribute greatly to this area; unfortunately, it is difficult to realistically simulate sugarcane data due to the complexity of its genome. Approximations can be made with software aimed at diploids and tetraploids, but it is unclear how well these results might be generalized. This is a rich area for future investigation and could provide considerable benefit to association study design in sugarcane.

The fourth consideration is the trait of interest. It has already been mentioned that prior knowledge of the trait, as to whether it is likely to be affected by one or many genes, will help to determine whether a candidate-gene or whole-genome study is more appropriate. The impact of the trait continues even after the selection of markers and samples, however, in the proper design of the phenotypic data collection. Field trials often stretch over multiple years and sites, and the collection of high quality data or use of data from existing plant breeding trials is an important aspect of plant association studies. Field trial design is a rich area of study and has been treated extensively elsewhere (e.g., Smith et al. 2006), so here we will not discuss the details further .

Along with study design, planning stages should encompass data analysis, since the method of data collection will often dictate possibilities for the analysis. The simplest form of association analysis compares the allele frequency between treatment groups (e.g., diseased and undiseased), or, for a quantitative trait, regresses the trait on each marker individually. This can straightforwardly incorporate other variables of interest and is computationally expedient, even for genome-wide studies. Multiple marker analyses have been proposed in humans that take into account the relationship between closely linked markers. These haplotype approaches have been shown to have greater power than the single marker approach for rare untyped causal variants (McCarthy et al. 2008). However, they are not a practical alternative in sugarcane without better knowledge of marker genome positions. Current studies are limited to single marker approaches until further progress can be made in map construction. Thus while maps are not a prerequisite for association mapping, they may prove useful in analysis. Tightly linked markers are likely to exhibit similar associations with the trait, so knowledge of map distance can both eliminate redundant associations and support evidence for association within a region.

For single marker approaches, the emphasis lies primarily on reducing the number of associations detected due to factors other than the markers. Environmental factors, polygenic effects and interactions may all need to be considered in order to fully dissect the sources of phenotypic variation. The number of tests performed must also be considered in selecting markers for further investigation. Box 4-5 outlines some methods of correcting for multiple

---

Box 4-5 Multiple test correction in association studies.

For a genome-wide association study, the sheer quantity of tests performed means that multiple testing is a critical issue. If a single test of marker-trait association produces a spurious result 1% of the time, then testing thousands of markers will produce hundreds of positive results simply by chance. One way to account for this is to reduce the significance level and make it more difficult to declare positive results. Below are listed some options for doing so. It should be noted, though, that none of these methods can distinguish between true and false positives. They simply provide a means to select loci for further investigation.

Method	Description	Disadvantages
Bonferroni	- Treats all tests as independent - Divide significance level by number of tests - Very simple to implement	- Most tests are not independent due to LD - Very conservative, so number of true findings is reduced
False Discovery Rate (FDR) (Benjamini and Hochberg 1995)	- Calculates the expected proportion of spurious associations	- Increased chance of type I errors - Does not account for LD between markers
Permutation	- Shuffles marker values to break up association with trait - Computes empirical significance level - Accounts for the LD between markers	- Many permutations (> 1,000) must be performed to get the significance level - Each permutation requires re-analysis of the data - Time-consuming and difficult - Does not account well for covariates

---

testing. In the end, though, it is only via replication or validation studies that researchers can gain more confidence in detected associations. Simulation studies have much to add in this area. They allow testing of novel approaches to assess efficiency, power, and the propensity for spurious associations. In silico approaches can thus lead to substantial savings in time and money by reducing the number of futile follow-up studies.

4.5.1 Population Structure

Since the advent of association-based methods it has been recognized that population structure can result in spurious associations. These Type I errors occur when allele frequencies differ between population subgroups (Lander and Schork 1994). The admixture of these subgroups within the study population can thus lead to the identification of markers associated with the population composition rather than the trait of interest. Box 4-6 gives an example of one such situation.

Several approaches have been proposed to deal with population structure in human studies, but not all are appropriate for studies in sugarcane. For example, the transmission-disequilibrium test (TDT)

Box 4-6 Example of association due to population structure.

Suppose a population is composed of a 50:50 mixture of two distinct populations. Further suppose that one population is composed primarily of cultivars highly resistant to smut, so that 80% of the population has a disease rating of 1 (vertical lines). In contrast, only 10% of the other population has the same disease rating. Consider a dominant marker genotyped in both populations, which is not associated with disease. If it is present in 50% of the first population and only 20% of the second population, then in the combined population that locus will appear to be associated with smut resistance, even though there is no association in either population considered separately. Conversely, combining populations can create a situation where markers associated with the trait in both populations fail to be detected in the combined sample. Spurious associations created by admixture are of far greater concern, though, given their high cost in terms of money and time spent in validation or replication studies. Thus removing or controlling population structure is critical in association studies.

Sample 1

Sample 2

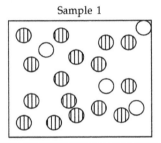
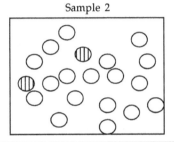

Population 1

	Marker Present	Marker Absent	
Smut rating = 1	40	40	80
Smut rating > 1	10	10	20
	50	50	100

Population 2

	Marker Present	Marker Absent	
Smut rating = 1	2	8	10
Smut rating > 1	18	72	90
	20	80	100

Population 2

	Marker Present	Marker Absent	
Smut rating = 1	42	48	90
Smut rating > 1	28	82	110
	70	130	200

considers the transmission of alleles only within families, and thus avoids the issue of structure in the population at large (Spielman et al. 1993). However, it requires a sample consisting of many diseased individuals and their parents and is not often applicable to plant studies.

Other methods developed in humans transfer better to plants. Pritchard et al. (2000a, b) proposed a method that uses information from unlinked markers to test for different population subgroups. This compares the allele frequency at independent loci throughout the genome between diseased and healthy individuals. If population structure (differences in the allele frequency) is detected, the analysis is stratified by the respective subpopulations. This reduces the Type I error rate and allows the detection of different associations in different populations. The method has been implemented in two freely available programs, STRUCTURE 2.2 and STRAT. Easy access to these computational resources has increased their usage in plants (Kraakman et al. 2004; Breseghello and Sorrels 2006b), and this, in turn, has led to new developments in the software which address some of the unique features of polyploids through the incorporation of methods to deal with dominant markers and null alleles (Falush et al. 2007).

In spite of these developments, it is important to remember that STRUCTURE was not developed with sugarcane in mind. Caution must be used in applying this method to general association mapping populations. STRUCTURE relies on certain assumptions to assign population subgroups, many of which may not be satisfied in sugarcane and other polyploids. The method is based on the assumption that genic associations between unlinked markers are the result of population substructure. Within each population, all loci are assumed to be in Hardy-Weinberg equilibrium with no linkage disequilibrium between loci if they are not tightly linked. These assumptions are violated in sugarcane, which has undergone generations of selection. It is unclear how robust the program is to these violations, so STRUCTURE may encounter difficulties with sugarcane data.

Table 4-1 summarizes two studies that applied STRUCTURE to sugarcane. The results of their population structure analysis are quite different, with one determining eight distinct subgroups and the other finding none. The differences in findings between studies may be due to differences between study populations, but may also reflect the difficulty of applying STRUCTURE to sugarcane. As sugarcane cultivars have been propagated worldwide, the population more closely resembles a very large interbred family than a mixture of distinct populations. STRUCTURE has had some success in maize (Yu et al. 2006) in distinguishing subgroups, but this may be a sign of differences in breeding practices and objectives, which are related to geographical location. Further investigation is necessary before general adoption of this methodology in sugarcane.

Two other methods, which have been proposed, also use unlinked background markers to assess the level of structure in the population. Genomic control (Devlin and Roeder 1999) estimates the effect of population structure at the background markers and then scales test statistics by this estimate. Although this procedure has performed well, it can be limited in

Table 4-1 Examples of sugarcane association mapping analysis.

STUDY	Wei et al. (2006)	Butterfield (2007)
PLAN	1. Detected association using simple regression analysis	1. Genetic diversity analysis to determine population structure
	2. STRUCTURE analysis revealed eight plausible subpopulations	2. Stepwise regression to select most significant marker associations
	3. Subpopulations incorporated into analysis a. Full model regressed disease rating on STRUCTURE group, marker, and interaction b. Significant interaction indicated differing effect of marker within subgroups c. Otherwise, reduced model regressed disease rating on STRUCTURE group and marker d. Significant main effect indicated association not related to population structure	
RESULT	- Very few significant interaction terms - Nearly half of original AFLP associations due to population structure - Follow-up study of population structure in different population was unable to distinguish clear number of subgroups (X. M. Wei, pers. comm.)	- No significant population structure found

applicability, conservative, and moreover assumes that the effect of population structure does not vary across the genome (Balding 2006). Principal component analysis (PCA) is a general statistical technique to decompose variation into a small number of components. It represents a different way of constructing subgroups to using STRUCTURE and is subject to some of the same problems when applied to sugarcane (Price et al. 2006).

Estimates of population structure can be incorporated into a variety of models to reduce the number of spurious marker associations detected. Mixed modeling has been used with some success (Yu et al. 2006; Zhao et al. 2007). It provides a more general and flexible framework than regression or cluster analysis in which to model both genetic and environmental variation. Considerations when constructing a model may include pedigree effects, environmental factors, and interactions—all of which can be accommodated in the mixed model framework. The use of pedigree information is common in animal breeding studies to account for the kinship between individuals, but this has not been a widespread practice in plant studies (Malosetti et al. 2007). However, the complicated web of interrelationships between cultivars in association studies lends itself to this approach. While STRUCTURE may have difficulty separating out distinct subpopulations due to the many generations of selection and interbreeding, these relationships will be reflected in the pedigree. In the situation of incomplete pedigree information, an empirical matrix of genetic/genotypic distances can be calculated from molecular marker information and used instead (Kraakman et al. 2004). This information can be used in cluster analysis to derive groups representing population structure (Simko et al. 2004).

The mixed model framework allows for more complex models to dissect the phenotypic variation into different genetic and environmental components. Yu et al. (2006) proposed a method for using pedigree information along with STRUCTURE group assignments to account for both obvious subpopulations and more complicated inbreeding. Through these two components, they attempted to simultaneously address broad and fine population structure. Malosetti et al. (2007) used a mixed model approach for association mapping in potato including pedigree information across multiple environments. The inclusion of pedigree information in the mixed model produced consistent significant associations across two independent studies. Aulchenko et al. (2007) proposed an approximation to this with a two-step approach, GRAMMAR. This first fits a mixed model for the familial correlations and population structure in the data and then regresses the residuals upon the marker data. This is a rapid method of analysis, which seems to perform well in humans/animals but has not yet been tested in plants. Further research into mixed model pedigree approaches is in progress in the association mapping study of Jackson and colleagues (P.A. Jackson, pers. comm.). The aim is to incorporate multiple marker associations into the model while simultaneously accounting for population structure.

4.5.2 *Software*

Software for association mapping has developed at the same pace as the studies themselves. A comprehensive listing of software for genetic studies, including websites for download and further details about packages, is available at *http://www.nslij-genetics.org/soft/*. Unfortunately, most software packages are developed for human studies. This means that they often cannot handle features specific to plants, e.g., dominant markers, polyploidy, and selfing. Even software oriented towards plants may not fully address the complexity of sugarcane. Awareness of inherent assumptions for a software package is critical before use, as violations of the assumptions may lead to poor or incorrect results. Although it is much more time-consuming to implement, personal code will often be a more foolproof method of analysis than using "off-the-shelf" software.

Table 4-2 lists several software packages, which can be used for association mapping in sugarcane. A similar table appears in Zhu et al.

(2008) with software applicable generally in plants. R (Ihaka and Gentleman 1996) and SAS (SAS Institute 2008) are general statistical programs that can be adapted to genetic analyses; however, they do require some familiarity with the language and are not particularly user-friendly. Modules are available for each program to perform specific tasks such as fitting mixed models, constructing pedigrees, or estimating linkage disequilibrium. The specific genetic modules were developed for human genetic analysis and should be utilized with caution. The mixed model capabilities are more generally applicable. They provided alternatives to ASReml (Gilmour et al. 2002), a commercial package for mixed modelling. ASReml was devised for

Table 4-2 Computing resources for use in association mapping in sugarcane.

Software	Website	Comment
General Statistical Platforms		
SAS	*www.sas.com*	Commercial, statistical analysis software. Built-in modules and interactive programming language.
R	*cran.r-project.org*	Free, statistical computing language and environment. Some programming skill required.
Mixed Model Platforms		
R/nlme	*cran.r-project.org/ web/packages/nlme*	Free, add-on package to R. Developed for general mixed (linear and nonlinear) modelling.
ASReml	*www.vsni.co.uk/ products/asreml*	Commercial, flexible and fast modelling for animal and plant breeding data.
Population Structure Estimation		
STRUCTURE	*pritch.bsd.uchicago. edu/structure.html*	Free, estimates percent membership in population subgroups.
EIGENSTRAT	*Genepath.med.harvard. edu/~reich/Software.htm*	Free, principal components analysis to detect population structure.
Kinship Estimation		
SPAGeDi	*www.ulb.ac.be/sciences /ecoevol/spagedi.html*	Free, calculates empirical kinship coefficients from marker data.
Association Analysis Package		
TASSEL	*www.maizegenetics.net*	Free, suite of programs to perform LD estimation, mixed modelling with kinship/ structure effects.

use in animal and plant breeding and the greater specialization results in efficient and fast algorithms for many applications.

The remaining packages listed in the table are specifically aimed at genetic analysis. The narrow focus of many packages often requires piecemeal analysis of data. For example, Yu et al. (2006) used SPAGeDi (Hardy and Vekemans 2002) to estimate kinship coefficients rather than using the observed pedigree; they used STRUCTURE (Pritchard et al. 2000a, b) to estimate population subgroups, and the two are incorporated into a mixed model, which is fit with SAS (SAS Institute 2008). EIGENSTRAT (Price et al. 2006) performs PCA analysis as an alternative to using STRUCTURE; generic forms of PCA can also be found within SAS and R. This piecewise analysis procedure was the impetus for Bradbury et al. (2007) to create TASSEL, which is a publically available suite of programs to perform association analysis. This is a Java platform to integrate the different steps of the analysis performed by Yu et al. (2006).

4.6 The Future of Association Studies

Association analysis in sugarcane is challenging. First, there are few published studies in sugarcane and the studies that have been completed are small in scale. Second, the complex genetic machinery underlying sugarcane has frustrated efforts to develop high-density marker maps (Rossi et al. 2003; Aitken et al. 2005, 2007). Third, there is evidence from LD studies in species, such as wheat, that LD could vary dramatically between populations with different evolutionary histories (Sorrells 2007). Hence, it is difficult to predict the marker density required for association analyses in sugarcane because the pattern of LD is population dependent. Fourth, population structure is very much an issue in sugarcane association studies. Many of the cultivars have ancestors in common. To reduce population structure, we could limit the study to progeny generated from a carefully chosen set of elite parents but this approach narrows the genetic base of the study.

Despite these challenges, the future of association analyses in sugarcane is bright. By following the footsteps of association studies in humans and animals, we can borrow from their achievements and learn from their mistakes. Not everything is transferrable due to the differences in population history and genetic structure, but much of the background for association in sugarcane has been laid in studies of other organisms. Consequently, progress should be more rapid than if the area was evolving *de novo*. Furthermore, in the next five years we expect sequence information to become increasingly available and consequently we will see great improvement in marker maps. These dense marker maps will give us the opportunity to significantly increase power of QTL detection and mapping resolution in our whole-genome association screens. In the not too distant future, we would expect whole-genome association analysis to become a key tool in the search for genomic regions housing genes influencing traits of interest.

References

Aitken KS, Jackson PA, McIntyre CL (2005) A combination of AFLP and SSR markers provides extensive map coverage and identification of homologous linkage groups in sugarcane. Theor Appl Genet 110: 789–801.

Aitken KS, Jackson PA, McIntyre CL (2007) Construction of a genetic linkage map for *Saccharum officinarum* incorporating both simplex and duplex markers to increase genome coverage. Genome 50: 742–756.

Ansari-Mahyari S, Berg P (2008) Combined use of phenotypic and genotypic information in sampling animals for genotyping in detection of quantitative trait loci. J Anim Breed Genet 125: 100–109.

Aranzana MJ, Kim S, Zhao K, Bakker E, Horton M, Jakob K, Lister C, Molitor J, Shindo C, Tang C, Toomajian C, Traw B, Zheng H, Bergelson J, Dean C, Marjoram P, Nordborg M (2005) Genome-wide association mapping in Arabidopsis identifies previously known flowering time and pathogen resistance genes. PLos Genet 1: e60 doi: 1371/journal.pgen.0010060.

Arceneaux G (1967) Cultivated sugarcane of the world and their botanical derivation. Proc Int Soc Sugar Cane Technol 12: 844–854.

Aulchenko YS, de Koning D-J, Haley C (2007) Genomewide rapid association using mixed model and regression: a fast and simple method for genomewide pedigree-based quantitative trait loci association analysis. Genetics 177: 577–585.

Balding DJ (2006) A tutorial on statistical methods for population association studies. Nat Rev Genet 7: 781–791.

Beló A, Zheng P, Luck S, Shen B, Meyer DJ, Li B, Tingey S, Rafalski A (2008) Whole genome scan detects an allelic variant of fad2 associated with increased oleic acid levels in maize. Mol Genet Genom 279: 1–10.

Benjamini Y, Hochberg Y (1995) Controlling the false discovery rate—a practical and powerful approach to multiple testing. J Roy Stat Soc B 57: 289–300.

Bradbury PJ, Zhang Z, Kroon DE, Casstevens TM, Ramdoss Y, Buckler ES (2007) TASSEL: Software for association mapping of complex traits in diverse samples. Bioinformatics 23: 2633–2635.

Breseghello F, Sorrells ME (2006a) Association analysis as a strategy for improvement of quantitative traits in plants. Crop Sci 46: 1323–1330.

Breseghello F, Sorrells ME (2006b) Association mapping of kernel size and milling quality in wheat (*Triticum aestivum* L.) cultivars. Genetics 172: 1165–1177.

Butterfield MK (2007) Marker Assisted Breeding in Sugarcane: A Complex Polyploidy. PhD Thesis, Univ Stellenbosch, Matieland, Stellenbosch, South Africa, pp 39–73.

Caldwell KS, Russell J, Langridge P, Powell W (2006) Extreme population-dependant linkage disequilibrium detected in an inbreeding plant species, *Hordeum vulgare*. Genetics 172: 557–567.

Cardon LR, Bell JI (2001) Association study designs for complex diseases. Nat Rev Genet 2: 91–99.

Ching A, Caldwell KS, Jung M, Dolan M, Smith OS, Tingey S, Morgante M, Rafalski AJ (2002) SNP frequency, haplotype structure and linkage disequilibrium in elite maize inbred lines. BMC Genet 3: 19.

Clark RM, Schweikert G, Toomajian C, Ossowski S, Zeller G, Shinn P, Warthmann N, Hu TT, Fu G, Hinds DA, Chen H, Frazer KA, Huson DH, Scholkopf B, Nordborg M, Ratsch G, Ecker JR, Weigel D (2007) Common sequence polymorphisms shaping genetic diversity in *Arabidopsis thaliana*. Science 317: 338–342.

Devlin B, Roeder K (1999) Genomic control for association studies. Biometrics 55: 997–1004.

Falconer DS, Mackay TF (1996) Introduction to Quantitative Genetics. Longmen Group, Essex, UK.

Falush D, Stephens M, Pritchard JK (2007) Inference of population structure using multilocus genotype data: dominant markers and null alleles. Mol Ecol Notes 7: 574–578.

Flint-Garcia SA, Thornsberry JM, Buckler ES (2003) Structure of linkage disequilibrium in plants. Annu Rev Plant Biol 54: 357–374.

Fusari CM, Lia VV, Hopp HE, Heinz RA, Paniego NB (2008) Identification of single nucleotide polymorphisms and analysis of linkage disequilibrium in sunflower elite inbred lines using the candidate gene approach. BMC Plant Biol 8: 7.

Garris AJ, McCouch SR, Kresovich S (2003) Population structure and its effect on haplotype diversity and linkage disequilibrium surrounding the *xa5* locus of rice (*Oryza sativa* L.). Genetics 165: 759–769.

Gaut BS, Long AD (2003) The lowdown on linkage disequilibrium. Plant Cell 15: 1502–1506.

Gebhardt C, Ballvora A, Walkemeier B, Oberhagemann P, Schüler K (2004) Assessing genetic potential in germplasm collections of crop plants by marker-trait association: a case study for potatoes with quantitative variation of resistance to late blight and maturity type. Mol Breed 13: 93–102.

Georges M (2007) Mapping, fine mapping, and molecular dissection of quantitative trait loci in domestic animals. Annu Rev Genom Hum Genet 8: 131–162.

Gilmour AR, Gogel BJ, Cullis BR, Thompson R (2006) ASReml User Guide Release 2.0. VSN Int, Hemel Hempstead, UK.

Goff SA (2002) A draft sequence of the rice genome (*Oryza sativa* L. ssp. *Japonica*). Science 309: 879.

Gupta PK, Rustgi S, Kulwal PL (2005) Linkage disequilibrium and association studies in higher plants: present status and future prospects. Plant Mol Biol 57: 461–485.

Hagenblad J, Nordborg M (2002) Sequence variation and haplotype structure surrounding the flowering time locus *FRI* in *Arabidopsis thaliana*. Genetics 161: 289–298.

Hamblin MT, Mitchell SE, White GM, Gallego J, Kukatla E, Wing RA, Paterson AH, Kresovich S (2004) Comparative population genetics of the Panicoid Grasses: sequence polymorphism, linkage disequilibrium and selection in a diverse sample of *Sorghum bicolor*. Genetics 167: 471–483.

Hardy OJ, Vekemans X (2002) SPAGeDi: A versatile computer program to analyse spatial genetic structure at the individual or population levels. Mol Ecol Notes 2: 618–620.

Harjes CE, Rocheford TR, Bai L, Brutnell TP, Kandianis KB, Sowinski SG, Stapleton AE, Vallabhaneni R, Williams M, Wurtzel ET, Yan J, Buckler ES (2008) Natural genetic variation in lycopene epsilon cyclase tapped for maize biofortification. Science 319: 330–333.

Heffner EL, Chomdej O, Williams KR, Sorrells ME (2008) Dominant male-sterile populations for association mapping and introgression of exotic wheat germplasm. Aust J Agri Res 59: 470–474.

Huang BE, Lin DY (2007) Efficient association mapping of quantitative trait loci with selective genotyping. Am J Hum Genet 80: 567–576.

Ihaka R, Gentleman R (1996) R: a language for data analysis and graphics. J Comp Graph Stat 5: 299–314.

The International HapMap Consortium (2005) A haplotype map of the human genome. Nature 437: 1299–1320.

Jaccoud D, Peng K, Feinstein D, Kilian A (2001) Diversity arrays: a solid state technology for sequence information independent genotyping. Nucl Acids Res 29: e25.

Jannoo N, Grivet G, Dookun A, D'Hont A, Glaszmann J-C (1999) Linkage disequilibrium among modern sugarcane cultivars. Theor Appl Genet 99: 1053–1060.

Jun T-H, Van K, Kim MY, Lee S-H, Walker DR, (2008) Association analysis using SSR markers to find QTL for seed protein content in soybean. Euphytica 162: 179–191.

Kim S, Zhao K, Jiang R, Molitor J, Borevitz JO, Nordborg M, Marjoram P (2006) Association mapping with single-feature polymorphisms. Genetics 173: 1125–1133.

Kraakman ATW, Niks RE, Van den Berg PMMM, Stam P, Van Eeuwijk FA (2004) Linkage equilibrium mapping of yield and yield stability in modern spring barley cultivars. Genetics 168: 435–446.

Lander ES, Schork NJ (1994) Genetic dissection of complex traits. Science 265: 2037–2048.

Maccaferri M, Sanguineti MC, Noli E, Tuberosa R (2005) Population structure and long-range linkage disequilibrium in a durum wheat elite collection. Mol Breed 15: 271–289.

Malosetti M, Van der Linden CG, Vosman B, Van Eeuwijk FA, (2007) A mixed-model approach to association mapping using pedigree information with an illustration of resistance to *Phytophthora infestans* in potato. Genetics 175: 879–889.

McCarthy MI, Abecasis GR, Cardon LR, Goldstein DB, Little J, Ioannidis JPA, Hirschhorn JN (2008) Genome-wide association studies for complex traits: consensus, uncertainty and challenges. Nat Rev Genet 9: 356–369.

McIntyre CL, Whan VA, Croft B, Magarey R, Smith GR (2005) Identification and validation of molecular markers associated with Pachymetra root rot and brown rust resistance in sugarcane using map- and association-based approaches. Mol Breed 16: 151–161.

Nordborg M (2000) Linkage disequilibrium, gene trees and selfing: an ancestral recombination graph with partial self-fertilisation. Genetics 154: 923–929.

Nordborg M, Borevitz JO, Bergelson J, Berry CC, Chory J, Hagenblad J, Kreitman M, Maloof JN, Noyes T, Oefner PJ, Stahl EA, Weigel D (2002) The extent of linkage disequilibrium in *Arabidopsis thaliana*. Nat Genet 30: 190–193.

Olsen KM, Caicedo AL, Polate N, McClung A, McCouch SR, Purugganan MD (2006) Selection under domestication: evidence for a sweep in the rice *Waxy* genomic region. Genetics 173: 975–983.

Palmer LJ, Cardon LR (2005) Shaking the tree: mapping complex disease genes with linkage disequilibrium. Lancet 366: 1223–1234.

Price AL, Patterson NJ, Plenge RM, Weinblatt ME, Shadick NA, Reich D (2006) Principal components analysis corrects for stratification in genome-wide association studies. Nat Genet 38: 904–909.

Pritchard JK, Rosenberg NA (1999) Use of unlinked genetic markers to detect population stratification in association studies. Am J Hum Genet 65: 220–228.

Pritchard JK, Przeworski M (2001) Linkage disequilibrium in Humans: Models and data. Am J Hum Genet 69: 1–14.

Pritchard JK, Stephens M, Donnelly P (2000a) Inference of population structure using multilocus genotype data. Genetics 155: 945–959.

Pritchard JK, Stephens M, Rosenberg NA, Donnelly P (2000b) Association mapping in structured populations. Am J Hum Genet 67: 170–181.

Raboin L-M, Pauquet J, Butterfield M, A. D'Hont, and J.C. Glaszmann. 2008. Analysis of genome-wide linkage disequilibrium in the highly polyploidy sugarcane. Theor Appl Genet 116: 701–714.

Rafalski A (2002) Applications of single nucleotide polymorphisms in crop genetics. Curr Opin Plant Biol 5: 94–100.

Remington DL, Thornsberry JM, Matsuoka Y, Wilson LM, Whitt SR, Doebley J, Kresovich S, Goodman MM, Buckler ES (2001) Structure of linkage disequilibrium and phenotypic associations in the maize genome. Proc Natl Acad Sci USA 98: 11479–11484.

Rossi M, Araujo PG, Paulet F, Garsmeur O, Dias VM, Chen H, Van Sluys M-A, D'Hont A (2003) Genomic distribution and characterization of EST-derived resistance gene analogs (RGAs) in sugarcane. Mol Genet Genom 269: 406–419.

SAS Institute (2008) SAS/STAT 9.2 user's guide. SAS Institute, Cary, NC, USA.

Slatkin M (2008) Linkage disequilibrium—understanding the evolutionary past and mapping the medical future. Nat Rev Genet 9: 477–485.

Simko I, Costanzo S, Haynes KG, Christ BJ, Jones RW (2004) Linkage disequilibrium mapping of a *Verticillium dahliae* resistance quantitative trait locus in tetraploid potato (*Solanum tuberosum*) through a candidate gene approach. Theor Appl Genet 108: 217–224.

Smith AB, Lim P, Cullis BR (2006) The design and analysis of multi-phase plant breeding experiments. J Agri Sci 144: 393–409.

Sorrells ME (2007) Application of new knowledge, technologies, and strategies to wheat improvement. Euphytica 157: 299–306.

Spielman RS, McGinnis RE, Ewens WJ (1993) Transmission test for linkage disequilibrium: The insulin gene region and insulin-dependent diabetes mellitus (IDDM). Am J Hum Genet 52: 506–516.

Tenaillon MI, Sawkins MC, Long AD, Gaut RL, Doebley JF, Gaut BS (2001) Patterns of DNA sequence polymorphism along chromosome 1 of maize (*Zea mays* ssp. *mays* L.). Proc Natl Acad Sci USA, 98: 9161–9166.

Thornsberry M, Goodman MM, Doebley J, Kresovich S, Nielsen D, Buckler ES (2001) *Dwarf8* polymorphisms associate with variation in flowering time. Nat Genet 28: 286–289.

Wang WYS, Barratt BJ, Clayton DG, Todd JA (2005) Genome-wide association studies: theoretical and practical concerns. Nat Rev Genet 6: 109–118.

Weber A, Clark RM, Vaughn L, Sanchez-Gonzalez JJ, Yu J, Yandell BS, Bradbury P, Doebley J (2007) Major regulatory genes in maize contribute to standing variation in Teosinte (*Zea mays* ssp. *parviglumis*). Genetics 177: 2349–2359.

Wei X, Jackson PA, McIntyre CL, Aitken KS, Croft B (2006) Associations between DNA markers and resistance to diseases in sugarcane and effects of population substructure. Theor Appl Genet 114: 155–164.

Weir BS (1996) Genetic Data Analysis II. Sinauer Associates, Sunderland MA, USA.

Yu J, Buckler ES (2006) Genetic association mapping and genome organization of maize. Curr Opin Biotechnol 17: 155–160.

Yu J, Pressoir G, Briggs WH, Vroh Bi I, Yamasaki M, Doebley JF, McMullen MD, Gaut BS, Nielsen DM, Holland JB, Kresovich S, Buckler ES (2006) A unified mixed-model method for association mapping that accounts for multiple levels of relatedness. Nat Genet 38: 203–208.

Yu J, Holland JB, McMullen MD, Bucker ES (2008) Genetic design and statistical power of nested association mapping in maize. Genetics 178: 539–551.

Zeggini E, Scott LJ, Saxena R, Voight BF (2008) Meta-analysis of genome-wide association data and large-scale replication identifies additional susceptibility loci for type 2 diabetes. Nat Genet 40: 638–645.

Zhao K, Aranzana MJ, Kim S, Lister C, Shindo C, Tang C, Toomajian C, Zheng H, Dean C, Marjoram P, Nordborg M (2007) An Arabidopsis example of association mapping in structured samples. PLoS Genet 3: e4.

Zhu C, Gore M, Buckler ES, Yu J (2008) Status and prospects of association mapping in plants. Plant Genome 1: 5–20.

5

Molecular Genetic Linkage Mapping in *Saccharum*: Strategies, Resources and Achievements

Sreedhar Alwala[†]* and Collins A. Kimbeng

ABSTRACT

Cultivated sugarcane (*Saccharum* spp. hybrids; $2n = 100$–130) is a complex aneu-polyploid with high levels of heterozygosity. Modern sugarcane cultivars are the derivatives of interspecific hybridization between the domesticated species *S. officinarum* and its wild relative *S. spontaneum* through a series of backcrosses to *S. officinarum*. The genome of cultivated sugarcane comprises 70–80% *S. officinarum*, 10–20% *S. spontaneum* and 5–17% of recombinant chromosomes. Genetic linkage maps have been difficult to construct in sugarcane due to the double genome structure, homologous and homoeologous chromosomes and aneuploid inheritance. However, with the advent and wide adoption of single dose markers, linkage mapping in sugarcane has witnessed several achievements in the last couple of decades. The rapid and concomitant advancements in PCR-based marker systems and high-throughput technologies have resulted in the development of several linkage maps in *S. officinarum*, *S. spontaneum*, and *S. robustum* species as well as in cultivars. Although incomplete, these genetic linkage maps not only improved our understanding of the sugarcane genome organization and its genetic

School of Plant, Environmental and Soil Sciences, M. B. Sturgis Hall, Louisiana State University Agricultural Center, Baton Rouge, LA 70803, USA.
[†]Current Address: Dow Agro Sciences, York, NE, USA.
*Corresponding author: *SAlwala@dow.com*

architecture but also facilitated quantitative trait loci (QTL) detection and identification, comparative mapping and map-based cloning. A futuristic multidisciplinary approach of integrating molecular genetics, cytogenetics and comparative genomic analyses involving maps of sorghum and other allied species would enable a comprehensive dissection of the sugarcane genome.

Keywords: linkage mapping, aneu-polyploid, pseudo test-cross, single-dose markers, multi-dose markers, quantitative trait loci, comparative mapping

5.1 Introduction

The genus *Saccharum* (family: *Poaceae*, tribe: *Andropogoneae*, subtribe: *Saccharinae*), collectively known as sugarcane, is comprised of six polyploid, outcrossing species namely *S. officinarum* Linnaeus ($2n = 80$), *S. barberi* Jeswiet ($2n = 81–124$), *S. sinense* Roxb. ($2n = 111–120$), *S. spontaneum* Linnaeus ($2n = 40–128$), *S. robustum* Brandes and Jeswiet ex Grassl ($2n = 60–80$), and *S. edule* Hassk. ($2n = T60, 70, 80$) (Brandes 1958, Sreenivasan et al. 1987). Modern sugarcane cultivars (*Saccharum* spp. hybrids; $2n = 100–130$) are the descendants of interspecific hybridization, that occurred in the early 1900s, between the domesticated species *S. officinarum*, and its wild relative *S. spontaneum*. The initial interspecific hybrids were repeatedly backcrossed to *S. officinarum* via a process termed "nobilization" mainly to recover the high sugar-producing ability of *S. officinarum* and to minimize the negative effects of *S. spontaneum* (Roach 1972; Sreenivasan et al. 1987). The nobilization process also resulted in improved cane yields, ratooning ability and increased resistance to biotic and abiotic stresses. During nobilization the progeny inherited $2n$ gametes from the *S. officinarum* parent (Bremer 1961; Bhat and Gill 1985). Modern sugarcane represents the extreme example of polyploidy, with a double genome structure in which the number of homologous and homoeologous chromosomes can vary among genotypes from the same cross. The genome composition of modern sugarcane is about 70–80% *S. officinarum*, 10–20% *S. spontaneum* and 5–17% of recombinant chromosomes (D'Hont et al. 1996; Piperidis and D'Hont 2001). Readers may refer to Chapter 3 for a detailed description of origin and history of cultivated sugarcane.

Saccharum officinarum is characterized by its thick stalks, high sucrose and low fiber content. Chromosome numbers have been reported consistently as $2n = 80$ with the basic chromosome number as $x = 10$ (Daniels and Roach 1987; Sreenivasan et al. 1987; D'Hont et al. 1998). *S. spontaneum*, on the other hand, is a wild species and its clones are characterized by thin stalks, low sucrose, high fiber, profuse flowering, good ratooning ability, and relatively high levels of disease and insect resistance. Chromosome numbers of *S.*

spontaneum have been reported to range from $2n = 40–128$ (Panje and Babu 1960) with $x = 8$ as the basic chromosome number (Al-Janabi et al. 1993; da Silva et al. 1993; D'Hont et al. 1998).

The last few decades have witnessed vast developments in the field of plant molecular genetics. The development and application of molecular (or DNA) markers has increased our knowledge of the genome constitution and genetic architecture of many plant species including several major crops. Molecular markers have also played a pivotal role in our understanding of population genetics with respect to genetic variability, evolutionary relationships and the dynamics of gene flow among plant populations. Molecular markers such as restriction fragment length polymorphism (RFLP), random amplified polymorphic DNA (RAPD), amplified fragment length polymorphism (AFLP), and simple sequence repeat (SSR) have been used to construct genetic linkage maps in several major diploid crop species such as rice, maize, tomato, and lettuce (Tanksley et al. 1992; Nagamura et al. 1997; Castiglioni et al. 1999; Syed et al. 2006). Genetic linkage maps, generated using molecular markers, have facilitated gene tagging (Sobral and Honeycutt 1994), map-based cloning (Dietrich et al. 1996), QTL mapping (Frewen et al. 2000) and supported marker-assisted selection in plant breeding programs (Dubcovsky 2004; Collard and Mackill 2008). For drop-wise details, readers are referred to the seven book volumes of the series on Genome Mapping and Molecular Breeding in Plants edited by C. Kole (2006-2007) published by Springer-Verlag.

Unfortunately, genetic linkage maps are inherently difficult to construct in polyploid species for several reasons: (1) the statistics are far more complicated for polyploids than for diploids, (2) a wide array of genotypes is expected in the segregating population, (3) there are several modes of gamete formation and random pairing of multiple homologous chromosomes, (4) the different gamete frequencies cannot be identified with certainty (Fisher 1949) because of the segregation of alleles with different dosage levels, and (5) the genome constitution of some polyploids, especially sugarcane, is still unclear and their inheritance pattern is difficult to determine (Wu et al. 1992; Milbourne et al. 2008) because the basic chromosome number is different in each of the *Saccharum* species and some have a mixtures of both allo- and auto-polyploid genomes. For example, the basic chromosome number is $x = 10$ in *S. officinarum* and $x = 8$ in *S. spontatenum* and both genomes occur disproportionately in the genomes of *S. barberi*, *S. sinense* and cultivated sugarcane. Maps of the diploid relatives have been exploited in some polyploid species such as wheat (Paillard et al. 2003), potato (Tanksley et al. 1992) and alfalfa (Kiss et al. 1993) to circumvent the complexity of the polyploid genome. Unfortunately, there are no known diploid relatives or progenitor species of sugarcane that could be exploited for this purpose. Another unique difficulty in sugarcane is the development of mapping populations. Because it is clonally propagated, highly heterozygous and suffers from inbreeding depression, traditional mapping populations, such as F_2, inbred-backcross and/or recombinant inbred lines, common in other crop species are difficult to construct in sugarcane (for details see Cordeiro et al. 2007).

In spite of these difficulties, quite a number of achievements have been made in the last decade in sugarcane molecular genetics. The development and wide adoption of the single-dose marker (or simplex markers) mapping approach has facilitated the development of genetic linkage maps as well as increased our understanding of the genetics of sugarcane and other polyploid species (Table 5-1). Genetic linkage maps have been constructed in several polyploid species including potato (Bonnierbale et al. 1988), eucalyptus (Yin et al. 2001), sweetpotato (Kriegner et al. 2003), and apple (Hemmat et al. 1994) using the single-dose molecular marker mapping approach (for details see Kole 2006, 2007).

Table 5-1 Segregation patterns of single-dose, double-dose, triple-dose and simple-duplex markers in an octoploid under polysomic inheritance.

Marker type	Ratio	Parent		Mapping population									
		P$_1$	P$_2$	M1	M2	M3	M4	M5	M6	M7	M8	M9	M10
Single-dose	1:1	–		–	–		–	–			–		
Double-dose	11:3		–	–	–	–		–	–		–	–	–
Triple-dose	13:1	–		–	–	–	–	–		–	–	–	–
Simple-duplex	3:1	–		–	–	–	–	–	–		–	–	

5.2 Mapping using Single-and Multi-dose Markers

A single-dose marker segregates in a 1:1 ratio for the presence or absence of a band in a mapping population derived from two heterozygous parents where, the band is present in one parent and absent in the other. For example, consider two heterozygous parents P_1 and P_2 and a F_1 mapping population derived from the cross $P_1 \times P_2$. A marker that is polymorphic among the parents, is present in one parent and absent in the other, and fits a χ^2 probability for a 1:1 segregation ratio in the F_1 mapping population is treated as a single-dose marker (Table 5-1). This segregation ratio is valid for allopolyploid, autopolyploid or even diploid species. The original theory of linkage mapping using single-dose markers was put forth by Ritter et al. (1990) and later refined by Wu et al. (1992) to suit polyploids. Sobral and Honeycutt (1993) and Al-Janabi et al. (1993) further generalized the theory of single-dose markers to any ploidy level involving two heterozygous parents. Grattapaglia and Sederoff (1994) named this mapping strategy as pseudo-testcross (if one parent is studied) or double pseudo-testcross (if both parents are studied). The pseudo-testcross strategy has virtually been extended to any biparental cross involving heterozygous parents in any species.

There are, however, two exceptions to the 1:1 single-dose marker segregation ratio. Although the single-dose segregation ratio is 1:1 in a F_1 mapping population, the ratio extends to 3:1 in populations derived from selfing a single heterozygous parent. Recently, Hoarau et al. (2001) used a mapping population derived by selfing the sugarcane cultivar "R570" and developed a genetic linkage map using a 3:1 single-dose (and 15: 1 double-dose) marker segregation ratio. The other exception in the F_1 mapping population is when certain markers that are monomorphic in both parents, segregate in a 3:1 ratio in the mapping population. Such markers are called bi-parental simplex (some authors term these markers as simple-duplex or double-simplex) markers. Bi-parental simplex markers have commonly been used as allelic bridges mainly to identify homo(eo)logous linkage groups (Edmé et al. 2006) and to infer conserved regions across different *Saccharum* species (Alwala et al. 2008).

Multi-dose markers on the other hand occur in two or more doses in the parents and segregate in ratios > 1:1 (Tables 5-1 and 5-2). The most common multi-dose markers are double-dose (DD) or duplex markers (present in two doses) and triple-dose (TD) or triplex markers (present in three doses). Whereas single-dose markers are expected to segregate in the same way regardless of the ploidy level and genome constitution, the expected segregation ratio of higher-dose markers, such as double-dose and triple-dose markers, varies according to the genome constitution (autopolyploidy *vs.* allopolyploidy), the ploidy level, and the pairing behavior (bivalents, trivalents, quadrivalents). Consider a F_1 mapping population consisting of 100 progeny derived from a polysomic octoploid (Table 5-2). A double-dose marker would segregate in a 79: 21 (11: 3) ratio for polysomic inheritance or in a 75: 25 (3:1) ratio for disomic inheritance and a triple-dose marker would segregate in a 93: 7 (13: 1) ratio for polysomic inheritance or in an 88: 12 (7: 1) fashion for disomic inheritance. Quadruple-dose (QD) or quadruplex markers segregating in a 93.75: 6.25 (15: 1) ratio for disomic inheritance and 98.5: 1.5 (69: 1) ratio for polysomic inheritance would be difficult to identify, especially for polysomic inheritance, as the population size required to detect the lower frequency genotypes would be prohibitively large.

Table 5-2 Segregation ratios for single-dose, double-dose, triple-dose and quadruple-dose markers in an F_1 mapping population derived from an octoploid species (modified from da Silva and Sobral 1996).

Marker dose	Parental marker cross	Segregation Ratio	
		Polysomic Inheritance	Disomic inheritance
Single-dose	*Aaaaaaaa* x *aaaaaaaa*	1:1	1:1
Double-dose	*AAaaaaaa* x *aaaaaaaa*	11:3	3:1
Triple-dose	*AAAaaaaa* x *aaaaaaaa*	13:1	7:1
Quadruple-dose	*AAAAaaaa* x *aaaaaaaa*	69:1	15:1

5.3 Marker Systems used for *Saccharum* Linkage Mapping

The earliest molecular genetic linkage maps in *Saccharum* were developed for progenitor species of modern sugarcane followed by cultivars such as "R570" and "Q164". Linkage maps have been developed in sugarcane using RFLP (da Silva et al. 1993,1995; Grivet et al. 1996; Ming et al. 1998), RAPD (Al-Janabi et al. 1993; Mudge et al. 1996), AFLP (Guimaráes et al. 1997; Hoarau et al. 2001; Aitken et al. 2005, 2007; Garcia et al. 2006; Alwala et al. 2008) and SSR (Aitken et al. 2005, 2007; Edmé et al. 2006; Garcia et al. 2006) markers. The RFLP, RAPD, AFLP and SSR markers are ideal for genetic fingerprinting and construction of linkage maps because they generate a large number of polymorphic markers, distributed anonymously across the genome and do not require prior gene sequence information. Because of their locus specificity, RFLP and SSR markers are also useful in identifying homo(eo)logous groups and in building consensus maps.

Recently, sequence-related amplified polymorphism (SRAP; Li and Quiros 2001) and target region amplification polymorphism (TRAP; Hu and Vick 2003) markers have been used in conjunction with AFLP markers to construct linkage maps in *S. officinarum* and *S. spontaneum* (Alwala et al. 2008). The PCR-based SRAP and TRAP markers detect intragenic polymorphisms. The SRAP primers are arbitrarily designed to contain AT- and GC-rich motifs that anneal to intron and exons, respectively (Li and Quiros 2001). Sequenced SRAP amplicons from *Brassica rapa* and *Brassica napus* (Li and Quiros 2001), *Cucurbita moschata* (Ferriol et al. 2003) and *S. officinarum* (Suman et al. 2008) when used in BLAST searches revealed significant similarities to reported gene sequences deposited in GenBank databases. The TRAP markers are generated with a fixed/forward primer designed from a gene or expressed sequence tag (EST) sequence and an arbitrary reverse primer that is similar to that of a SRAP primer. Using TRAP primers designed from resistance gene analogs, Miklas et al. (2006) reported that some of the polymorphisms produced on a pre-existing common bean (*Phaseolus vulgaris* L.) mapping population mapped to the vicinity of resistance gene QTLs. In sugarcane, BLAST searches of sequenced TRAP amplicons from a *S. spontaneum* clone revealed high homology with known gene sequences from other grass species (Alwala et al. 2006). Thus, SRAP and TRAP markers could be useful for QTL mapping because of their ability to target gene rich regions of the genome but this assertion remains to be seen in sugarcane. However, compared with the AFLP technique, the SRAP and TRAP techniques were less effective at generating the large number of genome-wide polymorphism required for linkage mapping in a genome as large as that of sugarcane. Alwala et al. (2008) observed an average of 12 polymorphic AFLP markers per primer combination whereas only five polymorphic markers per SRAP and/or TRAP primer combination. In the companion QTL study, several (relative to the proportion of polymorphism detected) SRAP and TRAP markers were found to be associated with sucrose related traits in sugarcane (Alwala et al. 2009).

The abundance of SSR sequences in expressed sequence tags (ESTs) has made EST-derived SSRs (EST-SSRs) an attractive source of polymorphisms (Kantley et al. 2002). SSR primers have been developed from EST sequences (EST-SSRs) in many crop species such as *Triticum* (Nicot et al. 2004), *Gossypium* (Han et al. 2006) and *Citrus* (Chen et al. 2006). EST-SSRs have also been developed for *Saccharum* species (Cordeiro et al. 2001; Pinto et al. 2004) using sugarcane EST sequences (Carson and Botha 2000; Casu et al. 2001). The Sugarcane Expressed Sequence Tag Project (SUCEST) from Brazil produced several thousands of EST sequences. Using the EST sequences in the SUCEST database, Oliveira et al. (2007) developed functionally associated EST-SSR and EST-RFLP markers and constructed an integrated genetic linkage map comprising previously mapped RFLP, AFLP and SSR markers (Garcia et al. 2006). Rossi et al. (2003) developed EST-derived resistance gene analog (RGA) markers and integrated them with RFLP and AFLP markers on the "R570" genetic linkage map.

Single nucleotide polymorphisms (SNPs) are naturally occurring point mutations in gene sequences. SNP markers provide a rich source of polymorphisms that could be useful to further elucidate genome architecture as well as detect marker-trait associations in crop species. The SNPs found in certain genes have been associated with the risks of developing diseases in humans (Mototani et al. 2005), and with fragrance (Bradbury et al. 2005) and gelatinization (Waters et al. 2006) traits in rice. The identification and development of SNP markers in sugarcane, from EST sequences adds one more tool to the sugarcane molecular geneticists' tool kit. In cultivated sugarcane, each homo(eo)logous group contains a multitude of chromosomes leading to several sets of alleles at each locus. Therefore, the frequency of SNPs in sugarcane is directly dependent on the number of chromosomes carrying the allele(s) rendering a complex detecting system (Storm et al. 2003). A combination of SNPs, rather than a single SNP, would likely be necessary to define an allele in sugarcane. Recently, McIntyre et al. (2006) detected several SNP markers based on eco-tilling and mapped single-dose SNP markers in a member of a sucrose phosphate synthase gene family. Cordeiro et al. (2006) identified several EST-derived SNPs (EST-SNPs) and accurately determined the quantitative presence of each base at 58 SNP loci using the pyrophosphate sequencing technology. However, not all of the potential SNPs could be used as single-dose markers. Only four of the 42 (9.5%) polymorphic SNPs were deemed single-dose in contrast to ~70% single-dose markers from other marker systems. Although the frequency of SNP-derived single-dose markers identified in sugarcane so far appears to be low compared to other marker systems, further research and improvement in genomic technologies might provide deeper insights into SNPs and their utility in mapping and tagging of traits of interest in sugarcane.

5.4 Types of Populations and Software used for Sugarcane Linkage Mapping

Traditional mapping populations such as F_2, inbred-backcross (BC), recombinant inbred lines (RILs) and/or doubled haploid (DH) lines are difficult to attain in *Saccharum* species. This is because *Saccharum* species are clonally propagated, highly heterozygous and suffer from inbreeding depression. Because of these difficulties, sugarcane geneticists have made use of segregating F_1 progeny from biparental crosses and/or selfed progeny from heterozygous parent as mapping populations to construct genetic linkage maps. The F_1 mapping population serves as a pseudo-testcross population from which linkage maps can be constructed for one or both parents. The markers pertaining to each parent are separated based on the co-segregation of alleles with that parent and this information can be used to construct a map for each of the parents using single- and possibly double-dose markers (Ming et al. 1998; Edmé et al. 2006; Aitken et al. 2007; Alwala et al. 2008).

The selfed progeny obtained from a clone, on the other hand, serves as a pseudo-F_2 mapping population because sugarcane clones are often *Saccharum* species hybrids. Unlike the F_1 mapping progeny where two parental maps can be constructed, the selfed population offers the development of a linkage map of the clone or the cultivar from which the population has been developed. Several genetic linkage maps have been developed using the selfed progeny of the cultivar "R570" (Grivet et al. 1996; Hoarau et al. 2001; Rossi et al. 2003). Efforts are underway in Louisiana to develop a linkage map of the popular cultivar "LCP85-384" using progeny derived from selfing.

The map generated from a biparental population is based not on pairing and recombination of chromosomes inherited from the two parents, but rather on heterozygosity and recombinations that occurred independently within each of the parents. Generally, a biparental population is expected to yield less informative markers for mapping compared to a selfed population for the same amount of genotyping effort. For example, when mapping with single-dose markers, those that occur in both parents are immediately rendered useless. Raboin et al. (2006) estimated that, when the cultivar "R570" was used as a parent in a biparental cross, about 40% of single-dose AFLP markers previously mapped in "R570" were common in the other parent of which, about 26% of them occurred as single-dose markers in both parents. The possibility to map with double-dose markers in a biparental cross is further reduced when markers occur in the single-dose in one parent and multiple-doses in the second parent. However, it must be noted here that, the appropriateness of a mapping population is best dictated by the overall goals of the research. For example, when the overall goal is to detect QTLs, a biparental population may offer the opportunity to combine several contrasting characteristics not present in a single selfed population.

The majority of sugarcane linkage maps (Hoarau et al. 2001; Aitken et al. 2005; Raboin et al. 2006) have been constructed using the MAPMAKER/ EXP software (Lander et al. 1987). The MAPMAKER/EXP software program performs multipoint linkage analysis between any number of marker loci and the marker order is presented based on the maximum likelihood approach. Indeed, this software package has been the choice for constructing linkage maps in a wide variety of crops. JoinMap ver 3.0 (Van Ooijen and Voorrips 2001), another software that has recently been used for sugarcane linkage mapping (Edmé et al. 2006; Garcia et al. 2006; Aitken et al. 2007; Alwala et al. 2008) utilizes a regression mapping algorithm. The latest version of JoinMap (ver 4.0) has an added mapping algorithm based on Monte Carlo Maximum Likelihood approach. JoinMap also has an option to converge two linkage maps developed using different mapping populations. Recently, Garcia et al. (2006) followed a method developed by Wu et al. (2002) that allows for the simultaneous estimation of recombination fractions and the linkage phases combining information from various types of molecular markers to construct a linkage map of sugarcane. In addition, the novel method of maximum likelihood proposed by Wu et al. (2002) determines linkage relationships of "ao x oo" (marker heterozygous in parent A and homozygous in parent B), "oo x ao" (marker homozygous in parent A and heterozygous in parent B) and "ao x ao" (double-simplex marker heterozygous in both parents) markers based on the linkage analysis of other marker types within the same co-segregation group.

5.5 Mapping Efforts in Sugarcane

5.5.1 *Saccharum spontaneum*

The first published genetic linkage map in a *Saccharum* species was constructed for the *S. spontaneum* clone "SES208" ($2n = 64$) using RFLP (da Silva et al. 1993, 1995) and RAPD markers (Al-Janabi et al. 1993) and a F_1 mapping progeny derived from ADP 85-0068 x SES 208 cross (ADP 85-0068 is a doubled haploid of SES 208; see Table 5-3 for details). The *S. spontaneum* map from da Silva et al. (1995) was based on 276 RFLP loci and 208 arbitrarily primed PCR loci spanning 64 linkage groups (LGs). The RAPD map by Al-Janabi et al. (1993) was based on 208 RAPD loci distributed across 42 LGs spanning approximately 1,500 cM in length. Relatively more saturated maps of two *S. spontaneum* clones were developed by Ming et al. (1998) based on RFLP markers. The linkage maps of the two *S. spontaneum* clones "IND81-146"

Table 5-3 Summary of mapping efforts in sugarcane.

Mapped species	Type of mapping population	Parents involved in creating mapping population	Mapped parent	Marker technique used	Reference
S. spontaneum	Ss x Ss F_1	ADP85-0068 x SES 208	SES 208 ($2n = 64$)	RFLP, AP-PCR, RAPD	da Silva et al. 1993, 1995
	So x Ss F_1	Green German x IND81-146	IND81-146 ($2n = 52$–56)	RFLP	Al-Janabi et al. 1993 Ming et al.1998, 2002
	So x Ss F_1	PIN84-1 x Muntok Java	PIN84-1 ($2n = 96$)	RFLP	Ming et al. 1998, 2002
	So x Ss F_1	Green German x IND81-146	IND81-146	SSR	Edmé et al. 2006
	So x Ss F_1	LA Striped x SES 147B	SES 147B ($2n = 64$)	AFLP, SRAP, TRAP	Alwala et al. 2008
S. officinarum	So x Sr F_1	LA Purple x Molokai 5829	LA Purple ($2n = 80$)	RAPD, RFLP, AP-PCR, AFLP	Mudge et al. 1996; Guimaráes et al. 1999
	So x Ss F_1	Green German x IND81-146	Green German ($2n = 97$–117)	RFLP	Ming et al. 1998, 2002
	So x Ss F_1	PIN84-1 x Muntok Java	Muntok Java ($2n = 140$)	RFLP	Ming et al. 1998, 2002
	So x Ss F_1	Green German x IND81-146	Green German	SSR	Edmé et al. 2006
	So x cultivar F_1	IJ76-514 x Q165	IJ76-514 ($2n = 80$)	AFLP, SSR	Aitken et al. 2007
	So x Ss F_1	LA Striped x SES 147B	LA Striped ($2n = 80$)	AFLP, SRAP, TRAP	Alwala et al. 2008
S. robustum	So x Sr F_1	LA Purple x Molokai 5829	Molokai 5829 ($2n = 80$)	RFLP, AP-PCR, AFLP	Guimaráes et al. 1999
Commercial Cultivar	Selfed Population	Selfed population of R570	R570 ($2n = 107$–117)	RFLP, AFLP, RGA, SSR	Grivet et al. 1996; Hoarau et al. 2001; Rossi et al. 2003

So x cultivar F_1	IJ76-514 x Q165	Q165 ($2n = 115$)	AFLP, SSR	Aitken et al. 2005
Cultivar x cultivar F_1	R570 x MQ76–53	R570 and MQ76–53	RFLP, SSR andAFLP,	Raboin et al. 2006
Cultivar x cultivar F_1	Q117 x MQ77-340	Q117 and MQ77-340	AFLP and SSR	Reffay et al. 2005
Selfed population	Selfed population of LCP85-384	LCP85-384	AFLP, SSR, SRAP	Pan et al. (Per. Comm.)
Commercial cultivars	Cultivar x cultivar F_1	SP80-180 x SP80-4966	AFLP, SSR, EST-RFLP, EST-SSR	Garcia et al. 2006; Oliveira et al. 2007

†ADP85-0068 is a doubled haploid derived from 'SES 208'

So = *S. officinarum*, Ss = *S. spontaneum*, Sr = *S. Robustum*

($2n$ = 52–56) and "PIN84-1" ($2n$ = 96) comprised of 257 and 194 linked loci across 69 and 72 LGs, respectively. The map length of "IND81-146" was 2,172 cM whereas that of "PIN84-1" was 1,395 cM. Edmé et al. (2006) constructed another linkage map of "IND81-146", using the same cross but a different set of progeny. The map was comprised of 46 SSR loci assembled into 10 LGs spanning across 614 cM. More recently, Alwala et al. (2008) constructed a linkage map of the *S. spontaneum* clone "SES 147B" ($2n$ = 64) using AFLP, SRAP and TRAP markers. This map comprising of 121 linked loci distributed across 45 LGs spans 1,491 cM.

5.5.2 Saccharum officinarum

Mudge et al. (1996) constructed the first published genetic linkage map of *S. officinarum* ("LA Purple", $2n$ = 80) using RAPD markers. The "LA Purple" map comprised 51 linkage groups with 160 loci and one morphological marker spanning 1,152 cM. Ming et al. (1998) produced maps of two clones, "Green German" ($2n$ = 97–117) and "Muntok Java" ($2n$ = 140) previously listed as *S. officinarum*, using RFLP markers. The "Green German" map comprised 289 linked single-dose markers distributed on 75 LGs and spanned 2,466 cM whereas that of "Muntok Java" comprised 214 markers assembled into 73 LGs and spanned 1,472 cM. The SSR based genetic linkage map of "Green German" (Edmé et al. 2006) comprised 91 loci covering 25 LGs and spanned 1,180 cM. The most extensive map of a *S. officinarum* clone, "IJ76-514", was developed by Aitken et al. (2007) using AFLP, SSR and randomly amplified DNA fingerprint (RAF) markers. The "IJ76-514" map comprised 534 (230 single-dose, 234 double-dose and 80 simple-duplex) markers across 123 LGs and covering 4,906 cM. The "IJ76-514" map was developed by incorporating both single-dose and double-dose markers based on statistical methodology suggested by Ripol et al. (1999) to increase the genome coverage. More recently, using AFLP markers and sequence targeted SRAP and TRAP markers Alwala et al. (2008) developed a genetic linkage map of "LA Striped" ($2n$ = 80) consisting of 49 LGs with 146 linked loci spanning 1,732 cM.

5.5.3 Saccharum robustum

A genetic linkage map of the *S. robustum* clone "Molokai 5829" ($2n$ = 80) was constructed using arbitrarily primed PCR, RFLP and AFLP markers (Guimaráes et al. 1999). The "Molokai 5829" map consisted of 301 single-dose loci covering 65 LGs and spanned 1,189 cM.

5.5.4 Cultivars and Commercial Crosses

A preliminary genetic linkage map of the cultivar "R570" was constructed using RFLP markers (Grivet et al. 1996) comprising 408 loci distributed across 96 LGs and grouped into 10 putative homologous groups. An extensive genetic linkage map of "R570" was later developed comprising 887 single-dose AFLP markers distributed on 120 LGs (Hoarau et al. 2001). The cumulative length of the "R570" AFLP map was 5,849 cM. Out of 120 LGs, 34 were grouped into 10 homologous groups previously obtained by Grivet et al. (1996). Raboin et al. (2006) produced another linkage map of "R570" using RFLP, SSR and AFLP markers comprising 424 linked loci distributed across 86 LGs and spanning 3,144 cM. Using a subset of individuals from the "R570" selfed population, Rossi et al. (2003) integrated 148 resistance gene analog (RGA) markers onto the AFLP map developed by Hoarau et al. (2001).

Aitken et al. (2005) developed an integrated genetic linkage map for the cultivar "Q165" using AFLP, SSR and RAF markers. The "Q165" map consisted of 1,069 loci distributed across 136 LGs. The cumulative length based on single-dose markers was 9,053 cM and 127 out of 136 LGs were grouped into eight homo(eo)logous groups. Reffay et al. (2005) constructed linkage maps of the cultivars "Q117" and "MQ77-340" using AFLP and SSR markers. The "Q117" map comprised 270 linked markers spanning 93 LGs (3,167 cM) whereas that of "MQ77-340" had 400 markers across 101 LGs (3,582 cM).

Garcia et al. (2006) developed an F_1 linkage map for a cross derived from two pre-commercial cultivars ("SP80-180" x "SP80-4966") using RFLP, SSR and AFLP markers. This map consisted of 357 linked loci assigned to 131 LGs spanning 2,602 cM. Later, this map was further refined and saturated by incorporating additional EST-SSR and EST-RFLP markers (Oliveira et al. 2007). The final map comprised of 664 markers scattered over 192 LGs with a cumulative length of 6,261 cM. Out of 192 LG, 120 were further formed into 14 homo(eo)logous groups. A summary of all genetic linkage maps generated thus far in sugarcane is presented in Table 5-3.

5.6 Segregation Distortion in *Saccharum* Interspecific Mapping Populations

Several sugarcane maps have been constructed using interspecific F_1 mapping populations. Segregation distortion is common in mapping populations derived from interspecific crosses because of disparities and asynchronous union between the two genomes. In a *S. officinarum* "LA Striped" x *S. spontaneum* "SES 147B" cross, Alwala et al. (2008) reported significant levels of segregation distortion in the *S. officinarum* (18%) and *S. spontaneum* (14%) parents. Using the same interspecific cross but a different sets of progeny, Edmé et al. (2006) reported 22% segregation distortion for the female parent (*S. officinarum* "Green German") whereas Ming et al. (1998) reported twice as much distortion (26%) for the male parent (*S. spontaneum* "IND81-146"). Differences in ploidy level, chromosome number, genome size and unequal transmission of gametes are all factors that can conspire to effect segregation distortion when interspecific populations are used for linkage mapping in sugarcane.

Non-biological factors such as small population sizes and fragment complexes consisting of co-migrating, non-allelic fragments can also affect segregation distortion. Nikaido et al. (2000) reported that in a conifer cross at least 30% of distorted AFLP alleles resulted from fragment complexes. Distorted markers would take on biological significance if they cluster around lethal genes or loci with high genetic load causing inbreeding depression. In pine (Kubisiak et al. 1995) and oak (Barreneche et al. 1998) tree maps, distorted markers tended to cluster in particular linkage groups. In sugarcane, Edmé et al. (2006) found evidence of clustering only for one linkage group and since the distorted markers were mostly in high dosage (non-SD), they postulated a possible role for double reduction in influencing distortion in that region of the genome. Also in sugarcane, Alwala et al. (2008) added distorted markers after the framework map (without distorted markers) had been constructed, and this resulted in the formation of new LGs made up exclusively of distorted markers. Some distorted markers also formed new LGs with previously unlinked markers and others mapped onto pre-existing LGs. Nevertheless, to accurately pinpoint loci with biological predisposition for segregation distortion in sugarcane would require population sizes larger than what has been used in these mapping studies.

5.7 Determining Ploidy Type using Mapping Populations

Classical and molecular cytogenetic studies have been instrumental in determining ploidy type in sugarcane (Bremer 1961; D'Hont et al. 1996). Ploidy type in sugarcane has also been inferred from sugarcane linkage mapping studies using two approaches: by detecting the proportion of coupling to repulsion phase linkages (Wu et al. 1992) and by χ^2 tests of the segregation ratios of single- versus multiple-dose markers (da Silva et al. 1993).

5.7.1 Proportion of Coupling to Repulsion Phase Linkages

A crude method to detect repulsion phase linkages in mapping studies is to invert the presence or absence ($0 \leftrightarrow 1$) score of the original single-dose markers. The inverted data matrix is appended to the original data set followed by two-point linkage analysis. The presence of repulsion phase linkage is indicated by co-localization of the original single-dose marker with its corresponding inverted marker. The proportion of coupling to repulsion phase linkages is used to distinguish between allopolyploids and autopolyploids in this way because repulsion phase linkages are theoretically much more difficult to detect in autopolyploids with polysomic inheritance than allopolyploids with disomic inheritance (Qu and Hancock 2001). In contrast, when chromosomes pair preferentially at meiosis (allopolyploids), the detection power of coupling and repulsion phase linkages is the same. Therefore, in allopolyploids, if the chromosomes pair preferentially, the original single-dose markers assigned to the co-segregation group will be in repulsion with a comparable number of inverted co-segregating single-dose markers, whereas their number will be much lower if the chromosomes pair randomly. This is usually accomplished by testing the observed ratio of coupling to repulsion phase linkages in a population against the theoretical ratio of 1:1 (for disomic pairing) and 1:0 (for random pairing) using a χ^2 test. Wu et al. (1992) observed that the proportion of repulsion phase linkages detected in polyploids is one-fourth that of coupling phase linkages and estimated that a minimum of about 300 individuals would be required to detect repulsion phase linkages in autopolyploids (with e.g., $2n = 8x$) with only about 75 individuals required for the corresponding allopolyploids.

It should be noted, however, that the observed repulsion phase linkages do not imply the actual recombination fraction (r) because in reality two alleles linked in repulsion phase would reside on two homologous chromosomes. The two alleles from the two homologous chromosomes can be brought together in the same gamete as a result of independent assortment. The theoretical threshold r-value in the two-point linkage analysis, to detect repulsion phase linkages, should be greater than the corresponding r-value arising from independent assortment. Qu and Hancock (2001) estimated a threshold r-value of ≥ 0.43 for an auto-octoploid. Generally the corresponding threshold value should be $r/2$ for the allopolyploid.

In sugarcane mapping studies, no repulsion phase linkages were detected in the *S. spontaneum* clones "SES 208" and "SES 147B" and it was suggested that these clones are autopolyploids undergoing random pairing (Al-Janabi et al. 1993; da Silva et al. 1993; Alwala et al. 2008). Conversely, repulsion phase linkages were detected for the *S. officinarum* clones "LA Purple", "Green German" and "IJ76-514" implying that *S. officinarum* behaves as an allopolyploid (Mudge et al. 1996; Edmé et al. 2006; Aitken et al. 2007). Detection of repulsion phase linkages would mean that a euploid map in *S. officinarum* could have less than $2n = 80$ linkage groups.

5.7.2 χ^2 Tests

The proportion of expected vs. observed single-dose and multi-dose markers provides additional information that could be used to bolster the results from repulsion phase linkage analysis. For allopolyploids, the expected ratio of DD, TD and QD markers is $1/4$, $1/8$ and $1/16$, respectively (da Silva et al. 1993). Therefore, the expected SD value is equal to $1-(1/4 + 1/8 + 1/16)$

= 0.56. Likewise, for autopolyploids, the expected SD value is estimated to be 0.70 (Table 5-4). The ploidy level could be determined based on the χ^2 tests of the proportion of SD versus multi-dose markers.

Table 5-4 χ^2 tests for determining the ploidy type based on observed vs. expected proportion of single-dose (SD) and multi-dose (MD) markers.

	Observed	Expected	
		Allo-octoploidy	Auto-octoploidy
SD markers	SD_{obs}	0.56	0.70
MD markers	MD_{obs}	0.44	0.30
χ^2		If NS = Allopolyploid	If NS = Autopolyploid
		If ** = Non-allopolyploid	If ** = Non-autoploid

NS = Non significant, ** = Significant at specified probability level.

Based on χ^2 tests it was determined that the *S. spontaneum* clones "SES 208", "IND81-146" and "SES 147B" were autopolyploids (Al-Janabi et al. 1993; da Silva et al. 1993; Edmé et al. 2006; Alwala et al. 2008) whereas, the *S. officinarum* clones "Green German", "IJ76-514" and "LA Striped" were determined to be allopolyploids (Edmé et al. 2006; Aitken et al. 2007; Alwala et al. 2008). Taken into consideration the combined results obtained from several mapping studies using these two techniques (the proportion of coupling to repulsion phase linkages and χ^2 tests) it is evident that *S. spontaneum* behaves as an autopolyploid with polysomic inheritance while *S. officinarum* behaves as an allopolyploid with partial (preferential) pairing.

5.8 Estimation of Genome Size and Coverage

The genome size of any species can be estimated by using the formula proposed by Hulbert et al. (1988):

$$G = \frac{n(n-1)}{2} \cdot \frac{2x}{y_x}$$

where n is the number of mapped markers, y_x is the number of two-point linkages at a distance equal to x cM. From the estimated genome size (G), the probability (P_n) that a randomly placed marker not covered by n randomly placed markers could be estimated as follows:

$$P_n = 2r/n+1 \left[(1 - x/2G)^{n+1} - (1 - x/G)^{n+1} \right] + (1 - rx/G)(1 - x/G)^n$$

where r is the number of chromosomes, x is the length of intervals in cM and G is the estimated genome size. Based on Bishop et al. (1983), the genome coverage (E) is further deduced as:

$$E = 1 - P_n$$

The genome coverage E denotes the % probability for any new marker to be placed within x cM of any existing linked marker on a map.

Based on the Hulbert et al. (1988) method, the genome size of the *S. spontaneum* clone, "SES 208", was estimated to be 4,214 cM. Since there were no repulsion phase linkages observed in the RFLP map, the estimated genome size was reduced by half and thus found to be 2,107 cM. The genome coverage (Bishop et al. 1983) was estimated to be 86% based on the 2,107 cM genome size, 64 chromosomes, 216 mapped markers at an estimated 25 cM interval (da Silva et al. 1993). (Note: The 86% genome coverage at a 25 cM interval implies that there is an 86% probability for any new marker to be placed within a 25 cM interval of an already existing linked marker). Comparatively similar results were obtained by Al-Janabi et al. (1993). The estimated genome size for "SES 208" based on RAPD markers was found to be 2.547 cM and the estimated genome coverage was found to be 85% based on 208 linked markers at a 30 cM interval.

Likewise, the estimated genome size of the *S. spontaneum* clone "SES 147B" ($2n = 64$) was found to be 3,232 cM with a genome coverage of 63% based on 121 linked markers at a 30 cM interval (Alwala et al. 2008). The estimated genome size of the *S. officinarum* clone "LA Striped" ($2n = 80$) was found to be 2,448 cM with 76% genome coverage based on 129 linked markers at a 30 cM interval. Since Alwala et al. (2008) used all of the mapped markers including duplex and other multi-dose markers there is a chance that genome size and genome coverage might have been overestimated. Using only single-dose markers, the estimated genome coverage of 63% for the *S. officinarum* clone "IJ76-514" (Aitken et al. 2007) is probably more realistic.

A different method has been adapted for estimating genome size and coverage in sugarcane cultivars because chromosome number is not consistent among cultivars ($2n = 100$ to 130). Hoarau et al. (2001) estimated the genome size in cultivar R570 based on the length of the longest LG. In their study, there were 112 LG and the length of the longest LG was found to be 150 cM. The genome size was estimated to be $112 \times 150 = 16,800$ cM (~17,000 cM). Likewise, Aitken et al. (2005) estimated the genome size of cultivar Q165 as $120 \times 150 = 18,000$ cM. It is clear from these estimates of genome size and coverage that, producing a saturated map in any of the *Saccharum* species or cultivars remains a gargantuan task. Although current sugarcane maps containing more than 1,000 linked markers (Hoarau et al. 2001; Aitken et al. 2005) may seem large relative to other crop species, they are still incomplete and unsaturated. Furthermore, in none of the *Saccharum* species maps, does the number of LGs reflect the basic chromosome number of the species and maps containing LGs with only two markers abound. Therefore, with all the mapping efforts so far, it could be estimated that only about one-half of the genome has been covered, leaving the prospects of producing a saturated sugarcane map to future advancements in molecular marker technology.

5.9 Comparative Mapping

Comparative mapping could simply be defined as the alignment of chromosomes of different genera based on genome-wide colinearity of common (orthologous) DNA markers (Bennetzen and Freeling 1993). Comparative and/or synteny mapping offers a repertoire of chromosomal and genomic information related especially to loci controlling agronomic traits identified in one species with potential usefulness in other species. From comparative genetic mapping studies, the conservation of gene order has been well established between divergent model species (such as *Arabidopsis*, rice, etc.) with smaller genomes across various plant families (Bennetzen 1996; Freeling 2001). Such an approach of a common genetic framework and correspondence of QTLs across divergent species can enable crop improvement via marker-assisted selection in recalcitrant species such as sugarcane.

In the early 1990s significant research was devoted to synteny mapping among members of the Poaceae (Doebley et al. 1990; Hulbert et al. 1990; Ahn and Tanksley 1993; Kurata et al. 1994), Solanaceae (Tanksley et al. 1992) and Brassicaceae (Song et al. 1988) families. Within the grass family, sorghum is the closest relative of sugarcane with their divergence dating back to only some five million years. Asnaghi et al. (2000) found up to 78% of cross-hybridization with sugarcane DNA using sorghum heterologous probes as compared to 68% and 45% using maize and rice heterologous probes, respectively. With high levels of taxonomical proximity and synteny between sugarcane and sorghum, the genomic information from sorghum could provide an excellent channel for either molecular mapping or positional cloning in sugarcane. Moreover, given the smaller genome size of sorghum (748–772 Mbp/1C) compared to sugarcane (*S. officinarum*: 2,547–3,605 Mbp/1C) (Arumuganathan and Earle 1991), a detailed comparative study between sorghum and sugarcane would benefit sugarcane geneticists as sugarcane has no known diploid progenitor species.

The most detailed comparative mapping effort involving *Saccharum* and sorghum was done by Ming et al. (1998, 2002) (Fig. 5-1). A consensus *Saccharum* map was first assembled consisting of 13 homologous groups (HGs) using the common markers from linkage groups of four parents (Table 5-3; Ming et al. 1998). Although the sugarcane consensus map contributed up to 70% of the sorghum genome coverage, a saturated map is still a requirement to confirm if any of the sugarcane groups are connected. Nevertheless, seven out of 10 basic *S. officinarum* groups shared high homology with sorghum LGs A, D, F, G, H, I and J while the other three had likely correspondence with LGs B, C and E. Likewise, six out of eight basic chromosomes of *S. spontaneum* are likely to have synteny with sorghum LGs A, D, F, H, I and J whereas the other two might have correspondence with LGs B, C, E and G. The high degree of colinearity between sugarcane and sorghum suggests that there might have been very few inter-chromosomal re-arrangements since the two species diverged. However, certain deviations from colinearity exist between sugarcane and sorghum that has been attributed to ancient genome duplications (Pereira et al. 1994). It was observed that a three-fold higher recombination rate exists in sugarcane compared to sorghum. The higher recombination rate across genomic regions places sugarcane in an advantageous position (fitness advantage) to evolve complex pattern of higher number of chromosomes.

Small segmental colinearity has also been confirmed between sugarcane and maize using maize probes (D'Hont et al. 1993; da Silva et al. 1993) although these results are complicated due to segmental polyploidy of the maize genome (Dufour et al. 1997). Comparison between *S. officinarum* and *S. spontaneum* revealed that 11 out of 13 HGs in sugarcane consensus map had significant homology between *S. officinarum* and *S. spontaneum* chromosomes. Also it was found that 11 major structural chromosomal arrangements distinguished *S. officinarum* from *S. spontaneum* although many more arrangements could potentially be discovered with further research (Ming et al. 2002)

As mentioned earlier, comparative mapping serves as a major tool to identify genomic regions associated with quantitative trait loci (QTL). Using probes from sorghum, maize and rice, Ashnagi et al. (2000) identified a sugarcane chromosomal region harboring a rust resistance gene. Another comparative mapping between sugarcane and sorghum (McIntyre et al. 2005) revealed that 17 out of 31 mapped sugarcane resistant gene analog (RGA) markers were syntenically mapped onto sorghum linkage groups. Interestingly three of the RGAs, which were shown to have significant association with rust resistance in sugarcane, were also associated with a major rust resistant QTL in sorghum. A near perfect marker colinearity between sorghum and three *Saccharum* species genomes was also observed by Guimaraes et al. (1997). In addition, a genomic region harboring flowering time locus in sugarcane was found to be homologous to one of the sorghum linkage groups (Guimaraes and Sobral 1998) and maize chromosome 9 (Lin et al. 1995), which in turn was found to be orthologous to the *Se1/Se3* region of rice chromosome 3 (Paterson et al. 1995). Jordan et al. (2004) detected significant associations between RFLP markers and stalk number and suckering in sugarcane and interestingly seven of those markers had been mapped in sorghum. The use of heterologous probes show tremendous promise in map-based cloning, however a complex polyploid like sugarcane would require further genomic micro-colinearity at the locus level. Micro-colinearity offers a better path to chromosome walking and thus would pave an easier way to identify genes of interest. Lastly, a comparative mapping approach even within a genus can also offer ways to validate quantitative trait loci (QTL) for agronomic traits of interest, for example, using similar markers and different mapping populations. Piperidis et al. (2008) identified several chromosomal regions/QTLs associated with sugar content-related traits by comparing the maps of the cultivar "R570" and "Q165" with those of Q117 and MQ77–340 using common SSR and AFLP markers.

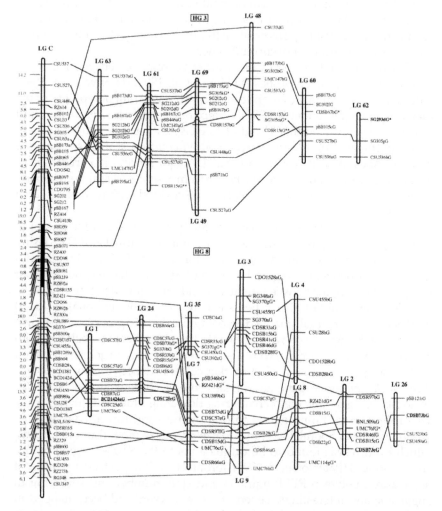

Figure 5-1 An example of synteny mapping between sugarcane and sorghum chromosomal regions. The lines connecting various linkage groups indicate that the loci were detected using the same probes in sugarcane as well as sorghum. For further details refer to Ming et al. (1998). The figure has been reproduced with kind permission from the Genetics Society of America.

5.10 Molecular Cytogenetics

Although genetic linkage maps have provided insights on the genetic recombinations and gene order, discrepancies still exist between genetic and physical linkages. A large scale development of cytogenetic markers is necessary to enable assigning linkage groups to various homologous groups, to allow comparisons between genetic and physical maps and to complement positional or map-based cloning. Advancements in cytogenetics, such as fluorescence in situ hybridization (FISH) and genomic in situ hybridization (GISH), have been extensively used in various crops to determine ploidy type, to track alien introgression chromosomes and to test genomic exchanges (Bennett et al. 1992; Schwarzacher et al. 1992; Kenton et al. 1993; Thomas et al. 1994).

In sugarcane, in situ hybridization methods ushered in new tools that facilitated a deeper understanding of the genomic architecture and genetic diversity (Glaszmann et al. 1990; Pan et al. 2000). Application of modern molecular cytogenetics have divulged the double genomic structure of modern sugarcane cultivars, the basic chromosome number of *S. officinarum* ($x = 10$) and *S. spontaneum* ($x = 8$) and ploidy type ($n = 8$) (D'Hont et al. 1996). These studies have also demonstrated the occurrence of recombination between *S. officinarum* and *S. spontaneum* chromosomes in modern sugarcane cultivars although the recombination events are infrequent. Furthermore, the first link between the genetic and physical maps was developed by physically localizing the *S. officinarum* and *S. spontaneum* rRNA genes onto the "R570" genetic linkage map (D' Hont et al. 1996).

A well established BAC-anchored molecular cytogenetic map of sorghum (Islam-Faridi et al. 2002; Kim et al. 2002) might also offer opportunities to decipher genetic/physical linkages in sugarcane. The genetic linkages in sorghum have been integrated with physical distances and all linkage groups have been associated with specific chromosomes. A multidisciplinary approach by integrating molecular genetics and cytogenetics will ultimately enable a comprehensive genetic dissection of the complex sugarcane genome. Readers may refer to Chapter 2 for a detailed description of sugarcane molecular cytogenetics.

5.11 Conclusion

The genus *Saccharum*, collectively known as sugarcane, contains species with the most complex genomes among plants. The species are usually vegetative propagated and characterized by high degree of polyploidy and heterozygosity. The cultivated sugarcane is even more complex with a double genome structure and varying numbers of homologous and homoeologous chromosomes. The complexity of the genome has complicated genetic linkage mapping efforts such that sugarcane lagged behind diploid or even some other polyploid crop species. The complexity stems from a combination of factors including the random pairing of multiple chromosomes, uncertain gametic frequencies, exceptionally wide range of marker segregation patterns, unknown genome constitution, and unclear inheritance patterns. Furthermore, diploid relatives of sugarcane that could be exploited in linkage mapping studies to circumvent some of the issues associated with polyploidy are either unknown or extinct.

It was not until the development of the single-dose markers mapping approach that it became possible to construct genetic linkage maps in sugarcane and other polyploid and highly heterozygous species. Single-dose markers are present in one parent, absent in the other and segregate in a 1:1 ratio in the progeny regardless of ploidy level, genome constitution or chromosome behavior. In populations derived from selfing a heterozygous parent, single-dose markers segregate in a 3:1 ratio. Single-dose markers are abundant in sugarcane representing about 70% of the detectable polymorphic loci and could easily be detected in a relatively small population of about 75 individuals.

The early sugarcane genetic linkage maps were constructed using the labor intensive RFLP marker technique. The power of the RFLP was its ability to detect locus specificity which is useful in identifying homo(eo)logous linkage groups. With the onset of PCR-based markers, mapping studies have transitioned to the use of RAPD, AFLP, SSR, EST-SSR, SRAP and TRAP markers with ongoing efforts to optimize and use SNP markers. The locus specificity of the PCR-based SSR is increasingly being used to group multiple homo(eo)logous linkage groups in sugarcane mapping studies.

Although current sugarcane maps containing more than 1,000 linked markers may seem large relative to other crop species, they are still incomplete and unsaturated. In none of the *Saccharum* species maps, for example, does the number of LGs reflect the basic chromosome number of the species and LGs assembled from two markers abound. Inspite of these shortcomings, genetic linkage mapping efforts in sugarcane have come a long way towards improving our understanding of its genetic constitution and architecture. With the rapid developments in high-throughput techniques and advancements in genomic technologies, it will not be surprising to obtain a saturated map in the near future. Comparative mapping between sugarcane and closely related grass species such as sorghum should help increase genome coverage of sugarcane maps while molecular cytogenetics should help bridge the gap between genetic and physical linkages.

Acknowledgements

The authors are grateful to Ms. Ruth A. Issacson, Genetics Society of America, for permitting the use of figures published in the journal "Genetics".

References

Ahn S, Tanksley SD (1993) Comparative linkage maps of the rice and maize genomes. Proc Natl Acad Sci USA, 90: 7980–7984.

Aitken KS, Jackson PA, McIntyre CL (2005) A combination of AFLP and SSR markers provide extensive map coverage and identification of homo(eo)logous linkage groups in a sugarcane cultivar. Theor Appl Genet 110: 789–801.

Aitken KS, Jackson PA, McIntyre CL (2007) Construction of genetic linkage map for *Saccharum officinarum* incorporating both simplex and duplex markers to increase genome coverage. Genome 50: 742–756.

Al-Janabi SM, Honeycutt RJ, McClelland M, Sobral BWS (1993) A genetic linkage map of *Saccharum spontaneum* L. 'SES 208'. Genetics 134: 1249–1260.

Alwala S, Suman A, Arro JA, Veremis JC, Kimbeng CA (2006) Target Region Amplification Polymorphism (TRAP) for assessing genetic diversity in sugarcane germplasm collections. Crop Sci 46: 448–455.

Alwala S, Kimbeng CA, Veremis JC, Gravois KA (2008) Linkage mapping and genome analysis in *Saccharum* interspecific cross using AFLP, SRAP and TRAP markers. Euphytica 164: 37–51.

Alwala S, Kimbeng CA, Veremis JC, Gravois KA (2009) Identification of molecular markers associated with sugar-related traits in a *Saccharum* interspecific cross. Euphytica 167: 127–142.

Arumuganathan K, Earle ED (1991) Nuclear DNA content of some important plant species. Plant Mol Biol Rep 9: 208–218.

Asnaghi C, Paulet F, Kaye C, Grivet L, Deu M, Glaszmann JC, D'Hont A (2000) Application of synteny across Poaceae to determine the map location of a sugarcane rust resistance gene. Theor Appl Genet 101: 962–969.

Barreneche T, Bodenes C, Lexer C, Trontin JF, Fluch S, Streiff R, Plomion C, Roussel G, Steinkellner H, Burg K, Favre JM, Glossl J, Kremer A (1998) A genetic linkage map of *Quercus robur* L. (pedunculate oak) based on RAPD, SCAR, microsatellite, minisatellite, isozyme and 55 rDNA markers. Theor Appl Genet 97: 1090–1103.

Bennett ST, Kenton AY, Bennett MD (1992) Genomic in situ hybridization reveals the alloploid nature of *Milium montianum* (Gramineae). Chromosoma 101: 420–424.

Bennetzen JL (1996) The use of comparative genome mapping in the identification, cloning and manipulation of important plant genes. In: BWS Sobral (ed) *The Impact of Plant Molecular Genetics*. Birkhauser, Boston, MA, USA, pp 71–85.

Bennetzen JL, Freeling M (1993) Grasses as single genetic system: genome composition, colinearity and compatibility. Trends Genet 9: 259–261.

Bhat SR, Gill BS (1985) The implication of 2n egg gametes in nobilisation and breeding of sugarcane. Euphytica 34: 377–384.

Bishop DT, Cannings C, Skolnick M, Williamson JA (1983) The number of polymorphic clones required to map the human genome. In: BS Weir (ed) *Statistical Analysis of DNA Sequence Data*. Marcel Dekker, NY, USA, pp 118–200.

Bonnierbale MW, Paisted RL, Tanksley SD (1988) RFLP maps based on a common set of clones reveal modes of chromosomal evolution in potato and tomato. Genetics 120: 1095–1103.

Bradbury LMT, Fitzgerald TL, Henry RJ, Jin QS, Waters DLE (2005) The gene for fragrance in rice. Plant Biotechnol J 3: 363–370.

Brandes EW (1958) Origin, classification and characteristics. In: E Artschwager, EW Brandes (eds) Sugarcane (*S. officinarum* L.). USDA Agri Handbook 122: 1–35, 260–262.

Bremer G (1961) Problems in breeding and cytology of sugarcane. Euphytica 10: 59–78.

Butterfield MK (2005) Molecular breeding of sugarcane using DNA markers and linkage disequilibrium mapping. Proc Int Soc Sugar Cane Technol 25: 481–486.

Carson DL, Botha FC (2000) Preliminary analysis of expressed sequence tags for sugarcane. Crop Sci 40: 1769–1779.

Castiglioni P, Ajmone-Marsan P, van Wijk R, Motto M (1999) AFLP markers in a molecular linkage map of maize: codominant scoring and linkage group distribution. Theor Appl Genet 99: 425–431.

Casu R, Dimmock C, Thomas M, Bower N, Knight D, Grof C, McIntyre CL, Jackson P, Jordan D, Whan V, Drenth J, Tao Y, Manners J (2001) Genetic and expression profiling in sugarcane. Proc Int Soc Sugar Cane Technol 24: 626–627.

Chen C, Zhou P, Choi YA, Huang S, Gmitter FG Jr (2006) Mining and characterizing microsatellites from citrus ESTs. Theor Appl Genet 112: 1248–1257.

Collard BC, Mackill DJ (2008) Marker-assisted selection: an approach for precision plant breeding in the twenty-first century. Phil Trans R Soc Lond B Biol Sci 363: 557–572.

Cordeiro GM, Casu R, McIntyre CL, Manners JM, Henry RJ (2001) Microsatellite markers from sugarcane (*Saccharum* spp) ESTs cross transferable to *Erianthus* and *Sorghum*. Plant Sci 160: 1115–1123.

Cordeiro GM, Eliott F, McIntyre CL, Casu RE, Henry RJ (2006) Characterization of single nucleotide polymorphisms in sugarcane ESTs. Theor Appl Genet 113: 331–343.

Cordeiro GM, Amouyal O, Eliott FG, Henry RJ (2007) Sugarcane. In: C Kole (ed) *Genome Mapping and Molecular Breeding in Plants, vol 3: Pulses, Sugar and Tuber Crops*. Springer, Heidelberg, Berlin, New York, pp 175–204.

D'Hont A, Lu YH, Feldmann P, Glaszmann JC (1993) Cytoplasmic diversity in sugar cane revealed by heterologous probes. Sugar Cane 1: 12–15.

D' Hont A, Grivet L, Feldmann P, Rao PS, Berding N, Glazmann JC (1996) Characterisation of the double genome structure of modern sugarcane cultivars (*Saccharum* spp.) by molecular cytogenetics. Mol Gen Genet 250: 405–413.

D' Hont A, Ison D, Alix K, Roux C, Glazmann JC (1998) Determination of basic chromosome numbers in the genus *Saccharum* by physical mapping of ribosomal RNA genes. Genome 41: 221–225.

Daniels J, Roach BT (1987) Taxonomy and evolution in sugarcane. In: DJ Heinz (ed) Sugarcane Improvement through Breeding. Elsevier, Amsterdam, The Netherlands, pp 7–84.

Dietrich WF, Miller J, Steen R, Merchant MA, Damron Boles D, et al. (1996) A comprehensive genetic map of the mouse genome. Nature (Lond) 380: 149–152.

Doebley J, Stec A, Wendel J, Edwards M (1990) Genetic and morphological analysis of a maize-teosinte F_2 population: implications for the origin of maize. Proc Natl Acad Sci USA 87: 9888–9892.

Dubcovsky J (2004) Marker-assisted selection in public breeding programs: The wheat experience. Crop Sci 44: 1895–1898.

Dufour P, Deu M, Grivet L, D'Hont A, Paulet F, Bouet A, Lanaud C, Glaszmann JC, Hamon P (1997) Construction of a composite sorghum genome map and comparison with sugarcane, a related complex polyploidy. Theor Appl Genet 94: 409–418.

Edmé SJ, Glynn NG, Comstock JC (2006) Genetic segregation of microsatellite markers in *Saccharum officinarum* and *S. spontaneum*. Heredity 97: 366–375.

Ferriol M, Picó B, Nuez F (2003) Genetic diversity of a germplasm collection of *Cucurbita pepo* using SRAP and AFLP markers. Theor Appl Genet 107: 271–282.

Fisher RA (1949) The linkage problem in a tetrasomic wild plant, *Lythrum salicaria*. In: Proc 8th Int Congr Genet, pp 225–233.

Freeling M (2001) Grasses as single genetic system: Reassessment. Plant Physiol 125: 1191–1197.

Frewen BE, Chen THH, Howe GT, Davis J, Rhode A, Boerjan W, Bradshaw HD (2000) Quantitative trait loci and candidate gene mapping of bud set and bud flush in *Populus*. Genetics 154: 837–845.

Garcia AAF, Kido EA, Meza AN, Souza HMB, Pinto LR, Pastina MM, Leite CS, da Silva JAG, Ulian EC, Figueira A, Souza AP (2006) Development of an integrated genetic map of a sugarcane (*Saccharum* spp.) commercial cross, based on a maximum-likelihood approach for estimation of linkage and linkage phases Theor Appl Genet 112: 298–314.

Glaszmann JC, Lu YH, Lanaud C (1990) Variation of nuclear ribosomal DNA in sugarcane. J Genet Breed 44: 191–198.

Grattapaglia D, Sederoff R (1994) Genetic linkage maps of *Eucalyptus grandis* and *Eucalyptus urophylla* using a pseudo-testcross: mapping strategy and RAPD markers. Genetics 137: 1121–1137.

Grivet L, D'Hont A, Roques D, Feldmann P, Lanaud C, Glaszmann J-C (1996) RFLP mapping in a highly polyploid and aneuploid interspecific hybrid. Genetics 142: 987–100.

Guimaráes CT, Sobral BWS (1998) The *Saccharum* complex: relation to other Andropogoneae. Plant Breed Rev 16: 269–288.

Guimaráes CT, Sills GR, Sobral BWS (1997) Comparative mapping of Andropogoneae: *Saccharum* (sugarcane) and its relation to sorghum and maize. Proc Natl Acad Sci USA 94: 14261–14266.

Guimaráes CT, Honeycutt RJ, Sills GR, Sobral BWS (1999) Genetic linkage maps of *Saccharum officinarum* L. and *Saccharum robustum* Brandes & Jew. Ex Grassl Genet Mol Biol 22: 125–132.

Han ZG, Wang C, Song X, Guo W, Gou J, Li C, Chen X, Zhang T (2006) Characteristics, development and mapping of *Gossypium hirsutum* derived EST-SSRs in allotetraploid cotton. Theor Appl Genet 112: 430–439.

Hemmat M, Weedon NF, Manganaris AG, Lawson DM (1994) Molecular marker linkage map for apple. J Hered 85: 4–11.

Hoarau JY, Offmann B, D'Hont A. Risterucci AM, Roques D, Glaszmann JC, Grivet L (2001) Genetic dissection of a modern sugarcane cultivar (*Saccharum* spp.). I. Genome mapping with AFLP markers. Theor Appl Genet 103: 84–97.

Hu JG, Vick BA (2003) Target region amplification polymorphism: A novel marker technique for plant genotyping. Plant Mol Biol Rep 21: 289–294.

Hulbert SH, Ilot TW, Egg EJL, Lincolne SE, Lander S, Michelmore RW (1988) Genetic analysis of the fungus *Bremia lactucae* using restriction fragment length polymorphisms. Genetics 120: 947–958.

Hulbert SH, Richter TE, Axtell JD, Bennetzen JL (1990) Genetic mapping and characterization of sorghum and related crops by means of maize DNA probes. Proc Natl Acad Sci USA 87: 4251–4255.

Islam-Faridi MN, Childs KL, Klein PE, Hodnett G, Menz MA, Klein RR, Rooney WL, Mullet JE, Stelly DM, Price HJ (2002) A molecular cytogenetic map of sorghum chromosome 1: fluorescence *in situ* hybridization analysis with mapped bacterial artificial chromosomes. Genetics 161: 345–353.

Jordan DR, Casu RE, Besse P, Carroll BC, Berding N, McIntyre CL (2004) Markers associated with stalk number and sugarcane collocate with tillering and rhizomatousness QTLs in sorghum. Genome 47: 988–993.

Kantley RV, Rota ML, Matthews DE, Sorrells ME (2002) Data mining for simple sequence repeats in expressed sequence tags from barley, maize, rice, sorghum and wheat. Plant Mol Biol 48: 501–510.

Kenton A, Parokonny A, Gleba Y, Bennett M (1993) Characterization of the *Nicotiana tabacum* L. genome by molecular cytogenetics. Mol Gen Genet 240: 159–169.

Kim J-S, Childs KL, Islam-Faridi MN, Menz MA, Klein RR, Klein PE, Price HJ, Mullet JE, Stelly DM (2002) Integrated karyotyping of sorghum by in situ hybridization of landed BACs. Genome 45: 402–412.

Kiss GB, Csanadi G, Kalman K, Kalo P, Okresz L (1993) Construction of a basic genetic map for alfalfa using RFLP, RAPD, isozyme and morphological markers. Mol Gen Genet 238: 129–137.

Kole C (ed) (2006-2007) Genome Mapping and Molecular Breeding in Plants, vol 1–7. Springer, Berlin, Heidelberg, New York.

Kriegner A, Cervantes JC, Burg K, Mwanga ROM, Zhang D (2003) A genetic linkage map of sweetpotato [*Ipomoea batatas* (L.) Lam.] based on AFLP markers. Mol Breed 11: 169–185.

Kubisiak TL, Nelson CD, Nance WL, Stine M (1995) RAPD linkage mapping in a longleaf pine x slash pine F1 family. Theor Appl Genet 90: 1119–1127.

Kurata N, Moore G, Nagamura Y, Foote T, Yano M, Minobe Y, Gale M (1994) Conservation of genome structure between rice and wheat. Bio/Technology 12: 270–278.

Lander ES, Green P, Abrahamson J, Barlow A, Daly MJ, Lincoln SE, Newburg L (1987) MAPMAKER: an interactive computer package for constructing primary genetic linkage maps of experimental and natural populations. Genomics 1: 174–181.

Li G, Quiros CF (2001) Sequence related amplified polymorphism (SRAP), a new marker system based on a simple PCR reaction: Its application to mapping and gene tagging in *Brassica*. Theor Appl Genet 103: 455–461.

Lin Y-R, Schertz KF, Paterson AH (1995) Comparative analysis of QTLs affecting plant height and maturity across the Poaceae, in reference to an interspecific sorghum population. Genetics 141: 391–411.

Manners J, McIntyre L, Casu R, Cordeiro G, Jackson M, Aitken K, Jackson P, Bonnett G, Lee S, Henry RJ (2004) Can genomics revolutionise genetics and breeding in sugarcane? Proc 4th Int Crop Sci Congr, Brisbane, Australia.

McIntyre CL, Casu RE, Drenth J, Knight D, Whan VA, Croft BJ, Jordan DR, Manners JM (2005) Resistance gene analogues in sugarcane and sorghum and their association with quantitative trait loci for rust resistance. Genome 48: 391–400.

McIntyre CL, Jackson M, Cordeiro G, Amouyal O, Eliott F, Henry RJ, Casu RE, Hermann S, Aitken KS, Bonnett GD (2006) The identification and characterization of alleles of sucrose phosphate synthase gene family III in sugarcane. Mol Breed 18: 39–50.

Miklas PN, Hu J, Grünwald NJ, Larsen KM (2006) Potential application of TRAP (Targeted Region Amplified Polymorphism) markers for mapping and tagging disease resistance traits in common bean. Crop Sci 46: 910–916.

Milbourne DM, Bradshaw JE, Hackett CA (2008) Molecular mapping and breeding in polyploid crop plants. In: C Kole, AG Abbott (eds) Principles *and* Practices *of* Plant Genomics, vol 2: Molecular Breeding. Science Publishers, New Hampshire, Plymouth, Jersey, USA, pp 355–394.

Ming R, Liu SC, Lin YR, Da Silva JAG, Wilson W, Braga D, van Devnze A, Wenslaff F, Wu KK, Moore PH, Burnquist W, Sorrells ME, Irvine JE, Paterson AH (1998) Detailed alignment of *Saccharum* and *Sorghum* chromosomes: comparative organization of closely related diploid and polyploid genomes. Genetics 150: 1663–1682.

Ming R, Liu S-C, Bowers JE, Moore PH, Irvine JE, Paterson AH (2002) Construction of *Saccharum* consensus genetic map from two interspecific crosses. Crop Sci 42: 570–583.

Mototani H, Mabuchi A, Saito S, Fujioka M, Iida A, Takatori Y, Kotani A, Kubo T, Nakamura K, Sekine A, Murakami Y, Tsunoda T, Notoya K, Nakamura Y, Ikegawa S (2005) A functional single nucleotide polymorphism in the core promoter region of CALM1 is associated with hip osteoarthritis in Japanese. Hum Mol Genet 14: 1009–1017.

Mudge J, Andersen WR, Kehrer RL, Fairbanks DJ (1996) A RAPD genetic map of *Saccharum officinarum*. Crop Sci 36: 1362–1366.

Nagamura Y, Antonio BA, Sasaki T (1997) Rice molecular genetic map using RFLPs and its applications. Plant Mol Biol 35: 79–87.

Nicot N, Chiquet V, Gandon B, Amilhat L, Legeai F, Leroy P, Bernard M, Sourdille P (2004) Study of simple sequence repeat (SSR) markers from wheat expressed sequence tags (ESTs). Theor Appl Genet 109: 800–805.

Nikaido AM, Ujino T, Iwata H, Yoshimura K, Yoshimura H, Suyama Y, Murai M, Nagasaka K, Tsumura Y (2000) AFLP and CAPS linkage maps of *Cryptomeria japonica*. Theor Appl Genet 100: 825–831.

Oliveira KM, Pinto LR, Marconi TG, Margarido GRA, Pastina MM, Teixeira LHM, Figueira AV, Ulian EC, Garcia AAF, Souza AP (2007) Functional integrated genetic linkage map based on EST-markers for a sugarcane (*Saccharum* spp.) commercial cross. Mol Breed 20: 189–208.

Paillard S, Schnurbusch T, Winzeler M, Messmer M, Sourdille P, Abderhalden O, Keller B, Schachermayr C (2003) An integrative genetic linkage map of winter wheat (*Triticum aestivum* L.). Theor Appl Genet 107: 1235–1242.

Pan YB, Burner DM, Legendre BL (2000) An assessment of the phylogenetic relationship among sugarcane and related taxa based on the nucleotide sequence of 5S rRNA intergenic spacers. Genetica 108: 285–295.

Panje RR, Babu CN (1960) Studies in *Saccharum spontaneum* distribution and geographical association of chromosome numbers. Cytologia 25: 152–172.

Paterson AH, Lin Y-R, Li Z, Schertz K, Doebley JF, Pinson SM, Liu S-C, Stansel JW, Irvine JE (1995) Convergent domestication of cereal crops by independent mutations at corresponding genetic loci. Science 269: 1714–1718.

Pereira MG, Lee M, Bramel-Cox P, Woodman W, Doebley J, Whitkus R (1994) Construction of an RFLP map in sorghum and comparative mapping in maize. Genome 37: 236–243.

Pinto LR, Oliveira KM, Ulian EC, Garcia AAF, Souza AP (2004) Survey in the expressed sequence tag database (SUCEST) for simple sequence repeats. Genome 47: 795–804.

Piperidis G, D'Hont A (2001) Chromosome composition analysis of various *Saccharum* interspecific hybrids by genomic in situ hybridization (GISH). Int Soc Sugar Cane Technol Congr 11: 565.

Piperidis N, Jackson PA, D'Hont A, Besse P, Hoarau J-Y, Courtis B, Aitken KS, McIntyre CL (2008) Comparative genetics in sugarcane enables structured map enhancement and validation of marker-trait associations. Mol Breed 21: 233–247.

Qu L, Hancock JF (2001) Detecting and mapping repulsion phase linkage in polyploids with polysomic inheritance. Theor Appl Genet 103: 136–143.

Raboin L-M, Oliveira KM, Lecunff L, Telismart H, Roques D, Butterfield M, Hoarau J-Y, D'Hont A (2006) Genetic mapping in sugarcane, a high polyploidy, using bi-parental progeny: identification of a gene controlling stalk colour and a new rust resistance gene. Theor Appl Genet 112: 1382–1391.

Reffay N, Jackson PA, Aitken KS, Hoarau J-Y, D'Hont A, Besse P, McIntyre CL (2005) Characterization of genome regions incorporated from an important wild relative into Australian sugarcane. Mol Breed 15: 367–381.

Ripol MI, Churchill GA, Da Silva JAG, Sorrells M (1999) Statistical aspects of genetic mapping in autopolyploids. Gene 235: 31–41.

Ritter E, Gebhardt C, Salamini F (1990) Estimation of recombination frequencies and construction of RFLP linkage maps in plants from between heterozygous parents. Genetics 125: 645–560.

Roach BT (1972) Nobilization of sugarcane. Proc Int Soc Sugar Cane Technol 14: 206–216.

Rossi M, Araujo GP, Paulet F, Garsmeur O, Dias VM, Chen H, Van Sluys Ma, D'Hont A (2003) Genomic distribution and characterization of EST-derived resistance gene analogs (RGAs) in sugarcane. Mol Genet Genom 269: 406–419.

Schwarzacher T, Anamthawat-Jonsson K, Harrisson G, Islam A, Jian J, Leitch A, Miller T, Rogers W, Heslop-Harrison JS (1992) Genomic in situ hybridization to identify alien chromosomes and chromosome segments in wheat. Theor Appl Genet 84: 778–786.

da Silva JAG, Sobral BWS (1996) Genetics of polyploids. In: BWS Sobral (ed) The Impact of Plant Molecular Genetics. Birkhauser, Boston, MA, USA, pp 3–37.

da Silva JAG, Sorrells ME, Burnquist W, Tanksley SD (1993) RFLP linkage map and genome analysis of *Saccharum spontaneum*. Genome 36: 782–791.

da Silva JAG, Honeycutt RJ, Burnquist W, Al-Janabi SM, Sorrells ME, Tanksley SD, Sobral BWS (1995) *Saccharum spontaneum* L. 'SES 208' genetic linkage map combining RFLP and PCR-based markers. Mol Breed 1: 165–179.

Sobral BWS, Honeycutt RJ (1993) High output genetic mapping in polyploids using PCR-generated markers. Theor Appl Genet 86: 105–112.

Sobral BWS, Honeycutt RJ (1994) Genetics, plants and the polymerase chain reaction. In: KB Mullis , F Ferré, A Gibbs (eds) The Polymerase Chain Reaction. Birkhauser, Boston, MA, USA, pp 304–320.

Song KM, Osborn TC, Williams PH (1988) Brassica taxonomy based on nuclear restriction fragment length polymorphisms (RFLPs). I. Genome evolution of diploid and amphidiploid species. Theor Appl Genet 75: 784–794.

Sreenivasan TV, Ahloowalia BS, Heinz DJ (1987) Cytogenetics. In: DJ Heinz (ed) Sugarcane Improvement through Breeding. Elsevier, Amsterdam, The Netherlands, pp 211–253.

Storm N, Darnhofer-Patel B, van den Boom D, Rodi CP (2003) MALDI-TOF mass spectrometry-based SNP genotyping. Meth Mol Biol 212: 241–262.

Suman A, Kimbeng CA, Edmé SJ, Veremis JC (2008) Sequence-related amplified polymorphism (SRAP) markers for assessing genetic relationships and diversity in sugarcane germplasm collections. Plant Genet Resour 6: 222–231.

Syed NH, Sørensen AP, Antonise R, van de Wiel C, van der Linden CG, van't Westende W, Hooftman DAP, den Nijs HCM, Flavell AJ (2006) A detailed linkage map of lettuce based on SSAP, AFLP and NBS markers. Theor Appl Genet 112: 517–527.

Tanksley SD, Ganal MW, Prince JP, deVicente MC, Bonierbale MW, Broun P, Fulton TM, Giovannoni JJ, Grandillo S, Martin GB, Messeguer R, Miller JC, Miller L, Paterson AH, Pineda O, Roder MS, Wing RA, Wu W, Young ND (1992) High density molecular linkage maps of tomato and potato genomes. Genetics 132: 1141–1160.

Thomas HM, Morgan W, Meredith M, Humphreys M, Thomas H, Leggett J (1994) Identification of parental and recombinant chromosomes in hybrid derivatives of *Lolium multiflorum* x *Festuca praatensis* by genomic in situ hybridization. Theor Appl Genet 88: 909–913.

Van Ooijen JW, Voorrips RE (2001) JOINMAP 3.0. Software for the calculation of genetic linkage maps. Plant Research International, Wageningen, The Netherlands.

Waters DLE, Henry RJ, Reinke RF, Fitzgerald MA (2006) Gelatinization temperature of rice explained by polymorphisms in starch synthase. Plant Biotechnol J 4: 115–122.

Wu KK, Burnquist W, Sorrells ME, Tew TL, Moore PH, Tanksley SD (1992) The detection and estimation of linkage in polyploids using single-dose restriction fragments. Theor Appl Genet 83: 294–300.

Wu R, Ma CX, Painter I, Zeng ZB (2002) Simultaneous maximum likelihood estimation of linkage and linkage phases in out-crossing species. Theor Popul Biol 61: 349–363.

Yin T, Huang M, Wang M, Zhu L-H, Zeng Z-B, Wu R (2001) Preliminary interspecific genetic maps of the *Populus* genome constructed from RAPD markers. Genome 44: 602–609.

6

Mapping, Tagging and Map-based Cloning of Simply Inherited Traits

Angélique D'Hont,[1]* Olivier Garsmeur[1] and Lynne McIntyre[2]

ABSTRACT

Sugarcane is one of the most genetically complex species. It is both aneuploid and polyploid with a large number of chromosomes. Consequently, simply inherited traits are rare; only five have been identified to date of which four are disease-related (including rust resistance) and one is a morphological trait. The genetic complexity and large genome size of sugarcane, together with the lack of a close diploid relative, make map-based gene cloning in sugarcane a challenge. This chapter outlines the strategies developed to overcome these difficulties and describes progress towards the cloning of a durable rust resistance gene. Using syntenic relationships with model crops (sorghum and rice), enriched BAC libraries and haplotype-specific chromosome walking, a high-density genetic map and a partial physical map of the region surrounding the rust resistance gene has been developed. A comparison with other homo(eo)logous chromosomes at this region reveals the presence of a large insertion—current efforts are focusd on characterizing and walking through the insertion.

Keywords: mapping, trait, BAC

[1] CIRAD, UMR 1096, TA40/03, Avenue Agropolis, 34 398 Montpellier, cedex 5, France.
[2] CSIRO Plant Industry, 306 Carmody Rd, St Lucia, QLD 4067, Australia.
*Corresponding author: *dhont@cirad.fr*

6.1 Introduction

Cultivated sugarcane cultivars (*Saccharum* spp. hybrid) are highly polyploid (> 10*x*) and aneuploid with large chromosome numbers (typically around 115) (D'Hont et al. 1996) making sugarcane one of the most genetically complex crop species. Cytogenetic studies of the two ancestral species, *S. officinarum* and *S. spontaneum*, using rDNA probes indicated that the basic chromosome number was different in the two species, with *x* = 10 and *x* = 8, respectively (D'Hont et al. 1998). The high level of ploidy combined with the large and irregular number of chromosomes makes genetic mapping in sugarcane very difficult. Despite this structural complexity, genetic maps have been constructed in sugarcane cultivars using simplex markers (markers that segregate 1:1 in progeny from a biparental cross or 3:1 in selfed progeny) from a variety of marker types (Hoarau et al. 2001; Rossi et al. 2003; Aitken et al. 2005; Garcia et al. 2006). The largest maps in cultivated sugarcane each contain more than 1,000 markers, mainly AFLP and SSR markers, distributed onto approximately 100 linkage groups (LGs) (Hoarau et al. 2001; Rossi et al. 2003; Aitken et al. 2005). The use of codominant markers such as those generated by restriction fragment length polymorphism (RFLP) probes, or simple sequence repeat (SSR) primers, have enabled these 100+ linkage groups (LG) to be assigned one of eight homology groups (HG) (Aitken et al. 2005; Grivet et al. 1996; Rossi et al. 2003), as *x* = 8 in *S. spontaneum* (D'Hont et al. 1998). Both maps have highly variable numbers of LGs in each HG. Some HGs currently contain less than five LGs while other HGs currently contain more than 20 (Hoarau et al. 2001; Rossi et al. 2003; Aitken et al. 2005). It is currently not known if this variation in number of LGs per HG reflects real variation in the number of homoeologous chromosomes among sugarcane HGs or reflects a sparse coverage of some chromosomes that are thus represented by more than one LG. For HG with large numbers of LGs, part of this variation must also reflect the different basic chromosome number in *S. officinarum* and *S. spontaneum*, where two *S. officinarum* LGs/chromosomes correspond to one *S. spontaneum* LG/chromosome or sparse coverage of chromosomes, with more than one LG per chromosome. If each homology group is evenly represented approximately 12 homoeologous chromosomes are expected in each homoeology group and thus potentially 12 alleles of every gene. Many alleles could thus have contributed to every trait.

6.2 Mapping of Simply Inherited Traits

Not surprisingly, most traits in sugarcane are quantitatively inherited (Hogarth et al. 1987), and QTL analysis for traits such as sugar content

(Aitken et al. 2006; Hoarau et al. 2002; Ming et al. 2002; Piperidis et al. 2008) and fiber content (Hoarau et al. 2002; Piperidis et al. 2008) has indicated many QTL of individually small effect. Nevertheless, there are a small number of examples of simply inherited traits. Four of the examples are disease-related and one is a morphological trait.

6.2.1 Major Genes for Rust Resistance

The first simply inherited trait reported in sugarcane was for resistance to brown rust (Daugrois et al. 1996). Brown rust, caused by *Puccinia melanocephala* H&P Sydow, occurs worldwide and, while once causing significant yield loss (Gilbert 2008), now tends to have less impact thanks to the selection of resistant cultivars. Rust resistance is generally considered to be a quantitatively inherited trait with high heritability (Hogarth et al. 1993; Tai et al. 1981). However, a major rust resistance gene was identified in the French cultivar, R570 (Daugrois et al. 1996). Using a population consisting of selfed progeny from R570, brown rust resistance was observed to clearly and significantly segregate in 3 (resistant): 1 (susceptible) ratio, which is indicative of a single dominant resistance gene. This gene was linked to a RFLP probe, CDSR29 that was initially not integrated in any linkage group in the R570 map. Subsequent additional mapping in R570 by Asnaghi et al. (2000) indicated that the rust gene, *Bru1*, was located on linkage group VII–1a in homology group (HG) VII (HGVII) of R570. More recently, a second major rust resistance gene was identified by Raboin et al. (2006) in MQ76–53 using a biparental cross between R570 and MQ76–53. The new rust resistance gene, *Bru2*, mapped to a linkage group in HGVIII, a different HG to the location of *Bru1*. Rust resistance has been analyzed in a third sugarcane population (McIntyre et al. 2005b) but in this population, rust resistance was quantitatively inherited with several QTLs identified that explained < 20% of the phenotypic variation.

Major genes and some quantitative resistance genes against rust have been reported in maize and sorghum, two close diploid relatives of sugarcane. In maize, many different major race-specific resistance genes conferring resistance to maize common rust (*Puccinia sorghi*), have been identified (Wilkinson and Hooker 1968) and most map to the *rp1* locus on the short arm of the maize chromosome 10 (Rhoades 1935; Saxena and Hooker 1968). Sugarcane homologs of *Rp1* genes have been mapped to unassigned LGs in R570 (Rossi et al. 2003), Q117 and 74C42 (McIntyre et al. 2005b). A minor QTL for rust resistance mapped in 74C42 to a region that may be orthologous to maize *rp1*. By contrast, the genetics of rust resistance in sorghum appears to be more complicated with varying numbers of genes at dispersed loci having differing effects and possible modes of action reported (Miller and Cruzado 1969; Patil-Kulkarni et al. 1972; Rana et al.

1976). Nevertheless, sorghum homologs of *Rp1* genes have been mapped to a major rust resistance locus in sorghum, which is syntenic with the maize chromosome 10 (Tao et al. 1998; Ramakrishna et al. 2002; McIntyre 2005a).

6.2.2 *Major Gene for Eyespot Resistance*

In the same year as the first report of a single gene for rust resistance by Daugrois et al. (1996), Mudge et al. (1996) published the first map of the sugarcane ancestral species, *S. officinarum*. Using a population developed from a biparental cross between *S. officinarum* and *S. robustum*, they constructed a map using RAPD markers and noted that a dominant allele for eyespot susceptibility, caused by *Bipolaris sacchari*, segregated in a 1:1 ratio in the population. Unfortunately, the use of RAPD markers in the construction of the map has meant that it has not been possible to determine in which sugarcane homology group the eyespot susceptibility allele is located. Eyespot has been reported as a disease of sugarcane in many sugarcane growing areas of the world and is usually well controlled by the use of resistant varieties (Gilbert 2008). However, large losses have been reported in some countries, such as Cuba (Ramos Leal et al. 1996). Eyespot has not been mapped in any other sugarcane populations or in sorghum. In maize, however, single gene (Chang and Peterson 1995) and multi-gene (Carson et al. 2004) resistance to *Bipolaris maydis*, the causal agent of southern corn leaf blight, has been reported.

6.2.3 *Major Gene for Yellow Spot Disease*

The most recent major gene identified in sugarcane is for yellow spot disease (Aljanabi et al. 2007). Yellow spot, caused by *Mycovellosiella koepkei*, is relatively unimportant in most sugarcane producing countries, except in areas where humid environmental conditions favor infection in susceptible varieties. Using a biparental cross between M134/75 and R570, a major gene was mapped to a linkage group in HG2, which is the equivalent to HGVIII in R570. There are no other reports of mapping yellow spot resistance in sugarcane or its relatives.

6.2.4 *Gene for a Morphological Trait, Stalk Color*

Using the same biparental cross population in which *Bru2* was identified (Raboin et al. 2006), the authors noted that red stalk color also segregated in a 1:1 ratio in the population. It, too, was mapped in MQ76-53 to an unassigned linkage group. Red stalk color is the only non-disease single gene identified in sugarcane to date. This trait has been mapped in a second biparental sugarcane population but did not segregate as a single major gene (K.S.

Aitken, pers. comm.). Both seedling and plant color have been mapped as single genes in sorghum (Tao et al. 1998; Menz et al. 2002) and numerous single plant color genes have been mapped in maize (*http://www.maizegdb.org/cgi-bin/displayrefresults.cgi?start=41&term=plant+color*).

6.2.5 Conclusion

Given the complexity of the sugarcane genome and the multiple copies of each chromosome, it is not surprising that major genes are uncommon and that the few major genes identified to date are for disease resistance and stalk color, traits for which single genes have been readily identified in other less genetically complex species, such as its close relatives maize and sorghum. Other major genes are likely to exist, but apparently appear not to be in economically valuable traits, such as the sugar-related and fiber-related traits studied to date; it is possible that major genes for these economically important traits have been selected and fixed in elite sugarcane germplasm.

In the following discussion, we describe how to identify a major gene conditioning a trait of interest base on its position on a genetic map by a process called map-based cloning. We will take as an example a current study aiming at cloning the resistance gene *Bru1* mentioned above.

6.3 Map-based Cloning in Sugarcane

6.3.1 Challenges of Map-based Cloning in Sugarcane

Map-based cloning is becoming increasingly efficient in model crops such as rice (Sun et al. 2004; Ueda et al. 2005; Xu et al. 2006) thanks to their simple diploid structure, their small genome size and tremendous molecular resources. However, it is still a major challenge in more complex crops in particular in high polyploid such as sugarcane ($2n\sim12x\sim120$). Modern sugarcane cultivars have a large genome of 10 Gb/2C (D'Hont 2005) with up to 12 homo(eo)logous highly heterozygous haplotypes on average that can coexist at a given locus. The size of its monoploid genome (i.e. basic set of chromosomes), ~900 Mb; (D'Hont and Glaszmann 2001; D'Hont et al. 2008) is similar to that of sorghum (730 Mb) and only 2-fold the rice genome (390 Mb). In addition, meiosis in modern sugarcane cultivars, which are hybrids between autopolyploid species, mainly involves bivalent pairing (Price 1963; Burner and Legendre 1994), but chromosome assortment results from a combination of polysomy and preferential pairing (Grivet et al. 1996; Hoarau et al. 2001; Jannoo et al. 2004).

This very high level of genetic redundancy (~12x) and the polysomic inheritance in sugarcane imply that: i) only single-dose markers (i.e., markers present on only one of the hom(oe)ologous haplotypes) can be used for high-resolution mapping; ii) it is very difficult to monitor specific loci in genetic and physical mapping approaches; and iii) a very large number of BAC clones is required to obtain a BAC library with genome coverage compatible with map-based cloning approaches. For example, although the R570 sugarcane cultivar BAC library made by Tomkins et al. (1999) includes as many as 103,296 BAC clones with an average size of 130 kb, this corresponded to a coverage of only 1.3-fold the total genome, and thus to a probability of only 73.9% of finding any particular DNA segment. By comparison, the SB_BBc sorghum (constructed by D. Begum; *http:// www.genome.clemson.edu*) library includes a similar number of BAC clones (110,592), with a similar average size but, since sorghum is diploid and homozygous, this represents a coverage of 17-fold the genome and a 99.9% probability of finding a given segment.

In addition, in sugarcane, there are no close diploid relatives, only polyploids are known in the *Saccharum* genus and thus no diploid genome can be used to facilitate map-based cloning as reported in wheat (Keller et al. 2005). Moreover, due to its genetic complexity, this species has received little research investment compared to its economic importance, and molecular resources are still limited (Grivet and Arruda 2002; D'Hont et al. 2008).

Despite this challenging complexity, the International Consortium for Sugarcane Biotechnology (ICSB) was interested in using the *Bru*1 rust resistance gene (Daugrois et al. 1996), at that time the only well characterized major gene identified in sugarcane, as a target to develop and test strategies for map-based cloning in sugarcane. This source of resistance is of particular interest since it is durable. Indeed, *Bru*1 resistance breakdown has never been observed despite intensive cultivation of R570 for over 20 years in various regions of the world. Moreover, tests under controlled conditions demonstrated that this gene provides resistance against diverse rust isolates collected in Africa and America (Asnaghi et al. 2001).

The following paragraphs describe the genomic strategies that we have developed to overcome constraints associated with high polyploidy in the successive steps of map-based cloning approaches and their application to the first and to-date unique attempt to isolate a gene from sugarcane by map-based cloning (Le Cunff et al. 2008). This approach targets the durable major rust resistance gene, *Bru*1.

6.3.2 Strategies Developed towards Map-based Cloning of the Resistance Gene (Bru1) in Cultivar R570

6.3.2.1 Development of a High-resolution Genetic Map of the Bru1 Region

The genetic map was developed using various strategies mainly exploiting the synteny relationship between sugarcane and model Poaceae. It was developed in several steps following the development of model Poaceae genomic resources.

6.3.2.1.1 Diploid/Polyploid Syntenic Shuttle Mapping using Maize, Sorghum and Rice as described above, the presence of the major resistance gene, *Bru1*, in the French sugarcane cultivar R570 was revealed by analyzing a selfed progeny from R570 including 147 individuals (Daugrois et al. 1996) and was further confirmed by analyzing segregation of rust resistance in a larger population of 658 individuals, also derived from selfing of R570 (Asnaghi et al. 2004). *Bru1* was originally linked to a single marker (CDSR29) on a genetic map built using selfed cv. R570 progeny (Daugrois et al. 1996; Grivet et al. 1996). Asnaghi et al. (2000) refined the genetic map around *Bru1* by exploiting existing rice, maize and sorghum genetic maps. This approach revealed that the targeted region is orthologous to one end of sorghum consensus linkage group (LG) 4, the end of the short arm of rice chromosome 2 and part of maize LG 4 and LG5. Probes from these regions were analyzed on the R570 mapping population. It enabled localization of *Bru1* at the end of one cosegregation group (CG VII.1) of the R570 homology group VII. However, it did not enable marker saturation of the region due to the distal position of the gene and the poor density of markers in the distal orthologous map area of rice and sorghum at that time.

6.3.2.1.2 Bulked Segregant Analysis with AFLP Markers

In parallel, Asnaghi et al. (2004) screened 443 AFLP primer pairs by bulked segregant analysis (BSA) using bulk of resistant and susceptible R570 progenies. Eight AFLP markers surrounding the resistance gene were identified in an interval of 10 cM, with the closest markers located at 2 cM on both sides of the gene.

6.3.2.1.3 Diploid/Polyploid Syntenic Shuttle High-resolution Mapping using New Genomic Resources from Sorghum

New genomic resources from sorghum including high-density genetic maps and partially ordered BAC library became progressively available. To exploit them, we refined the location of the sorghum region orthologous to the R570 target region. For that we cloned the two closest AFLP markers (*aaccac6* and *attcag*) distally flanking *Bru1* (Fig. 6-1, Step 1) and analyzed by RFLP on our

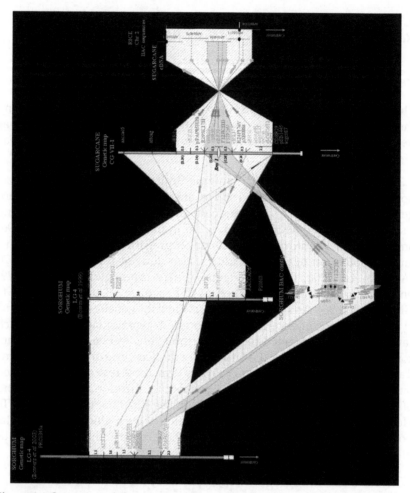

Figure 6-1 Saturation of the sugarcane target genetic region using various genomic resources.

Step 1: Three markers originally surrounding *Bru*1 in sugarcane (orange arrows) were mapped on the sorghum genetic map. Double green arrows link common markers between the two sorghum genetics maps (Boivin et al. 1999 and Bowers et al. 2003). The green arrows point to the 4 new sorghum markers that could be mapped on sugarcane.

Step 2: Four sorghum probes (indicated in italic in the Bowers et al. 2003 sorghum genetic map) were used to construct a sorghum BAC contig. Four BAC-ends markers derived from this contig could be mapped in the sugarcane target region (pointed by green arrows).

Step 3: Two probes were used to identify the corresponding rice orthologous physical map. Orange arrows point to the rice BACs identified. Seven sugarcane cDNA with homology to the rice orthologous sequence were mapped on the sugarcane genetic map (red arrows). Markers used to build the sugarcane physical map are indicated in bold. Figure from Le Cunff et al. (2008).

Color image of this figure appears in the color plate section at the end of the book.

sorghum mapping population. Both loci were mapped on sorghum LG4 proximally from *CDSR29*. These results allowed us to refine the sorghum region orthologous to the target sugarcane region to an 8.8 cM interval. They also revealed a local inversion on the R570 genetic map as compared to the sorghum map.

We aligned our sorghum LG4 (previously LG D, Boivin et al. 1999, and unpublished data) with LG4 (previously LGF) of the high-density RFLP map of Bowers et al. (2003). Nine RFLP loci located in the target sorghum region over the two maps at that stage were analyzed on the R570 progeny (Fig. 6-1, Step 1). This allowed us to surround *Bru*1 with four new markers at 0.3 cM and 0.6 cM on each side of the gene and to reduce the target region to a 0.9 cM interval. The position of the mapped markers confirmed the local inversion between sugarcane R570 CG VII.1 and sorghum LG4 and was in agreement with colinearity inside the inversion.

At that stage, because we anticipated that physical mapping in the highly polyploid sugarcane was going to be complicated and because a partially ordered physical map of sorghum was available (*http://www.stardaddy.uga.edu/fpc/WebAGCoL/bicolor/WebFPC/*), we decided to exploit this physical map to further refine the sugarcane genetic map. For that, we built an orthologous sorghum physical map and then derived new probes from BAC clones of this physical map (Fig. 6-1, Step 2). Four loci located in the 3.1 cM orthologous target region in sorghum were used to screen a partially contiged sorghum BAC library (SB_BBc, *http://www.genome.arizona.edu/genome/sorghum.html*). The identified BAC clones were searched for on the on-line sorghum physical map (*http://www.stardaddy.uga.edu/fpc/bicolor/WebAGCoL/WebFPC*) to find the corresponding contig. Four contigs were identified that were further shown to correspond to one unique contig of 91 BAC clones representing a region of around 350–400 kb (Fig. 6-1, Step 2).

A total of 57 BAC-ends, from this contig, were tested by Southern analysis on sugarcane DNA. Out of them, 14% did not hybridize well and around 56% corresponded to repeated sequences in sugarcane. From the remaining17, four revealed a marker closely linked to *Bru*1. Two were located at 0.3 cM proximally from the gene, one at 0.3 cM distally from the gene, and one cosegregated with the gene. Finally, this step allowed reducing the target region from 4.1 cM to 0.6 cM.

6.3.2.1.4 Diploid/Polyploid Syntenic Shuttle High-resolution Mapping using the Rice Genome Sequence

The rice genome sequence became available (International Rice Genome Sequencing Project 2005) and was exploited to continue refining the sugarcane genetic map of the *Bru*1 one region.

To identify the rice region orthologous to the R570 target region, probes revealing a marker surrounding *Bru*1 were tested for hybridization with rice. Two probes hybridized to rice DNA and thus were used to screen the OSJNBa rice BAC library (constructed by M.A. Budiman, *http:// www.genome.clemson.edu*). The identified BAC clones were located on the short arm of chromosome 2 (*www.genome.arizona.edu*). This allowed us to restrict the orthologous physical target region in rice to around 600 kb on the short arm of chromosome 2.

The sequence of the orthologous region was compared with available sugarcane ESTs (255,635 sugarcane ESTs assembled in 43,000 clusters, SUCEST Data Based, (Vettore et al. 2003)). A total of 90 EST clusters having high homologies to the orthologous rice BAC clones were identified. The blast of these clusters did not reveal homology to any cloned resistance gene or resistant gene analogue (RGA). The cDNAs corresponding to 34 of the sugarcane EST clusters were analyzed on the R570 mapping population. From the 34 cDNAs tested, seven generated markers that were mapped on the cosegregation group bearing *Bru*1 from which two were located at 0.3 cM on the each sides of *Bru*1 and two cosegregated with *Bru*1.

6.3.2.1.5 RGA Mapping

In parallel to these approaches, 50 RGAs that were selected from the SUCEST Data Base were mapped on R570. Two RGAs mapped on the homology group VII, which contains *Bru*1 but no marker was found closely linked to *Bru*1 (Rossi et al. 2003).

6.3.2.1.6 Development of a New Segregating Population of Self-progeny of R570 to Increase the Resolution of the Map

A new population of 1,600 individuals was developed by selfing R570 to increase the resolution of the genetic map around *Bru*1. The 400 susceptible individuals identified in this population were analyzed with four AFLP markers surrounding the gene to identify individuals locally-recombinant in the target area. We used only susceptible individuals because the susceptible phenotype is easier to establish with certainty and since susceptible individuals represent a quarter of the population but encompass half of the detectable recombinant individuals in the region. The nine markers closest to *Bru*1 were analyzed on these individuals. Two new recombinations were identified in the 0.6 cM target area but they did not allow separation of previous cosegregating markers. The new high-resolution map of the target region included three markers mapped at 0.28 cM on both sides of *Bru*1 proximally and three markers cosegregating with *Bru*1 (Fig. 6-1).

At that, stage we decided that the resolution of the genetic map was good enough to start building a sugarcane physical map of the target region.

6.3.2.2 Construction of a Physical Map of Region Bearing Bru1

6.3.2.2.1 Screening of the R570 BAC Library

A BAC library of the sugarcane cultivar R570 was constructed by Tomkins et al. (1999) with financial support of ICSB to allow cloning of *Bru1*. The R570 BAC library contains 103,296 BAC clones with an average size of 130 kb. We estimated the size of the genome of R570 by flow cytometry to 11 pg, which represents around 10 Gb (D'Hont et al. 1995; D'Hont and Glaszmann 2001). The total genome, that means any particular chromosome segment, is thus statistically represented only 1.3 times in the BAC library.

We screened the whole R570 BAC library with probes surrounding *Bru1*. To build the physical map, we used BAC clones that were detected by at least two probes. We were thus able to build a physical map of 32 BAC clones covering the *Bru1* region. The physical map, at that stage, contained one gap. To complete this map, several BAC-ends were isolated and analyzed on the R570 progeny. One BAC-end was mapped at 0.14 cM on the distal side of *Bru1*. This allowed us to reduce the target area and to eliminate the gap from the target area. The BAC-end was used to screen the R570 BAC library and allowed us to identify an additional BAC that partially covered the target area.

Finally, at this stage, the entire target region was contained in three R570 BAC clones and seven R570 BAC clones contained part of the target area (Fig. 6-2).

6.3.2.2.2 Identification of the BAC Corresponding to the Target Haplotype

because sugarcane is polyploid and heterozygous, the resulting physical map did not correspond to a unique BAC contig but to several hom(oe)ologous BAC contigs corresponding to hom(oe)ologous chromosomes in HG VII. To differentiate the haplotypes, we compared the restriction profiles of all the BAC clones belonging to the target region. Seven hom(oe)ologous haplotypes (1 to 7) were identified among the 10 BAC clones covering the target area (Fig. 6-2).

To identify the haplotype bearing *Bru1* (target haplotype), we had to determine which BAC clones were bearing the markers (alleles) linked to *Bru1*. We thus analyzed the BAC clones by RFLP with the probe/enzyme combinations that revealed the markers linked to *Bru1* and compared their RFLP profiles with that of R570 (Fig. 6-3). The RFLP patterns of BAC clones corresponding to the haplotype bearing *Bru1* should include the bands/

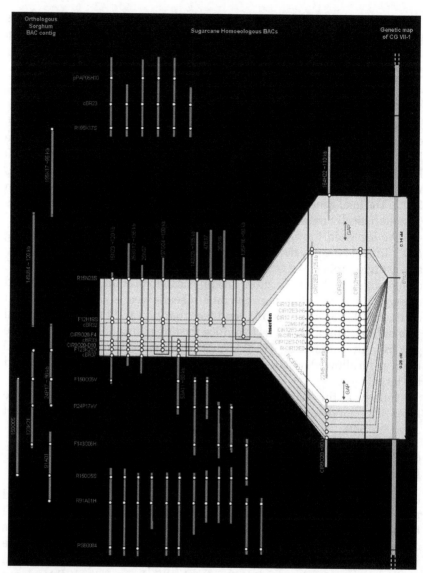

Figure 6-2 Physical map of the *Bru*1 region.

BAC clones are represented by vertical lines: orange for the target haplotype, brown for the hom(e)ologous haplotypes and green for sorghum. Dotted-lines and white rounds indicate the localization of probes used on the sorghum or/and sugarcane physical map or/and on the genetic map of the *Bru*1 region in sugarcane cv R570. Boxes assemble BAC clones for the same haplotype. Probes in green represent those from the sorghum genetic or physical map, those in red are from the sugarcane cDNA library, and those in orange are BAC-ends or subclones of sugarcane BACs. Figure from Le Cunff et al. (2008).

Color image of this figure appears in the color plate section at the end of the book.

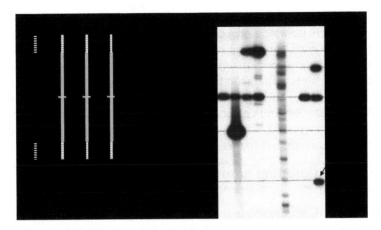

Figure 6-3 Method for identifying the BAC corresponding to the target haplotype versus hom(oe)ologous BACs.

A: Schematic representation of hom(oe)ologous chromosome segments (haplotypes) bearing four allelic RFLP markers with probeA/*Hind*III (a1 to a4). The target haplotype bearing allele a1 is represented in black.

B: RFLP profiles of R570 and six BACs analyzed with probe A (R15N23) in combination with *Hind*III. Asterics indicate BACs selected with probe A. BAC 164H22 contains marker a1 (pointed by an arrow) and thus belongs to the target haplotype. Figure from Le Cunff et al. (2008).

markers linked to *Bru*1. Using this procedure, we were able to identify only one BAC clone corresponding to the target haplotype, BAC 164H22. This BAC clone only partially covered the target (Fig. 6.2).

6.3.2.2.3 Haplotype-specific Chromosome Walking using a *Bru*1-enriched BAC Library in order to complete the physical map of the target haplotype, a new BAC library was built. We used a mix of DNA from four resistant selfed R570 progeny that contained two copies of *Bru*1 with the aim of increasing the proportion of the target haplotype in this library. The new library contained 110,592 clones of 130 kb in average and covers 2.8-fold the target haplotype and 1.4-fold the total genome. Added to the 1.3 genome coverage of the R570 BAC library of Tomkins et al. (1999), we now have BAC libraries covering 4-fold (4x) the target area overall, with a 98.5% probability of finding the target segment. The new library was screened with five probes mapped in a 0.42 cM region around *Bru*1. Two new BAC clones of the target haplotype were recovered. However, two gaps still remain in the physical map (Fig. 6-2).

BAC-ends and BAC-subclones from the three BAC clones from the target haplotype were isolated and a chromosome walk was undertaken in order to fill the two gaps remaining on the target haplotype. Due to polyploidy, only mapped probes can be used for this chromosome walking step to ensure that this walking is actually on the target chromosome. We thus first had to map the sub-clones and select the ones revealing a marker linked to *Bru*1.

Finally, after several chromosome-walking steps, the physical map of the target haplotype contained six BAC clones but two gaps remained. The genetic map included 15 markers that cosegregated with *Bru*1 (Fig. 6-2).

6.3.2.2.4 Identification of an Insertion—Impact on Recombination

During the BAC library screening steps with the BAC subclones, we noticed that several of the BAC subclones that revealed a marker cosegregating with *Bru*1 and thus belong to the target haplotype, did not hybridize with the other hom(oe)ologous BAC clones (Fig. 6-2). This highlighted the presence of an insertion specific to the target haplotype. This type of discovery has already complicated several map-based cloning projects in diploid species (Stirling et al. 2001; Barker et al. 2005). This is due to the fact that these insertions generally induce severe repression of recombination (Wei et al. 1999; Stirling et al. 2001; Neu et al. 2002; Barker et al. 2005).We indeed showed that the insertion induced a reduction of recombination on the target haplotype.

6.4 Conclusions and Perspectives

In this example, we report on the first attempt to isolate a gene from sugarcane by map-based cloning, targeting the durable major rust resistance gene (*Bru*1). We developed strategies to overcome the difficulties associated with high ploidy level for map-based cloning in sugarcane. These strategies included i) the use of the good syntenic relationships between sugarcane and two model species, i.e. sorghum and rice; ii) haplotype-specific chromosome walking; and iii) the development of specific BAC libraries.

The use of synteny relationships between sugarcane and two model diploid Poaceae species, i.e. sorghum and rice (Grivet et al. 1994; Dufour et al. 1997; Glaszmann et al. 1997; Guimaraes et al. 1997; Ming et al. 1998; Jannoo et al. 2007), that are estimated to have diverged from sugarcane around 8 and 50 MYA, respectively (Wolfe et al. 1989; Jannoo et al. 2007) was very successful since it allowed us to reduce the size of the genetic interval containing the *Bru*1 gene from 4 cM to 0.56 cM and to identify nine markers within this genetic interval. This strategy was developed in several steps following the development of rice and sorghum genomic resources in becoming more and more efficient. The availability of the complete sorghum genome sequence (Paterson et al. 2009) will greatly accelerate the high-resolution mapping steps in future studies of this type in sugarcane.

Since sugarcane is polyploid and highly heterozygous, to develop the physical map, we had to develop a haplotype-specific chromosome walking procedure. To identify among hom(oe)ologous BAC clones the ones corresponding to the haplotype bearing *Bru*1, we had to determine which ones encompass the markers (alleles) linked to *Bru*1. This meant that to develop the physical map, and for further steps of chromosome walking toward the gene, only probes revealing a marker linked to *Bru*1 could be used. Therefore physical and genetic mapping had to be tightly associated throughout the physical mapping process.

The other difficulty, due to the large size of the sugarcane genome, was that to obtain a BAC library with reasonable genome coverage, a very large number of BAC clones are required. One way to overcome this problem is to build a much larger library, which then becomes very tedious to manipulate. We thus developed an alternative strategy that involved building a library with a genotype comprising more copies of the target haplotype. This material allowed us to specifically double the proportion of the target haplotype in the new BAC library (as compared to the R750 library). Genotypes with even more copies of the target haplotype could be used in order to continue increasing the proportion of the target haplotype.

Two BAC subclones from the target haplotype showed homology with the S/T kinase barley rust resistance gene *Rpg*1. This suggests the possible presence of an S/T kinase cluster in the target region. These clones are being further investigated as they represent potential candidate genes for *Bru*1. In addition, several BAC clones representing the different homologous haplotypes are currently being sequenced. Their sequences will be used to continue the characterization of this resistance locus and the complete map-based cloning of the gene.

During this work, we developed strategies to overcome the difficulties associated with high ploidy level for map-based cloning in sugarcane (Le Cunff et al. 2008). We demonstrated their efficiency since we were able to i) develop a high density-map including markers at 0.28 and 0.14 cM on both sides and 13 markers cosegregating with *Bru*1; and ii) develop a physical map of the target haplotype that still includes two gaps at this stage due to the discovery of an insertion specific to this haplotype. These strategies will pave the way for future map-based cloning approaches in sugarcane and other complex polyploid species.

References

Aitken KS, Jackson PA, McIntyre CL (2005) A combination of AFLP and SSR markers provides extensive map coverage and identification of homo(eo)logous linkage groups in a sugarcane cultivar. Theor Appl Genet 110: 789–801.

Aitken KS, Jackson PA, McIntyre CL (2006) Quantitative trait loci identified for sugar related traits in a sugarcane (*Saccharum* spp.) cultivar x *Saccharum officinarum* population. Theor Appl Genet 112: 1306–1317.

Aljanabi SM, Parmessur Y, Kross H, Dhayan S, Saumtally S, Ramdoyal K, Autrey LJC, Dookun-Saumtally A (2007) Identification of a major quantitative trait locus (QTL) for yellow spot (*Mycovellosiella koepkei*) disease resistance in sugarcane. Mol Breed 19: 1–14.

Asnaghi C, Paulet F, Kaye C, Grivet L, Deu M, Glaszmann JC, D'Hont A (2000) Application of synteny across Poaceae to determine the map location of a sugarcane rust resitance gene. Theor Appl Genet 101: 962–969.

Asnaghi C, D'Hont A, Glaszmann JC, Rott P (2001) Resistance of sugarcane cultivar R570 to *Puccinia melanocephala* isolates from different geographic locations. Plant Dis 85: 82–286.

Asnaghi C, Roques D, Ruffel S, Kaye C, Hoarau JY, Telismart H, Girard JC, Raboin LM, Risterucci AM, Grivet L, D'Hont A (2004) Targeted mapping of a sugarcane rust resistance gene (*Bru1*) using bulked segregant analysis and AFLP markers. Theor Appl Genet 108: 759–764.

Barker CL, Donald T, Pauquet J, Ratnaparkhe MB, Bouquet A, Adam-Blondon AF, Thomas MR, Dry I (2005) Genetic and physical mapping of the grapevine powdery mildew resistance gene, *Run1*, using a bacterial artificial chromosome library. Theor Appl Genet 111: 370–377.

Boivin K, Deu M, Rami JF, Trouche G, Hamon P (1999) Towards a saturated sorghum map using RFLP and AFLP markers. Theor Appl Genet 98: 320–328.

Bowers JE, Abbey C, Anderson S, Chang C, Draye X, Hoppe AH, Jessup R, Lemke C, Lennington J, Lin Z, Lin YR, Liu SC, Luo L, Marler BS, Ming R, Mitchell SE, Qiang D, Reischmann K, Schulze SR, Skinner DN, Wang YW, Kresovich S, Schertz KF, Paterson AH (2003) A high-density genetic recombination map of sequence-tagged sites for sorghum, as a framework for comparative structural and evolutionary genomics of tropical grains and grasses. Genetics 165: 367–386.

Burner DM, Legendre BL (1994) Cytogenetic and fertility characteristics of elite sugarcane clones. Sugar Cane 1: 6–10.

Carson ML, Stuber CW, Senior ML (2004) Identification and mapping of quantitative trait loci conditioning resistance to southern leaf blight of maize caused by *Cochliobolus heterostrophus* race O. Phytopathology 94: 862–867.

Chang RY, Peterson PA (1995) Genetic-control of resistance to *Bipolaris maydis*—one gene or 2 genes. J Hered 86: 94–97.

D'Hont A (2005) Unraveling the genome structure of polyploids using FISH and GISH; examples of sugarcane and banana. Cytogenet Genome Res 109: 27–33.

D'Hont A, Glaszmann JC (2001) Sugarcane genome analysis with molecular markers, a first decade of research. Proc Int Soc Sugarcane Technol 24: 556–559.

D'Hont A, Rao PS, Feldmann P, Grivet L, Islam-Faridi N, Taylor P, Glaszmann JC (1995) Identification and characterisation of intergeneric hybrids, *S. officinarum* X *Erianthus arundinaceus*, with molecular markers and in situ hybridization. Theor Appl Genet 91: 320–326.

D'Hont A, Grivet L, Feldmann P, Rao S, Berding N, Glaszmann JC (1996) Characterisation of the double genome structure of modern sugarcane cultivars (*Saccharum* spp.) by molecular cytogenetics. Mol Gen Genet 250: 405–413.

D'Hont A, Ison D, Alix K, Roux C, Glaszmann JC (1998) Determination of basic chromosome numbers in the genus *Saccharum* by physical mapping of RNA genes. Genome 41: 221–225.

D'Hont A, Souza GM, Menossi M, Vincentz M, van Sluys MA, Glaszmann JC (2008) Sugarcane: a major source of sweetness, alcohol, and bio-energy. In: PH Moore , R Ming (eds) Genomics of Tropical Crop Plants. Springer, New York, USA, pp 483–513.

Daugrois JH, Grivet L, Roques D, Hoarau JY, Lombard H, Glaszmann JC, D'Hont A (1996) A putative major gene for rust resistant linked with an RFLP marker in sugarcane cultivar R570. Theor Appl Genet 92: 1059–1064.

Dufour P, Deu M, Grivet L, D'Hont A, Paulet F, Glaszmann JC (1997) Construction of a composite sorghum genome map and comparison with sugarcane, a related complex polyploid. Theor Appl Genet 94: 409–418.

Garcia AA, Kido EA, Meza AN, Souza HM, Pinto LR, Pastina MM, Leite CS, Silva JA, Ulian EC, Figueira A, Souza AP (2006) Development of an integrated genetic map of a sugarcane (*Saccharum* spp.) commercial cross, based on a maximum-likelihood approach for estimation of linkage and linkage phases. Theor Appl Genet 112: 298–314.

Gilbert RA (2008) Florida Sugarcane Handbook. Department, University of Florida, USA: *http://edis.ifas.ufl.edu/TOPIC_BOOK_Sugarcane_Handbook* (Retrieved 02/12/08).

Glaszmann JC, Dufour P, Grivet L, D'Hont A, Deu M, Paulet F, Hamon P (1997) Comparative genome analysis between several tropical grasses. Euphytica 96: 13–21.

Grivet L, Arruda P (2002) Sugarcane genomics: depicting the complex genome of an important tropical crop. Curr Opin Plant Biol 5: 122–127.

Grivet L, D'Hont A, Dufour P, Hamon P, Roques D, Glaszmann JC (1994) Comparative genome mapping of sugarcane with other species within the Andropogoneae tribe. Heredity 73: 500–508.

Grivet L, D'Hont A, Roques D, Feldmann P, Lanaud C, Glaszmann JC (1996) RFLP mapping in cultivated sugarcane (*Saccharum* spp.): genome organization in a highly polyploid and aneuploid interspecific hybrid. Genetics 142: 987–1000.

Guimaraes CT, Sills GR, Sobral BWS (1997) Comparative mapping of andropogoneae: *Saccharum* L. (sugarcane) and its relation to sorghum and maize. Proc Natl Acad Sci USA 94: 14261–14266.

Hoarau JY, Offmann B, D'Hont A, Risterucci AM, Roques D, Glaszmann JC, Grivet L (2001) Genetic dissection of a modern cultivar (*Saccharum* spp.). I. Genome mapping with AFLP. Theor Appl Genet 103: 84–97.

Hoarau JY, Grivet L, Offmann B, Raboin LM, Diorflar JP, Payet J, Hellmann M, D'Hont A, Glaszmann JC (2002) Genetic dissection of a modern sugarcane cultivar (*Saccharum* spp.).II. Detection of QTLs for yield components. Theor Appl Genet 105: 1027–1037.

Hogarth DM, Wu KK, Heinz DJ (1987) Genetics of sugarcane. In: DJ Heinz (ed) Sugarcane Improvement through Breeding. Elsevier, Amsterdam, The Netherlands, pp 255–271

Hogarth DM, Ryan CC, Taylor PWJ (1993) Quantitative inheritance of rust resistance in sugarcane. Field Crops Res 34: 187–193.

International Rice Genome Sequencing Project (2005) The map-based sequence of the rice genome. Nature 436: 793–800.

Jannoo N, Grivet L, David J, D'Hont A, Glaszmann JC (2004) Differential chromosome pairing affinities at meiosis in polyploid sugarcane revealed by molecular markers. Heredity 93: 460–467.

Jannoo N, Grivet L, Chantret N, Garsmeur O, Glaszmann JC, Arruda P, D'Hont A (2007) Orthologous comparison in a gene-rich region among grasses reveals stability in the sugarcane polyploid genome. Plant J 50: 574–585.

Keller B, Feuillet C, Yahiaoui N (2005) Map-based isolation of disease resistance genes from bread wheat: cloning in a supersize genome. Genet Res 85: 93–100.

Le Cunff L, Garsmeur O, Raboin LM, Pauquet J, Telismart H, Selvi A, Grivet L, Philippe R, Begum D, Deu M, Costet L, Wing R, Glaszmann JC, D'Hont A (2008) Diploid/polyploid syntenic shuttle mapping and haplotype-specific chromosome walking toward a rust resistance gene (*Bru*1) in highly polyploid sugarcane (2n approximately 12x approximately 115). Genetics 180: 649–660.

McIntyre CL, Casu RE, Drenth J, Knight D, Whan VA, Croft BJ, Jordan DR, Manners JM (2005a) Resistance gene analogues in sugarcane and sorghum and their association with quantitative trait loci for rust resistance. Genome 48: 391–400.

McIntyre CL, Whan VA, Croft B, Magarey R, Smith GR (2005b) Identification and validation of molecular markers associated with Pachymetra root rot and brown rust resistance in sugarcane using map- and association-based approches. Mol Breed 16: 151–161.

Menz MA, Klein RR, Mullet JE, Obert JA, Unruh NC, Klein PE (2002) A high-density genetic map of *Sorghum bicolor* (L.) Moench based on 2926 AFLP, RFLP and SSR markers. Plant Mol Biol 48: 483–499.

Miller RR, Cruzado HJ (1969) Allelic interactions at the Pu locus in *Sorghum bicolor* (L.) Moench. Crop Sci 9: 336–338.

Ming R, Liu SC, Lin YR, da Silva J, Wilson W, Braga D, van Deynze A, Wenslaff TF, Wu KK, Moore PH, Burnquist W, Sorrells ME, Irvine JE, Paterson AH (1998) Detailed alignment of *Saccharum* and sorghum chromosomes: comparative organization of closely related diploid and polyploid genomes. Genetics 150: 1663–1682.

Ming R, Wang Y-W, Draye X, Moore PH, IrvineJE, Paterson AH (2002) Molecular dissection of complex traits in autopolyploids: mapping QTLs affecting sugar yield and related traits in sugarcane. Theor Appl Genet 105: 332–345.

Mudge J, Andersen WR, Kehrer RL, Fairbanks DJ (1996) A RAPD genetic map of *Saccharum officinarum*. Crop Sci 36: 1362–1366.

Neu C, Stein N, Keller B (2002) Genetic mapping of the *Lr20-Pm1* resistance locus reveals suppressed recombination on chromosome arm 7AL in hexaploid wheat. Genome 45: 737–744.

Paterson AH, Bowers JE, Bruggmann R, Dubchak I, Grimwood J, Gundlach H, Haberer G, Hellsten U, Mitros T, Poliakov A, Schmutz J, Spannagl M, Tang H, Wang X, Wicker T, Bharti AK, Chapman J, Feltus FA, Gowik U, Grigoriev IV, Lyons E, Maher CA, Martis M, Narechania A, Otillar RP, Penning BW, Salamov AA, Wang Y, Zhang L, Carpita NC, Freeling M, Gingle AR, Hash CT, Keller B, Klein P, Kresovich S, McCann MC, Ming R, Peterson DG, Mehboob ur R, Ware D, Westhoff P, Mayer KF, Messing J, Rokhsar DS (2009) The *Sorghum bicolor* genome and the diversification of grasses. Nature 457: 551–556.

Patil-Kulkarni BG, Puttarudrappa A, Kajjari NB, Goud JV (1972) Breeding for rust resistance in sorghum. Indian physiopathol 25: 166–168.

Piperidis N, Jackson PA, D'Hont A, Besse P, Hoarau JY, Courtois B, Aitken KS, McIntyre CL (2008) Comparative genetics in sugarcane enables structured map enhancement and validation of marker-trait associations. Mol Breed 21(2): 233–247.

Price S (1963) Cytogenetics of modern sugar canes. Econ Bot 17: 97–105.

Raboin LM, Oliveira KM, Lecunff L, Telismart H, Roques D, Butterfield M, Hoarau JY, D'Hont A (2006) Genetic mapping in sugarcane, a high polyploid, using bi-parental progeny: identification of a gene controlling stalk colour and a new rust resistance gene. Theor Appl Genet 112: 1382–1391.

Ramakrishna W, Emberton J, SanMiguel P, Ogden M, Llaca V, Messing J, Bennetzen JL (2002) Comparative sequence analysis of the sorghum *Rph* region and the maize *Rp1* resistance gene complex. Plant Physiol 130: 1728–1738.

Ramos Leal M, Maribona RH, Ruiz A, Korneva S, Canales E, Dinkova TD, Izquierdo F, Goto O, Rizo D (1996) Somaclonal variation as a source of resistance to eyespot disease of sugarcane. Plant Breed 115: 37–42.

Rana BS, Tripathi DP, Rao NG (1976) Genetic analysis of some exotic x Indian crosses in sorghum. XV. Inheritance of resistance to sorghum rust. Indian J Genet 36: 244–249.

Rhoades VH (1935) The location of a gene for disease resistance in maize. Proc Natl Acad Sci USA, 21: 243–246.

Rossi M, Araujo PG, Paulet F, Garsmeur O, Dias VM, Chen H, Van Sluys MA, D'Hont A (2003) Genomic distribution and characterization of EST-derived resistance gene analogs (RGAs) in sugarcane. Mol Genet Genom 269: 406–419.

Saxena KMS, Hooker AL (1968) On the structure of a gene for disease resistance in maize. Proc Natl Acad Sci USA, 4: 261–268.

Stirling B, Newcombe G, Vrebalov J, Bosdet I, Bradshaw HD (2001) Suppressed recombination around the *MXC3* locus, a major gene for resistance to poplar leaf rust. Theor Appl Genet 103: 1129–1137.

Sun X, Cao Y, Yang Z, Xu C, Li X, Wang S, Zhang Q (2004) *Xa26*, a gene conferring resistance to *Xanthomonas oryzae* pv. *oryzae* in rice, encodes an LRR receptor kinase-like protein. Plant J 37: 517–527.

Tai PYP, Miller JD, Dean JL (1981) Inheritance of resistance to rust in sugarcane. Field Crops Res 4: 261–268.

Tao Y, Jordan DR, Henzell RG, McIntyre CL (1998) Identification of genomic regions for rust resistance in sorghum. Euphytica 103: 287–292.

Tomkins JP, Yu Y, Miller-Smith H, Frisch DA, Woo SS, Wing RA (1999) A bacterial artificial chromosome library for sugarcane. Theor Appl Genet 99: 419–424.

Ueda T, Sato T, Hidema J, Hirouchi T, Yamamoto K, Kumagai T, Yano M (2005) qUVR-10, a major quantitative trait locus for ultraviolet-B resistance in rice, encodes cyclobutane pyrimidine dimer photolyase. Genetics 171: 1941–1950.

Vettore AL, da Silva FR, Kemper EL, Souza GM, da Silva AM, Ferro MI, Henrique-Silva F, Giglioti EA, Lemos MV, Coutinho LL, Nobrega MP, Carrer H, Franca SC, Bacci Junior M, Goldman MH, Gomes SL, Nunes LR, Camargo LE, Siqueira WJ, Van Sluys MA, Thiemann OH, Kuramae EE, Santelli RV, Marino CL, Targon ML, Ferro JA, Silveira HC, Marini DC, Lemos EG, Monteiro-Vitorello CB, Tambor JH, Carraro DM, Roberto PG, Martins VG, Goldman GH, de Oliveira RC, Truffi D, Colombo CA, Rossi M, de Araujo PG, Sculaccio SA, Angella A, Lima MM, de Rosa Junior VE, Siviero F, Coscrato VE, Machado MA, Grivet L, Di Mauro SM, Nobrega FG, Menck CF, Braga MD, Telles GP, Cara FA, Pedrosa G, Meidanis J, Arruda P (2003) Analysis and functional annotation of an expressed sequence tag collection for tropical crop sugarcane. Genome Res 13: 2725–2735.

Wei F, Gobelman-Werner K, Morroll SM, Kurth J, Mao L, Wing R, Leister D, Schulze-Lefert P, Wise RP (1999) The Mla (powdery mildew) resistance cluster is associated with three NBS-LRR gene families and suppressed recombination within a 240-kb DNA interval on chromosome 5S (1HS) of barley. Genetics 153: 1929–1948.

Wilkinson DR, Hooker AL (1968) Genetics of reaction to *Puccinia sorghi* in ten corn inbred lines from Africa and Europe. Phytopathology 58: 605–608.

Wolfe KH, Gouy M, Yang YW, Sharp PM, Li WH (1989) Date of the monocot-dicot divergence estimated from chloroplast DNA sequence data. Proc Natl Acad Sci USA, 86: 6201–6205.

Xu K, Xu X, Fukao T, Canlas P, Maghirang-Rodriguez R, Heuer S, Ismail AM, Bailey-Serres J, Ronald PC, Mackill DJ (2006) Sub1A is an ethylene-response-factor-like gene that confers submergence tolerance to rice. Nature 442: 705–708.

Molecular Mapping of Complex Traits

Maria Marta Pastina,[1] Luciana Rossini Pinto,[2]
Karine Miranda Oliveira,[3] Anete Pereira de Souza[4] and
*Antonio Augusto Franco Garcia[1]**

ABSTRACT

This chapter presents details of the main strategies, populations, molecular markers, traits, statistical methods and software used for QTL mapping in sugarcane. Most of the sugarcane linkage maps are based on first generation progenies derived from biparental crosses, using the pseudo-testcross strategy and markers such as AFLP, isozyme, RAPD, RFLP, SSRs and EST-SSRs. For QTL mapping, single marker and (composite) interval mapping are used for individual maps obtained for each parent, for traits related to resistance to disease (e.g., smut, brown rust, leaf scald, yellow spot) and sugar yield and its components (POL, Brix, tonnes of cane per hectare, tonnes of sugar per hectare, fiber content, stalk number, stalk diameter, stalk length and stalk weight). The percentage of the phenotypic variation explained by each QTL was in general low, ranging from 2 to 22.6%. A synthesis of the results is presented and discussed.

Keywords: polyploids, single dose markers, genetic linkage maps, Quantitative Trait Alleles, marker-assisted selection

[1]Departamento de Genética, Escola Superior de Agricultura Luiz de Queiroz (ESALQ), Universidade de São Paulo (USP), CP 83, CEP 13400-970, Piracicaba-SP, Brazil.
[2]Centro Avançado da Pesquisa Tecnológica do Agronegócio de Cana—IAC/Apta, Anel Viário Contorno Sul, Km 321, CP 206, CEP 14.001-970, Ribeirão Preto-SP, Brazil.
[3]Centro de Tecnologia Canavieira—CTC, Caixa Postal 162, 13400-970, Piracicaba, São Paulo, Brazil.
[4]Centro de Biologia Molecular e Engenharia Genética (CBMEG), Instituto de Biologia, Departamento. de Biologia Vegetal—Universidade Estadual de Campinas (UNICAMP), Cidade Universitária Zeferino Vaz, CP 6010, CEP 13083-875, Campinas-SP, Brasil.
*Corresponding author: *aafgarci@esalq.usp.br*

7.1 Introduction

Plant breeding has significantly contributed to the increase of productivity of several species, making evident its importance to agricultural development. This is achieved through selection schemes that rely on scientific criteria, using phenotypic measures that reflect the genotypic value of the individuals for the genes that control the trait under selection. However, most of the selected traits, such as sugarcane production, fiber content, stalk diameter, stalk length, stalk weight, among several others, are of quantitative nature, i.e., controlled by many genes whose expression is highly influenced by environmental action.

The recent development of techniques to detect and use molecular markers allowed a better understanding of the breeding process, since molecular markers provide information of the genetic architecture of the quantitative traits at the DNA level. Molecular markers have several applications in basic studies and in applied ones (Andersson 2001; Mackay 2001; Dekkers and Hospital 2002; Maurício 2001; Morgante and Salamini 2003; Schlötterer 2004; Takeda and Matsuoka 2008 among others).

In the first studies, molecular markers were based on the products of gene expression, being named morphological or isoenzymatic markers. Due to their rare occurrence and lack of abundant distribution across the genome, they are of limited use. With the development of DNA-based molecular markers, these problems were circumvented resulting in the wide use of these types of genetic markers. Nowadays, restriction fragment length polymorphisms (RFLP; Botstein et al. 1980), random amplified polymorphic DNA (RAPD; Williams et al. 1990), amplified fragment length polymorphism (AFLP; Zabeau and Vos 1993) and simple sequence repeat (SSR; Tautz 1989) are among the most used ones. With the development of the expressed sequence databases, expressed sequence tag (EST) based markers are receiving considerable attention, especially in sugarcane, where they have being used to construct a functional map (Oliveira et al. 2007). Single nucleotide polymorphism (SNP) marker has also recently emerged and shows a great potential for application in genetic studies (Syvänen 2001).

Information provided by molecular markers can be useful in several ways for plant breeding, especially through quantitative trait loci (QTL) mapping that should in theory lead to marker assisted selection (MAS; Mohan et al. 1997; Morgante and Salamini 2003; Charcosset and Moreau 2004; Takeda and Matsuoka 2008). QTL mapping is useful to have a better knowledge of the genetic architecture of quantitative traits, which are difficult to handle and are of great importance for breeding. Eventually, QTL mapping could provide the means to develop breeding methods that incorporate marker information as well as phenotypic information. QTL mapping means finding genomic regions that are associated with phenotypic expression, estimate their effects, gene action, incorporate number of loci and interaction among them and with the environment (Zeng et al. 1999). For mapping studies, a population with genetic variability and highly linkage disequilibrium (LD) is needed. Initially, a genetic map is built to serve as a basis to locate QTLs. QTL mapping has several applications. For example, in breeding programs occurrence of genetic correlations between traits is common. As a consequence, their selection responses are not independent. The causes of correlation are attributed to pleiotropy and/or genetic linkage (Falconer and MacKay 1996; Malosetti et al. 2008). Thus, using only phenotypic information it is possible to estimate only the magnitude of these correlations for all loci simultaneously. However, once QTLs are mapped for several traits, it is possible to verify which of them are linked or have a pleiotropic effect. This allows to develop breeding schemes in order to break or maintain the correlation. Another example is related to genotype-environment interaction. Once QTLs are mapped, it is possible to determinate which ones are more stable, and which environments are more favorable to each QTL (Malosetti et al. 2004; van Eeuwijk et al. 2005, 2007; Boer et al. 2007). Based on QTL mapping, it is also possible to obtain a better understanding of the genetic basis of heterosis (Garcia et al. 2008), which is a phenomenon of great importance for breeding.

In sugarcane, molecular markers have been used to construct genetic linkage maps (Al-Janabi et al. 1993; da Silva et al. 1993, 1995; D'Hont et al. 1994; Grivet et al. 1996; Mudge et al. 1996; Ming et al. 1998, 2002a; Guimarães et al. 1997, 1999; Hoarau et al. 2001; Aitken et al. 2005, 2007; Raboin et al. 2006; Garcia et al. 2006; Oliveira et al. 2007; presented in Chapter 5) and for QTL mapping (Sills et al. 1995; Daugrois et al. 1996; Ming et al. 2001, 2002b, c; Hoarau et al. 2002; Jordan et al. 2004; da Silva and Bressiani 2005; McIntyre et al. 2005a, b, 2006; Reffay et al. 2005; Aitken et al. 2006, 2008; Raboin et al. 2006; Wei et al. 2006; Al-Janabi et al. 2007; Piperidis et al. 2008). In this chapter, we will present details of the main strategies, methods and softwares used for QTL mapping in sugarcane, as well as types of populations, molecular markers and commonly measured traits. A synthesis of the results will be presented and discussed.

7.2 Framework Maps and Markers

A range of DNA marker-based technologies have been established to explore various DNA polymorphisms and they can be classified into three broad types, namely polymerase chain reaction (PCR)-based markers, hybridization-based markers and sequencing-based markers. A number of issues have to be considered in selecting one of these types. The most desirable criteria are: high polymorphism, frequent occurrence and even distribution throughout the genome, low cost and high-throughput. However, it is extremely difficult to find a molecular marker type, which would meet all these criteria. Depending on the study to be undertaken, a marker system can be identified that would fulfill at least a few of them. Hence, the choice will depend on the aim of the researchers, since they develop and apply molecular marker technology for diverse purposes, including assessment of genetic biodiversity, mapping studies and marker-assisted breeding.

The construction of genetic linkage maps is influenced by the genetic structure of the mapping population and the size and complexity of the genome. Due to the large size of the sugarcane genome and its high ploidy level, none of the genetic maps constructed for sugarcane until now is complete. Contrary to other crops, such maize and soybean, in sugarcane it is not possible to generate inbred lines for mapping. However, as sugarcane is highly heterozygous and polyploid, segregation can be observed already at the first generation of a cross. Therefore, conventional mapping in sugarcane generally relies on first generation progenies derived from biparental crosses.

Genetic mapping in sugarcane started after Wu et al. (1992) proposed a general strategy for mapping in highly polyploids with bivalent pairing, based on segregation of single-dose restriction fragments—single dose markers (SDM). In this method, only markers present at one copy (dose) in one of the mapping parents or in one copy in both parents segregating into 1:1 for presence:absence or 3:1 for presence:absence ratio in the progeny are able to be mapped. The single-dose marker approach opened the opportunity to start the construction of genetic linkage maps not only in sugarcane but also in several other crops, including diploids. The SDM approach can be used to any cross between heterozygous individuals and is also known as pseudo-testcross (Grattapaglia and Sederoff 1994; Porceddu et al. 2002; Shepherd et al. 2003; Carlier et al. 2004; Cavalcanti and Wilkinson 2007; Chen et al. 2008) as the testcross mating configuration is not known *a priori*, but inferred *a posteriori* from segregation analysis on the progeny. Two separated maps are obtained, one for each parent. Although single-dose markers (or simplex markers) are present at high frequencies in the sugarcane genome (approximately 70%), duplex markers (present at two copies in only one parent) can also be used to map regions with low levels of simplex markers, and thus increase map coverage (Aitken et al. 2007).

The first sugarcane genetic maps were developed in order to understand the organization of sugarcane genome and also to investigate on the best mapping population type needed to maximize the acquisition of SDM (da Silva et al. 1993; D'Hont 1994; Grivet et al. 1996) rather than to map QTLs. Indeed, these maps used progeny sizes less than one hundred of individuals. Although that was larger than 75 individuals, which is required to detect SDM (Wu et al. 1992), it was not large enough to provide good statistical power to detect QTL (Table 7-1).

Assuming that haploid mapping population derived from highly heterozygous plant could maximize SDM detection (Wu et al. 1992; da Silva et al. 1993), the first sugarcane genetic maps were published by da Silva et al. (1993) and Al-Janabi et al. (1993) using, respectively RFLP and RAPD markers. Both markers were detected in progenies from the cross between the doubled-haploid ADP85-0068 (female parent) derived through anther culture from the *Saccharum spontaneum* clone SES208, which was also used as the male parent. Later, the data of both types of markers were joined into a single map (da Silva et al. 1995).

A mapping study was undertaken on the progeny of the selfed cultivar R570 (Grivet et al. 1996). This map was entirely based on isozyme and RFLP markers and comprised 408 linked markers placed onto 96 cosegregation groups, assembled into the 10 basic linkage groups. The RFLP markers derived from maize probes allowed the investigation of synteny and colinearity between homo(eo)logous co-segregation groups and species origin (*S. officinarum* or *S. spontaneum*). Also, the distribution of some of the markers in the R570 map provided insights about the genome organization.

As R570 is resistant to brown rust (*Puccinia melanocephala*) its selfed progeny was evaluated for rust resistance segregation (Daugrois et al. 1996) as also for tagging of a major gene responsible for resistance (Asnaghi et al. 2001). The R570 selfed progeny was extended to 295 individuals and used to construct a reference genetic map based on AFLP markers. This map covered 5,849 cM, representing approximately 1/3rd of the sugarcane genome length, and was considered as the most saturated sugarcane map of that time (Hoarau et al. 2001). Further, 148 and 134 single-dose resistance gene analogous (RGAs) RFLP and SSR, respectively, derived markers were added to this map (Rossi et al. 2003) making it possible to investigate the distribution and organization of the RGAs along the sugarcane genome.

The variety R570 was also used in crosses with MQ76-53, an old Australian clone derived from a cross between Trojan and SES528 (Raboin et al. 2006), and with the yellow spot (*Mycovellosiella koepkei*) resistant sugarcane variety M134/75 (Al-Janabi et al. 2007). The map obtained for MQ76-53 (R570 x MQ76-53) was also used to identify a gene controlling the red stalk color and a new brown rust resistance gene (Raboin et al. 2006) while the M134/75 map (M134/75 x R570) was used to determine the number and location of QTL for resistance to yellow spot (Al-Janabi et al. 2007).

Aside the R570 genetic map, interespecific crosses involving *Saccharum officinarum* (La Purple) and its supposed progenitor species *S. robustum* (Mol 5829) (Mudge et al. 1996; Guimarães et al. 1999) as also between *S. officinarum* (Green German) and *S. spontaneum* (IND 81-146), and between *S. spontaneum* (PIN84-1) and *S. officinarum* (Muntok Java) (Ming et al. 2002a) allowed the construction of six genetic linkage maps, one for each of the parents (Table

Table 7-1 Genetic maps available for sugarcane.

Mapping population	Progeny Size	Marker Type	SD (1:1)	SD (3:1)	DP	Linked markers	Nº CG	Map coverage*	Software	Reference
				Dosage						
ADP068 x SES208	90	RFLP	216	–	–	188	44	1,361	MAPMAKER	da Silva et al. 1993
ADP068 x SES208	88	RAPD	208	–	–	176	42	1,500	MAPMAKER v1.0	Al-Janabi et al. 1993
SP701006	32	RFLP	253	–	–	94	25	–	MAPMAKER v1.0	D'Hont et al. 1994
R570	77	RFLP, Isozyme	505	–	–	408	96	2,008	MAPMAKER v3.0	Grivet et al. 1996
La Purple x Mol 5829	84	RAPD	279	–	–	161	50	1,152	MAPMAKER v2.0	Mudge et al. 1996
La Purple[a] x Mol 5829[b]	100	RAPD, RFLP, AFLP	341[a] 301[b]	–	–	283[a]208[b]	74[a]65[b]	1,881[a] 1,189[b]	MAPMAKER v2.0	Guimarães et al. 1999
R570	295	AFLP	939	–	–	887	120	5,849	MAPMAKER v3.0	Hoarau et al. 2001
Green German[a] x IND81-145[b]	85	RFLP	434[a]395[b] 308[c]	132[a] 54[b]86[c]	–	289[a]257[b] 194[c]	75[a]70[b] 71[c]	2,466[a]2, 172[b]1,395[c]		Ming et al. 2002a
PIN84-1[c] x Muntok Java[d]			359[d]	159[d]		214[d]	73[d]	1,472[d]		
R570	112	RFLP, SSR	347	–	–	282	128	–	MAPMAKER v3.0	Rossi et al. 2003
IJ76-514 x Q165	227	AFLP, RAF, SSR	967	36	123	1074	136	9,058.3	MAPMAKER v2.0	Aitken et al. 2005
IJ76-514 x Q165	227	AFLP, SSR	240	–	234	534	123	4,906.4	JoinMap v3.0	Aitken et al. 2007
Q117 x MQ77-340	232	AFLP, SSR	395	58	–	342	101	3,582	MAPMAKER v3.0	Reffay et al. 2005
SP80-180 x SP80-4966	100	RFLP, SSR AFLP	441	677	–	357[1]217[2]	131[1]98[2]	2,602.4[1], 340[2]	OneMap[1]JoinMap[2] v3.0	Garcia et al. 2006

R570[a] x MQ76-53[b]	198	AFLP, SSR, RFLP	1057	–	424[a]536[b]	86[a]105[b]	3,144[a] 4,329[b]	MAPMAKER v3.0	Raboin et al. 2006
M134/75 x R570	227	AFLP, SSR	557	–	474	95	6,200	MAPMAKER v3.0	Al-Janabi et al. 2007
SP80-180 x SP80-4966	100	AFLP, gSSRs, EST-SSRs, EST-RFLP, g-RFLP	800	869	664	192	6,261.1	OneMap	Oliveira et al. 2007
La Striped[i] x SES 147B[j]	100	AFLP, SRAP, TRAP	247[i] 221[j]	33[i] 43[j]	146[i] 121[j]	49[i] 45[j]	1,732[i] 1,491[j]	JoinMap v3.0	Alwala et al. 2008

*Cumulative length in centi-Morgan (cM). SD (1:1): single-dose markers, marker present only once in the genome segregating in a 1:1 ratio. SD (3:1): single dose marker, marker present in one copy in both parental genomes, segregating in a 3:1 ratio. DP: duplex marker, marker present twice in one parental genome segregating in 11:3 ratio ($x = 8$) or in a 7:2 ratio ($x = 10$). CG: co-segregation group. a,b,c,d,e,f,g,h,i,j: refers to the map information of the respective parental. Bold parents correspond to the parental map constructed.

7-1). These maps were focused on comparative mapping studies among the Andropogonae tribe, mainly among sugarcane, maize and sorghum, and this was of great contribution to sugarcane, as they illustrated the usefulness of sorghum linkage maps to understand the complex sugarcane genome (Ming et al. 2002a).

Another interspecific cross involving the progenitor species of cultivated sugarcane, *S. officinarum* (La Striped) and *S. spontaneum* (SES147B) were used to construct framework genetic linkage maps for each progenitor species using either simplex and duplex markers through JoinMap software (Van Ooijen and Voorrips 2001). These maps were constructed with AFLP, SRAP (sequence related amplified polymorphism) and TRAP (target region amplification polymorphism) the later ones being derived from gene sequence information (Alwala et al. 2008).

Interesting results were obtained in the Australian sugarcane population derived from the cross between Q117 x MQ77-340 (Reffay et al. 2005). MQ77-340 is a direct descendent of Mandalay, an important *Saccharum spontaneum* clone used in the ancestry of many important Australian varieties and elite parents. This cross was used for the construction of a genetic map for MQ77-340 clone and identification of QTL with positive or negative effect on important traits, and also allowed the identification of specific genomic regions contributed by Mandalay to the Australian commercial varieties and elite parental materials (Reffay et al. 2005).

Another linkage mapping was carried out by Aitken et al. (2005) in a population in which the *Saccharum officinarum* clone IJ76-514 was used as the female in a cross with the variety Q165. The main goal was to take advantage of the recent release of a large number of SSRs for sugarcane to construct a high coverage genetic map of the variety Q165. This map combined AFLP and SSR markers and highlighted the utility of SSRs to allocate the linkage groups to homology groups and to compare genetic maps. As pointed by Aitken et al. (2005), genetic maps from different cultivars can reveal different chromosome arrangements, having a great impact on the use of molecular markers for sugarcane breeding. Due to the important contribution of *S. officinarum* genome to the commercial sugarcane varieties, the cross between IJ76-514 and Q165 was also used to construct a map for IJ76-514, integrating simplex (1:1 and 3:1 segregation ratio) and duplex (11:3 segregation ratio) markers (Aitken et al. 2007).

Usually, in sugarcane mapping using biparental crosses, single dose polymorphisms identified in each parent allow the construction of individual parental maps . Garcia et al. (2006) used an approach based on the simultaneous maximum-likelihood estimation of linkage and linkage phases (Wu et al. 2002) to construct a single (integrated) map of a full-sib family derived from the cross between the pre-commercial Brazilian sugarcane varieties SP80-180 and SP80-4966. The integrated genetic map obtained with this approach gave rise to 357 linked markers (RFLP, SSP and AFLP) assigned to 131 co-segregation groups and had a total length of 2,602.4 cM. With the development of expressed sequence tag (EST) markers derived from the SUCEST database, EST-SSRs and EST-RFLPs were added to the SP80-180 and SP80-4966 map. This genetic linkage map containing function-associated markers had 664 single dose markers distributed into 192 co-segregation groups and a total length of 6,261.1 cM (Oliveira et al. 2007).

Over the past decade, there has been an exponential increase in the availability of DNA sequence data from a wide variety of taxa. ESTs are typically single-pass sequences produced from cDNAs, provide one way of overcoming the problem of cost and to become a promising alternative to the development of traditional "anonymous" markers following standard methods. The ESTs come from the partial sequencing of the 3' or 5' cDNA extremities of a gene that was expressed in a tissue, at a certain stage (Sterky and Lundeberg 2000). The identification of ESTs has proceeded rapidly, with over 6 million ESTs now available in computerized databases. ESTs were originally intended as a way to identify gene transcripts, but have since been instrumental in gene discovery, for obtaining data on gene expression and regulation, sequence determination, and for developing highly valuable molecular markers, such as EST-based RFLPs, SSRs, SNPs and CAPS (Semagn et al. 2006). RFLP markers developed from ESTs (EST-RFLP) have been extensively used for the construction of high-density genetic linkage maps and physical maps.

The more completed maps contain about 2,000 markers scattered in about 100 linkage groups and were constructed using mapping populations of about 200 individuals. Nevertheless, they have not already achieved good genome coverage. It is estimated that only 33 to 50% of the genome is represented by the markers in these maps. This means that more than half of the sugarcane genetic constitution is not being sampled in these studies. The necessity of a high number of markers for the complete analysis of the sugarcane genome is justified by the complex genome of this species, as well as its polyploid structure and relatively narrow genetic base, if cultivars are taken into consideration.

The abundance of SSRs and SNPs in transcribed regions of the genome and the high level of polymorphism of these markers make EST libraries a valuable resource for the supply of markers for sugarcane genetic studies. These EST-derived markers make a substantial contribution to the understanding of the structure and function of the sugarcane genome, which can lead to improvements in sugarcane production and quality.

7.3 Strategies for QTL Mapping in Sugarcane

7.3.1 Population Type

QTL mapping is based on the establishment of relations between the phenotype (expression of quantitative traits) and the genotype (evaluated with molecular markers). For doing this, sophisticated statistical methodologies are commonly used, relying on a strong computational support due to the complexity of the analysis. Such models include single marker analysis (for example, Soller et al. 1976; Weller 1986; Edwards et al. 1987; Stuber et al. 1987), interval mapping (Lander and Botstein 1989), composite interval mapping (Zeng 1993, 1994; Jansen and Stam 1994), Bayesian inference (Satagopan et al. 1996; Health 1997; Sillanpaa and Arjas 1998; Yi and Xu 2001; Yi et al. 2005, 2007a, b), multiple interval mapping (Kao and Zeng 1997; Kao et al. 1999) and the mixed models approach (Malosetti et al. 2004, 2008; Van Eeuwijk et al. 2005, 2007; Boer et al. 2007). In the case of sugarcane, as will be presented, due to its polyploid nature and high genomic complexity, single marker analysis is widely used, as well as, interval mapping or composite interval mapping, based on genetic maps of each parent (pseudo-testcross strategy).

The majority of the experimental crosses used for the construction of genetic maps and QTL mapping in plants are based on populations derived from crosses between inbreed lines, such as RILs, backcross and F_2. The statistical methods are well established and implemented in several softwares, such as MAPMAKER/EXP (Lander et al. 1987; Lincoln et al. 1992a) for genetic map construction and QTL-Cartographer (Basten et al. 2005) and WinQTL-Cartographer (Wang et al. 2007) for QTL mapping.

However, for sugarcane, it is impractical or even impossible to obtain inbreed lines, mainly due to the high inbreeding depression that occurs when plants are selfed. In this case, none of the sophisticated models could be directly used, making it difficult to obtain good mapping results. A commonly used type of population used in sugarcane is obtained from the crosses between non-inbred parents. Thus, a single locus could present several segregating alleles having different patterns of segregations for the markers (Garcia et al. 2006) and QTLs (Lin et al. 2003). Moreover, the linkage phases between markers and QTLs are unknown, making QTL mapping more complicated and challenging. Specifically for sugarcane, the high level of polyploidy brings additional problems, due to the complex pattern of chromosomal segregation at meiosis (Heinz and Tew 1987). Therefore, an approach that is being used is based on the mapping of quantitative trait allele (QTA). In this case, one tries to find significant associations between the phenotypic variation observed for the trait of interest and the different alleles that can be segregated for a specific locus. However, the effects of these QTAs might be smaller than those observed for diploid species, mainly due to the high number of segregating alleles per locus for the target trait (Aitken et al. 2008).

7.3.2 Statistical Models and Softwares

Single marker (SM) analysis, widely used for QTL mapping in sugarcane, is based on the comparison between trait means of different marker genotype classes. This can be easily implemented through t-tests, analysis of variance, simple and multiple linear regression, maximum likelihood analysis, among others. Such analyses test the null hypothesis (H_0) that the observed phenotypic values are independent of the genotype at a particular marker (if the marker is unlinked to the putative QTL). If the null hypothesis is rejected, it is assumed that there is a putative QTL associated with the marker for the quantitative trait (Doerge et al. 1997; Liu 1998; Lynch and Walsh 1997; Doerge 2002). The results of this type of analysis can be presented in tables with the values of the p-values from the statistical test for each marker, or using plots with the LOD (logarithm of odds) score or LRT (likelihood ratio test) values obtained for each marker of a given linkage group, when the genetic map is available (Fig. 7-1). The main advantages of this method are: simplicity and fast speed of execution; it can be carried out through widely used statistical softwares, such as SAS (SAS Institute 1989a) and R (R Development Core Team 2009). In addition, it can be easily extended to multiple loci models and does not need a genetic linkage map established for the population, enabling the inclusion of unlinked markers (which is common in the sugarcane linkage maps). However, the statistical tests are biased, because there is a confounding between the QTL effect and its distance from the marker, resulting in a low power to detect QTL when the markers do not completely cover the genome and/or when a small sample size is considered (Doerge et al. 1997; Doerge 2002). Furthermore, it is not possible to infer about the location of the QTL.

Interval mapping (IM), proposed by Lander and Botstein (1989), uses information from a pair of adjacent molecular markers to make inferences, about the existence of a putative QTL at each position within the interval between these markers. For this, an estimated genetic map is needed as a framework for the localization of QTL. The results of IM, usually expressed as LOD scores, reflect the difference between the value of the likelihood function under the null hypothesis (no QTL) and the value of the likelihood function under the alternative hypothesis (QTL at the position tested). In other words, the greater the LOD value, more likely is the presence of a putative QTL at the testing position. Figure 7-2 presents the results of the application of IM for QTL detection in sugarcane (red line). The LOD values obtained at each position in a linkage group are shown, making clear that

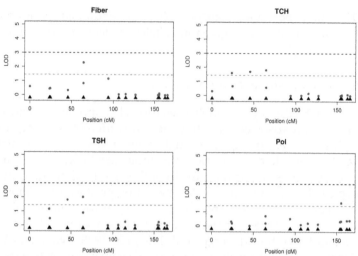

Figure 7-1 Graphic representation of the results of the Single marker analysis for sugarcane data (Garcia et al. 2009, unpub results).

Fiber: Fiber percent; TCH: tons of cane per hectare; TSH: tons of sugar per hectare; Pol: g of sucrose per kg per 100 g of fresh cane; single marker method: blue points; molecular markers: black triangle; dashed red line: LOD value considering 1% of significance level as a threshold; dashed black line: LOD = 3.

Color image of this figure appears in the color plate section at the end of the book.

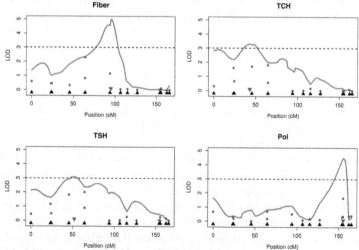

Figure 7-2 Graphic representation comparing the results of Single marker analysis and Interval Mapping for QTL detection in sugarcane (Garcia et al. 2009, unpublished results).

Fiber: Fiber percent; TCH: tonnes of cane per hectare; TSH: tonnes of sugar per hectare; Pol: g of sucrose per kg per 100 g of fresh cane; single marker method: blue points; interval mapping: continued red line; molecular markers: black triangle; green triangles indicate the position of QTL mapped based on IM; and dashed black line: LOD = 3 as the threshold.

Color image of this figure appears in the color plate section at the end of the book.

IM has higher statistical power than SM. However, although the IM allows estimating the effects and positions of putative QTLs, it also has some limitations since this method uses information only from two markers at a time, not considering the interference of other QTLs located outside the mapping interval, which can result in the detection of false positives (Doerge 2002).

Zeng (1993, 1994) and Jansen and Stam (1994) extended IM proposing a procedure named composite interval mapping (CIM), which is a model that also includes markers outside the mapping interval as cofactors. This makes possible to control the effects of other QTL(s) outside the mapping interval. In comparison with SM analysis, IM and CIM have several advantages, such as the lack of confounding between effect and position of the putative QTL, and higher statistical power (Doerge et al. 1997). However, IM and CIM also have limitations, overcome by other methods, such as multiple interval mapping (MIM) (Kao and Zeng 1997; Kao et al. 1999), which in addition to more accuracy to estimate genetic parameters, allows the identification of interactions between QTLs (epistasis). To our knowledge, MIM have not been used in sugarcane.

Considering that for sugarcane in general a genetic map for each parent is estimated based on markers segregating in a 1:1 ratio, QTL mapping is done through statistical methods developed for backcross, such as single marker analysis and simple or composite interval mapping. Such analyses are implemented in softwares developed for experimental populations, such as MAPMAKER/QTL (Paterson et al. 1988; Lincoln et al. 1992b), QTLCartographer (Basten et al. 2005), WinQTL-Cartographer (Wang et al. 2007), R/qtl (Broman et al. 2003), QTL Express (Seaton et al. 2002), PLABQTL (Utz and Melchinger 2003), QTX (Manly et al. 2001) among others. Some authors were able to map QTLs using these approaches for different traits, such as sugarcane brown rust resistance (Daugrois et al. 1996; McIntyre et al. 2005a, b; Raboin et al. 2006), flourish, plant height (Ming et al. 2002b), sugar production, weight and stalk number, fiber content (Ming et al. 2001, 2002a; Hoarau et al. 2002; Jordan et al. 2002; da Silva and Bressiani 2005; Aitken et al. 2006).

In spite of the relative success obtained with the pseudo-testcross strategy in sugarcane, the use of integrated maps (Garcia et al. 2006; Oliveira et al. 2007) presents several advantages, as they allow better saturation and the characterization of the polymorphic variation in the whole genome, which could provide better results for QTL mapping. Although several authors presented alternatives for the construction of integrated genetic maps in populations derived from crosses between non-inbred parents (Ritter and Salamini 1996; Maliepaard et al. 1997;Ridout et al. 1998; Wu et al. 2002; Garcia et al. 2006; Margarido et al. 2007; Oliveira et al. 2007), results for sugarcane are still scarce.

7.4 Results of QTL Mapping

As pointed out before, QTL mapping in sugarcane is done usually considering the same populations used to construct the linkage maps. These populations were generally obtained from biparental interspecific crosses between *S. officinarum* and *S. robustum*, *S. officinarum* and *S. spontaneum*, and self-fertilization or the biparental crosses between commercial materials. The most studied traits are: brown rust (*Puccinia melanocephala*) resistance; smut (*Ustilago scitaminea*) resistance; yellow spot (*Mycovellosiella koepkey*) resistance; flowering time; sugar yield; stalk length; stalk weight; stalk diameter; stalk number; and fiber content. Table 7-2 presents published results for QTL mapping in sugarcane, including the progeny type, molecular markers used and traits addressed. It is clear that in the early studies only markers segregating in a 1:1 ratio were used. More recently, there was inclusion of markers segregating in a 3:1 ratio or even multiplexes. However, the mapping strategy remains basically the same for the majority of the studies: pseudo-testcross for the construction of genetic maps, taking into account only information from markers segregating in a 1:1 ratio, and single marker analysis to detect QTLs associated with different traits of agronomic importance in sugarcane.

Sills et al. (1995) presented the first results of QTL mapping in sugarcane, using RAPD markers and a progeny obtained from the cross between *S. officinarum* ("LA Purple") and *S. robustum* ("Mol 5829"). SM analyses were performed using ANOVA and multiple regression, resulting in the identification of 24 significant associations ($p \leq 0.05$) for stalk number, stalk diameter, tasseled stalks, POL (g of sucrose per 100 grams of fresh cane), Fiber content and smut (*Ustilago scitaminea*) resistance. Three significant epistatic interactions were also identified for the traits under evaluation, considering $p \leq 0.05$. However, only the epistatic interaction between the markers P10.516 and V15.397, associated with stalk diameter, remained significant after multiple regression analysis. These analyses were performed using the program JMP v2.0.2 (SAS Institute 1989b). Later, for the same progeny, Mudge et al. (1996) and Guimarães et al. (1997) cited the association of molecular markers with eyespot disease and flowering time, respectively, but not providing detailed information.

Through the study of the inheritance of rust resistance, Daugrois et al. (1996) identified an RFLP marker linked at 10 cM with a possible dominant resistance gene, which showed a 3 (resistant): 1 (susceptible) segregation in the self progeny of the sugarcane modern cultivar "R570" (resistant to rust). For this, single marker analyses were performed through the software MapQTL (van Ooijen 2004) using the nonparametric test of Kruskal-Wallis, as the data did not follow normal distribution. Hoarau et al. (2002), using the same progeny expanded to 295 individuals, identified 73 putative QTAs, consistent across both years of evaluation, i.e. with significant associations between the observed phenotypic variation for a given trait and a particular allele of a locus, considering 5% significance level. In addition, 41 epistatic interactions were identified considering $p \leq (0.005)^2$, for the different years and traits (stalk number, stalk diameter, stalk length and brix). Tests were carried out based on the SM approach, through the analysis of variance using the program SAS (SAS Institute 1989a) for AFLP in single dose (segregating in 1:1 and 3:1 ratio) and multiple dose markers.

Ming et al. (2001) through the evaluation of two different full-sib progenies derived from heterozygous parents, one obtained from a cross between *S. officinarum* "Green German" and *S. spontaneum* "IND 81-146", and another from the cross between *S. spontaneum* "PIN 84-1" and *S.officinarunm* "Muntok Java", identified 36 associations for sugar content, of which 14 were detected for the first progeny and 22 for the second ($p \leq 0.01$). These tests were performed employing the IM approach, using the software MAPMAKER/QTL, in the case of markers placed on individual maps for each parent (constructed based on pseudo-testcross strategy), or SM analysis (ANOVA SAS program) for unlinked markers. Later, Ming et al. (2002b), using the same full-sib progenies, identified 102 significant associations for sugar yield, POL (sucrose content), stalk weight, stalk number, fiber content and ash content. Of these 102 marker-QTL associations, 61 were identified for markers placed on the genetic linkage map and 41 for unlinked markers. The tests were carried out by IM, using the MAPMAKER/QTL program for markers included in the linkage groups, and SM analysis conducting ANOVA using the SAS program, for unlinked markers. Still using the same full-sib progenies, Ming et al. (2002c) detected 65 significant associations for plant height and flowering time, of which 30 were identified for markers placed on linkage groups and 35 for unlinked ones. The analyses were performed based on SM and IM approaches. The three studies reported were carried out using RFLPs.

Considering the population obtained from a cross between two Australian elite clones, "Q117" (female) and "74C42" (male), evaluated in two different sites and years, Jordan et al. (2004) identified seven and six RFLP markers associated with stalk number and sucker number, respectively. These associations were consistent across sites and years, and three of these markers were identified to be related to the two traits simultaneously. The tests were carried out based on SM analysis through the MapQTL program. Comparative mapping with sorghum suggested that there are unlinked regions of the genome associated with stalk number and sucker number. This result leads one to conclude that it should be possible to select concurrently for high stalk number and low sucker types.

McIntyre et al. (2005a) evaluated the same progeny combining information from different marker types (AFLP, RFLP and SSR). For QTL

Table 7-2 QTL mapping results for sugarcane.

Reference	Mapping population	Marker type	Strategy (genetic map)	Statistical analysis	Years/ Phisiological state	Traits	QTLs*	Effects**	R^2 (min/ max)***	Threshold
Sills et al. (1995)	LA Purple x Mol 5829	RAPD (1:1)	–	SM (JMP v2.0.2)	1/plant crop	stalk number	2	2.99 and 4.52	34	$p \le 0.05$
						tasseled stalks (%)	4	−16.94 to 13.81	50	
						smutted stalks (%)	4	−0.99 to 1.08	43	
						stalk diameter (cm)	4	−0.14 to 0.20	49	
						POL	4	−1.07 to 0.77	43	
						fiber content	4	−1.37 to 1.28	58	
						plot weight (kg)	2	−5.02 and 6.79	23	
Daugrois et al. (1996)	self-fertilization of R570	Isozyme and RFLP (1:1)	pseudo-testcross	SM (MapQTL)	3/plant and ratoon crop	brown rust resistance (3 R :1 S, in the progeny)	1	–	26	$p=10^{-4}$
Ming et al. (2001)	Green German x IND81-146	RFLP (1:1)	pseudo-testcross	SM (SAS) and (MAPMAKER/ IM 1/QTL)	plant crop(field and green house)	sugar-content (pounds of sugar per ton of cane)	14	−23.40 to 20.00	65.5	$p \le 0.01$
Ming et al. (2001)	PIN84-1 x Muntok Java	RFLP (1:1)	pseudo-testcross	SM (SAS) and IM (MAPMAKER/ QTL)	1/plant crop(field and green house)	sugar-content (pounds of sugar per ton of cane)	22	−9.0 to 18.8	68.3	

Reference	Cross	Marker		Method	Crop	Trait		Range		LOD ≤ 2.5 and P ≤ 0.003
Ming et al. (2002b)	Green German x IND81-146'	RFLP (1:1)	pseudo-testcross	SM and IM (MAPMAKER /QTL)	1/plant crop	sugar yield (tonnes per and hectare)	3	-0.46 to 0.55	18.4	
						fiber content	19	-1.61 to 2.67	60.6	
						POL	2	-1.27 and 0.92	18.5	
						Ash (mMhos/cm)	7	-1.05 to 1.04	39.1	
						stalk number	2	-5.05 and 7.38	13.9	
						stalk weight (kg)	10	-4.45 to 2.66	62.7	
Ming et al. (2002b)	PIN84-1 x Muntok Java	RFLP (1:1)	pseudo-testcross	SM and IM (MAPMAKER /QTL)	1/plant crop	sugar yield (tonnes per hectare)	7	0.08 to 0.14	30.2	
						fiber content	1	-1.28	7.0	
						POL	12	-0.57 to 0.98	39.9	
						Ash (mMhos/cm)	13	-0.78 to 0.71	41.4	
						stalk number	1	6.08	6.1	
						stalk weight (kg)	24	-1.91 to 2.36	37.8	

Table 7-2 contd....

Table 7-2 contd....

Reference	Mapping population	Marker type	Strategy (genetic map)	Statistical analysis	Years/Phisiological state	Traits	QTLs*	Effects**	R² (min/max)***	Threshold
Ming et al. (2002c)	Green German x IND81-146	RFLP (1:1)	pseudo-testcross	SM and IM (MAPMAKER/QTL)	1/plant crop	plant height (cm)	3	-24.42 to 29.38	16.3	$LOD \leq 2.5$ (para IM) and $P \leq 0.003$ (para SM)
						flowering data(0, no flower and 1, any stage after flowering)	9	-0.31 to 0.31	(5.0/11.9)	
Ming et al. (2002c)	PIN84-1 x Muntok Java	RFLP (1:1)	pseudo-testcross	SM and IM (MAPMAKER/QTL)	1/plant crop	plant height (cm)	53	-19.11 to 37.79	(4.6/22.6)	
						flowering data (0, no flower and 1, any stage after flowering)	0	–	–	
Hoarau et al. (2002)	atofecundação de R570	AFLP(1:1, 3:1; and multiplex)	pseudo-testcross	SM (SAS)	2/plant and ratoon crop	Brix	10	-0.57 to 0.51	(3/7)	$p \leq 0.05$
						stalk diameter (cm)	21	-0.109 to 0.112	(3/7)	
						stalk lenght (cm)	20	-8.96 to 11.13	(3/7)	
						number of millable stalks	22	-2.96 to 2.50	(3/7)	

Reference	Cross	Marker	Method	Crop	Trait	No.	Range	(fraction)	p
Jordan et al. (2004)	Q117 x 74C42	RFLP and RAF (Radio Labelled Amplified Fragment) pseudo-testcross	SM (MapQTL)	2/plant and ratoon crop(2 sites)	stalk number	7	-1.87 to 1.45	–	p ≤ 0.05
					sucker number	6	-1.22 to 1.25	–	
McIntyre et al. (2005a)	Q117 x 74C42	AFLP, RFLP and SSR (1:1 and 3:1) pseudo-testcross	SM (Map Manager QTXb16)	2/plant and ratoon crop	*Pachimetra Root Rot* (PRR) resistance	7	-1.3 to 2.0	(6/13)	p ≤ 0.01
					brown rust resistance	4	-1.7 to 1.2	(5/18)	
da Silva and Bressiani (2005)	SP80-180 x SP80-4966	EST-RFLP –	SM	1/plant crop	POL	1	9.45	–	p ≤ 0.05
					tonnes of cane per hectare (TCH)	–	–	–	–
					fiber content	–	–	–	–
Reffay et al. (2005)	Q117 x MQ77-340	AFLP and SSR (1:1 and 3:1) pseudo-testcross	SM	3/plant and ratoon crop	Commercial Cane Sugar (CCS)	18	–	(2/7)	p ≤ 0.01
					Brix		–		
					POL		–		
					fiber content	13	–		
					stalk weight (kg)			–	
					tonnes of cane per hectare (TCH)		–	(2/7)	p ≤ 0.01
					tonnes of sugar per hectare (TSH)		–		

Table 7-2 contd.....

Table 7-2 contd....

Reference	Mapping population	Marker type	Strategy (genetic map)	Statistical analysis	Years/ Phisiological state	Traits	QTLs*	Effects**	R^2 (min/ max)***	Threshold
McIntyre et al. (2005b)	Q117 x 74C42	AFLP, RFLP and SSR (1:1 and 3:1)	pseudo-testcross	SM	2/plant crop	brown rust resistance	4	–	(4/10)	$p \leq 0.01$
Aitken et al. (2006)	IJ76-514 x Q165^A	AFLP and SSR (1:1, 3.1 and multiplex)	pseudo-testcross	SM	2/plant crop (early and mid season)	Brix	5	-0.66 to 0.56	(2/9)	$p \leq 0.05$
						POL	6	-3.5 to 2.99	(2/8)	
Raboin et al. (2006)	R570 x MQ76-53	AFLP, RFLP and SSR (1:1 and 3:1)	pseudo-testcross	SM	2/plant crop	brown rust resistance(1 R:1 S in the progeny)	1	–	13	$p = 3\times10^{-5}$
					3/plant crop	stalk colour (1 red : 1 non-red)	1	–	–	$p = 1.6 \times10^{-34}$
McIntyre et al. (2006)	IJ76-514 x Q165^A	SNP	pseudo-testcross	SM	2/plant crop	Commercial Cane Sugar (CCS)	–	–	–	–
Wei et al. (2006)	154 clones sampled from germplasm collections	AFLP and SSR	–	SM	–	*Pachymetra Root Rot (PRR)* resistance	5	–	–	$p \leq 0.001$
						leaf scald resistance	6	–	–	
						Fiji leaf gall resistance	7	–	–	
						smut resistance	21	–	–	

Reference	Cross	Marker type	Method	Crop/progeny	Trait		Effect		Significance
Aljanabi et al. (2007)	M134/75 x R570	AFLP and pseudo-testcross SSR (1:1 and 3:1 and multiplex)	SM, IM and CIM (QTLCartographer)	2/plant crop and first ratoon crop (2 sites) the progeny	yellow spot resistance 3 R : 1 S in	2	-10.05% and-4.7%	23.8 and 7.7	LOD=8.7 and LOD=3.5
Piperidis et al. (2008)	Q117 x MQ 77-340	AFLP and pseudo-testcross SSR (1:1 and 3:1)	SM and IM (QTLCartographer)	3/plant crop and first ratoon crop	Brix	1	-0.57 to 0.50	(3/5)	p ≤ 0.01
					POL	1	-0.68 to -0.55	(3/5)	
					Commercial Cane Sugar (CCS)	2	-0.60 to -0.42	(3/6)	
Aitken et al. (2008)	IJ76-514 x Q165[A]	AFLP and SSR (1:1 and 3:1)	pseudo-testcross	SM and CIM (QTLCartographer)	3/plant crop, first and second ratoon crops (2 sites)				
					stalk weight	18	-0.88 to 0.79	(2/9)	p ≤ 0.05
					stalk height	8	-12.28 to 11.92	(2/7)	
					stalk number	7	-1.28 to 1.27	(2/9)	
					stalk diameter	19	-1.53 to 1.47	(2/7)	
					tonnes of cane per hectare (TCH)	6	-8.58 to 12.78	(2/6)	

*: QTLs consistent across sites (if this occurs), years and physiological stage, with effects measured for each situation. **: if more than one value is showed for a single QTL, it corresponds to the effects of different years and sites of evaluation. ***: total R^2 (fitted model with all QTLs simultaneously); value in brackets indicates the minimum and maximum R^2 value observed among the QTLs. SM: single marker; IM: interval mapping; CIM: composite interval mapping.

mapping, the tests were performed based on SM approach, considering markers segregating for each parent (1:1), used to construct the genetic linkage map through the double pseudo-testcross strategy, and markers segregating for both parents (3:1). Thus, in addition to an unlinked marker segregating in a 3:1 ratio, two genomic regions, one in each parent, were identified to be associated with Pachymetra root rot (*Pachymetra chaunorhiza*) resistance. For brown rust, significant associations were identified in two genomic regions, one in each parent. Moreover, through the assessment of 154 elite clones, it was found that some of these markers remained associated with these diseases (Wei et al. 2006). Such results suggest that these markers can be used for marker-assisted selection (MAS) in breeding programs that are trying to select for envisage resistance to both diseases. Still in the same progeny, McIntyre et al. (2005b) used similar sequences of genes for brown rust resistance to identify QTLs associated with RFLP, AFLP and SSR markers. Through the comparative mapping with sorghum, it was found that markers obtained from one of these genes were associated with a chromosomal region of sorghum, in which it had already been identified as a major QTL for brown rust resistance. QTL analyses were performed based on SM approach through the program MapManager QTXb16 (Meer et al. 2002).

Da Silva and Bressiani (2005), considering a progeny derived from the cross between two *Saccharum* spp. elite clones, "SP80-180" (female) and "SP80-4966" (male), identified an EST-RFLP marker inversely associated with sucrose content (POL) at 1% significance level, indicating that plants without this marker would have a higher POL value than plants with the marker. However, this marker was present in the parent "SP80-4966" (high sugar yield) and absent in the parent "SP80-180" (low sugar yield), indicating a transgressive segregation of such a marker in the progeny. This suggests that parent "SP80-4966" has other alleles that contribute to POL increase, compensating the negative effect of the pA35 (EST-RFLP) marker. Single marker analyses were carried out based on linear regression.

To study the genetic contribution of the "Mandalay" (*S. spontaneum*) clone for Australian elite varieties and parents, Reffay et al. (2005) performed QTL mapping in a progeny derived from the cross between the clones "Q117" and "MQ77-340", the latter being a direct descendant of "Mandalay" (obtained from the cross between the *S. officinarum* "Korpi" clone and "Mandalay"). For this, AFLP and SSR markers were used, segregating for each parent (with 1:1 ratio) and for both parents (with 3:1 ratio). From a total of 101 linkage groups, 65 had markers originated from the parent "Mandalay" and/or "Korpi". QTL analyses were performed for single markers, considering sugar related traits (POL, brix and Commercial Cane Sugar), fiber content, stalk number, tons of cane per hectare (TCH) and tons of sugar per hectare (TSH). Markers from the two parents of "MQ77-340",

"Mandalay" and "Korpi" , were identified to be associated with different traits, expressing both positive and negative phenotypic effects. Recently, Piperidis et al. (2008), through the comparative mapping between individual maps obtained for each parent of the same progeny and maps of cultivars "R570" (French origin) and "Q165^A" (Australian origin), identified significant marker-QTL associations for brix, in linkage groups consistent across two or three maps of each parent. The analyses to detect associations between unlinked markers and the phenotypic variation of traits (POL, brix or Commercial Cane Sugar) were performed using ANOVA, considering single markers. For markers placed on individual maps for each parent, QTL mapping was carried out based on IM approach with the software QTLCartographer.

Aitken et al. (2006) identified 37 marker-QTL associations for brix and POL in a progeny derived from the cross between the clone *S. officinarum* "IJ76-514" and the cultivar "Q165^A". Each QTL explained from 3 to 9% of the observed phenotypic variation, showing both positive and negative effects. When all the detected QTLs were included simultaneously in a multiple regression model, from 37 to 66% of the observed phenotypic variation was explained, depending on the trait. In addition, using models fitted for the effect of two markers and their interaction, 97 epistatic interactions were identified, of which nine were consistent across two years (eight affecting more than one trait simultaneously), considering $p < 0.00001$. Such epistatic interactions explained individually between 11 and 13% of the observed phenotypic variation, and between 24 and 36% when considered jointly. The QTL detection was performed based on the SM approach, using ANOVA. In this study, AFLP and SSR markers were used. McIntyre et al. (2006), considering the same progeny and SNP (polymorphisms in the sucrose phosphate synthase—SPS—gene family III) markers, identified by SM analysis (t-test) that there is no significant difference in the SNP frequency among individuals with low and high sucrose content in the progeny. However, using an ecotilling approach, two of the SPS gene family III haplotypes were mapped to two different linkage groups from homology group 1 in "Q165^A". Both haplotypes mapped near QTLs for increased sucrose content, but none of them were associated with any sugar related traits. Recently, Aitken et al. (2008), using AFLP and SSR markers and the same progeny, identified 27 genomic regions significantly associated (5% significance level, using permutation test) with at least one of the traits including: cane yield (tons of cane per hectare, TCH), stalk weight, stalk number, stalk length, stalk diameter; 46% of the marker-QTL associations were consistent across different years of evaluation. In addition, using SNPs, two alleles of the TEOSINTE BRANCHED 1 gene (*TB1*, a major gene controlling branching in maize) showed some association with stalk number in two years of evaluation, but unlike maize, *TB1* has a small effect in sugarcane (explaining from 6 to 8% of the observed phenotypic variation). Moreover, 195 epistatic interactions were identified ($p \leq 0.00001$) considering all the traits and years under study.

Based on the progeny obtained from the cross between the modern cultivar "R570" and the clone "MQ76-53", Raboin et al. (2006) identified an AFLP marker linked at 6.5 cM to a gene controlling the red stalk color (segregating in a 1:1 ratio) and another AFLP marker linked at 23 cM to a new brown rust resistance gene (segregating in a 1:1 ratio), both genes placed on the "MQ76-53" genetic linkage map. A Fisher's exact test was performed to assess associations between each trait and the segregating markers under evaluation.

Wei et al. (2006), using association mapping, identified an AFLP and an SSR marker associated with some major diseases affecting Australian sugarcane crops: Pachymetra root rot (*Pachymetra chanorhiza*), leaf scald (*Xanthomonas albilineans*), Fiji leaf gall (*Fiji Disease Virus*) and smut (*Ustilago scitaminea*). For this, a collection of 154 sugarcane clones (maintained by BSES LTDA. and CSR LTDA) were used that consists of important ancestors or parents, and cultivars used in breeding programs in Australia. The analyses to identify associations between disease resistance and markers were performed based on simple linear regression. However, to minimize the population structure effect in the detection of false positives, a population structure analysis was carried out using the software Structure (Pritchard et al. 2000).

Although the yellow spot disease, caused by the imperfect fungus *Mycovellosiella koepkey*, is not of great importance for many sugarcane producing countries, it can be severe and can cause low juice purity, high reducing sugar/sucrose ratio, sucrose losses at early harvest and it can also affect cane yield at late harvest, when grown under high relative humidity. Al-Janabi et al. (2007) used a progeny derived from a biparental cross between "M134/75" (resistant cultivar) and "R570" (susceptible cultivar) as the mapping population and performed QTL mapping based on IM for markers segregating in a 1:1 ratio and placed on the individual maps of each parent, using the programs QTLCartographer and MAPMAKER/QTL. For unlinked markers, the tests were carried out based on SM method using ANOVA. A major QTL was found to be associated with a resistance gene (segregating in a 3 resistant :1 susceptible in the progeny), placed on the linkage group LG87, between the "actctc10" (AFLP) and "CIR12284" (SSR) markers.

7.5 Retrospect on QTL Mapping in Sugarcane

Some interesting points can be considered for further evaluation of the results generated for QTL mapping in sugarcane and their implications in future planning as discussed below.

7.5.1 Genetic Linkage Maps

Several genetic linkage maps were constructed, using in general the first generation progenies derived from biparental crosses with sample size ranging between 32 and 295 individuals. These maps were obtained using information from only one type of molecular marker (AFLP, RFLP or RAPD) or by combining information from different types (AFLP, Isozyme, RAPD, RFLP, SSRs and EST-SSRs). Pseudo-testcross is the most used strategy.

7.5.2 QTL Analysis

For QTL mapping, single marker analysis (SM), interval mapping (IM) and / or composite interval mapping (CIM) are used, considering two individual maps, one for each parent. Statistical analyses are carried out by using well established models for backcross, using software appropriate for this experimental population. It is important to mention that the pseudo-testcross strategy, has some disadvantages including: a) reduction of the genome coverage, because normally only markers segregating in a 1:1 ratio are used to build the map (other segregation types such as 3:1 are sometimes included, but not making usage of modern multipoint features for map construction); b) as a consequence, low-density genetic linkage maps are obtained; c) reduced statistical power; d) difficulty to interpret the results, since the mapping should refer to the mapping population, rather than for each parent; e) use of non-appropriate statistical models for QTL mapping, since additive and dominance effects cannot be estimated separately using backcross models.

7.5.3 Traits

The most studied traits can be grouped into two categories: disease resistance (smut, brown rust, leaf scald, Fiji leaf gall, yellow spot and Pachymetra root rot) and yield components (POL, Brix, tons of cane per hectare, tons of sugar per hectare, fiber content, stalk number, stalk diameter, stalk length and stalk weight).

7.5.4 R^2

The percentages of the phenotypic variation explained by QTLs, measured by the coefficient of determination (R^2), were in general low, ranging from 4 to 26% and from 2 to 22.6% for disease resistance and yield components, respectively. In turn, the global coefficient of determination (R^2), i.e., considering all mapped QTLs simultaneously, ranged from 16.3 (3 QTLs) to 68.3% (22 QTLs) for yield components. Based on these observations, it seems clear that the QTLs in general have small effects. This may be one of the reasons that no reports were found showing the effective use of MAS in sugarcane.

7.5.5 *Statistical Models*

In terms of statistical analysis, sophisticated methods could produce more accurate and reliable results, allowing that more realistic assumptions were made. For example, Lin et al. (2003) presented a statistical method that allows QTL mapping based on IM approach, considering information from molecular markers with different segregation types and an integrated genetic linkage map. However, this model and other similar ones were not developed for polyploid species, such as sugarcane. Thus, for integrated genetic linkage maps, it is necessary to develop appropriate statistical models, including the possibility of precisely using information from molecular markers with different patterns of segregation that exists in polyploids (such as 1:1, 3:1, 7:1, 7:2, 11:1, 11:3, 13:1, 15:1, 64:1 and 69:1) (Edmé et al. 2006) and that also allow more powerful QTL mapping, producing results with greater possibility of application for MAS. Also, to guarantee that QTL mapping results may be useful for breeding purposes, it is also important to identify QTLs that have consistent expression under different environmental conditions, years and ratoons. Information about the genetic base of correlated traits (pleiotropy and/or linked QTLs) may as well help the breeder in the selection process. Therefore, it is important to develop mapping strategies considering multiple traits and environments simultaneously. Combining these new approaches with the more saturated maps that certainly will be produced in a near future could lead to more reliable QTL identification.

References

Aitken KS, Jackson PA, McIntyre CL (2005) A combination of AFLP and SSR markers provides extensive map coverage and identification of homo(eo)logous linkage groups in a Sugarcane cultivar. Theor Appl Genet 110: 789–801.

Aitken KS, Jackson PA, McIntyre CL (2006) Quantitative trait loci identified for sugar related traits in a sugarcane (*Saccharum* spp.) cultivar x *Saccharum officinarum* population. Theor Appl Genet 112: 1306–1317.

Aitken KS, Jackson PA, McIntyre CL (2007) Construction of a genetic linkage map for Saccharum officinarum incorporating both simplex and duplex markers to increase genome coverage. Genome 50: 742–756.

Aitken KS, Hermann S, Karno K, Bonnett GD, McIntyre CL, Jackson PA (2008) Genetic control of yield related stalk traits in sugarcane. Theor Appl Genet 117: 1191–1203.

Alwala S, Kimberg CA, Veremis JC, Gravois KA (2008) Linkage mapping and genome analysis in a Saccharum interspecific cross using AFLP, SRAP and TRAP markers. Euphytica 164: 37–51.

Al-Janabi SM, Honeycutt RJ, McClelland M, Sobral BWS (1993) A genetic linkage map of *Saccharum spontaneum* (L.) 'SES 208'. Genetics 134: 1249–1260.

Al-Janabi SM, Parmessur Y, Kross H, Dhayan S, Saumtally S, Ramdoyal K, Autrey LJC, Dookun-Saumtally A (2007) Identification of a major quantitative trait locus (QTL) for yellow spot (*Mycovellosiella koepkei*) disease resistance in sugarcane. Mol Breed 19: 1–14.

Andersson L (2001) Genetic dissection of phenotypic diversity in farm animals. Nat Rev Genet 2: 130–138.

Asnaghi C, D'HontA, Glaszmann JC, Rott P (2001) Resistance of sugarcane cultivar R 570 to *Puccinia melanocephala* isolates from different geographic locations. Plant Dis 85: 282–286.

Basten CJ, Weir BS, Zeng ZB (2005) QTL-Cartographer: A Reference Manual and Tutorial for QTL Mapping. Center for Quantitative Genetics, NCSU, USA: *statgen.ncsu.edu/qtlcart*.

Boer M, Wright D, Feng L, Podlich D, Luo L, Cooper M van Eeuwijk F (2007) A mixed model QTL analysis for multiple environment trial data using environmental covariables for QTLxE, with an example in maize. Genetics 177: 1801–1813.

Botstein D, White RL, Skolnick M, Davis RW (1980) Construction of a genetic linkage map in man using restriction fragment length polymorphisms. Am J Hum Genet 32: 314–331.

Broman KW, Wu H, Sen S, Churchill GA (2003) R/qtl: QTL mapping in experimental crosses. Bioinformatics 19: 889–890.

Carlier JD, Reis A, Duval MF, D'Eeckenbrugge GC, Leitão M (2004) Genetic maps of RAPD, AFLP and ISSR markers in *Ananas bracteatus* and *A. comosus* using the pseudotestcross strategy. Plant Breed 123: 186–192.

Cavalcanti JJV, Wilkinson MJ (2007) The first genetic maps of cashew (*Anacardium occidentale* L.). Euphytica 157: 131–143.

Charcosset A, Moreau L (2004) Use of molecular markers for the development of new cultivars and the evaluation of genetic diversity. Euphytica 137: 81–94.

Chen C, Bowman KD, Choi YA, Dang PM, Rao MN, Huang S, Soneji JR, McCollum TG, Gmitter Jr FG (2008) EST-SSR genetic maps for Citrus sinensis and Poncirus trifoliata. Tree Genet Genom 4: 1–10.

Daugrois JH, Grivet L, Roques D, Hoarau JY, Lombardi H, Glaszmann JC, D'Hont A (1996) A putative major gene for rust resistence linked with a RFLP marker in Sugarcane cultivar 'R570'. Theor Appl Genet 92: 1059–1064.

Dekkers JCM, Hospital F (2002) The use of molecular genetics in the improvment of agricultural populations. Nat Rev Genet 3: 22–32.

D'Hont A, Lu YH, Gonzáles de Leon D, Grivet L, Feldmann P, Lanaud C, Glaszmann JC (1994) A molecular approach to unravelling the genetics of sugarcane, a complex polyploid of the andropogoneae. Genome 37: 222–230.

Doerge RW (2002) Mapping and analysis of quantitative trait loci in experimental populations. Nat Rev Genet 3: 43–52.

Doerge RW, Zeng ZB, Weir BS (1997) Statistical issues in the search for genes affecting quantitative traits in experimental populations. Stat Sci 12: 195–219.

Edmé SJ, Glynn NG, Comstock JC (2006) Genetic segregation of microsatellite markers in Saccharum officinarum and S. spontaneum. Heredity 97: 366–375.

Edwards MD, Stuber CW, Wendel JF (1987) Molecular-marker-facilitated investigations of quantitative trait loci in maize. I. Numbers, genomic distribution and types of gene action. Genetics 116: 113–125.

Falconer DS, MacKay TFC (1996) Introduction to Quantitative Genetics. 4th Edn. Longman, New York, USA.

Garcia AAF, Kido EA, Meza AN, Souza HMB, Pinto LR, Pastina MM, Leite CS, da Silva JAG, Ulian EC, Figueira A, Souza AP (2006) Development of an integrated genetic map of a sugarcane (*Saccharum* spp.) commercial cross, based on a maximum-likelihood approach for estimation of linkage and linkage phases. Theor Appl Genet 112: 298–314.

Garcia AAF, Wang S, Melchinger AE, Zeng ZB (2008) Quantitative trait loci mapping and the genetic basis of heterosis in maize and rice. Genetics 180: 1707–1724.

Grattapaglia D, Sederoff R (1994) Genetic linkage maps of Eucalyptus grandis and Eucalyptus urophylla using a pseudo-testcross mapping strategy and RAPD markers. Genetics 137: 1121–1137.

Grivet L, D'Hont A, Roques D, Feldmann P, Lanaud CE, Glaszmann JC (1996) RFLP mapping in cultivated sugarcane (*Saccharum* spp.): Genome organization in a highly polyploid and aneuploid interespecific hybrid. Genetics 142: 987–1000.

Guimarães CT, Sills GR, Sobral BWS (1997) Comparative mapping of Andropogoneae: Saccharum L. (sugarcane) and its relation to sorghum and maize. Proc Natl Acad Sci 94: 14261–14266.

Guimarães CT, Honeycutt RJ, Sills GR, Sobral BWS (1999) Genetic maps of *Saccharum officinarum* L. and *Saccharum robustum* Brandes and Jew. Ex. Grassl. Genet Mol Biol 22: 125–132.

Health SC (1997) Markov chain Monte Carlo segregation and linkage analysis for oligogenic models. Am J Hum Genet 61: 748–760.

Heinz DJ, Tew TL (1987) Hybridization procedures. In: DJ Heinz (ed) Sugarcane Improvement through Breeding. Elsevier, Amsterdam, The Netherlands, pp 313–342.

Hoarau JY, Offman B, D'Hont A, Risterucci AM, Roques D, Glaszmann JC, Grivet L (2001) Genetic dissection of a modern sugarcane cultivar (*Saccharum* spp.). I. Genome mapping with AFLP markers. Theor Appl Genet 103: 84–97.

Hoarau JY, Grivet L, Offman B, Raboin LM, Diorflar JP, Payet J, Hellman M, D'Hont A, Glaszmann JC (2002) Genetic dissection of a modern sugarcane cultivar (*Saccharum* spp.). II. Detection of QTL's for yield components. Theor Appl Genet 105: 1027–1037.

Jansen RC, Stam P (1994) High resolution of quantitative traits into multiple loci via interval mapping. Genetics 136: 1447–1455.

Jordan DR, Casu RE, Besse P, Carroll BC, Berding N, McIntyre CL (2004) Markers associated with stalk number and suckering in sugarcane colocate with tillering and rhizomatousness QTLs in sorghum. Genome 47: 988–993.

Kao CH, Zeng ZB (1997) General formulae for obtaining the MLEs and the asymptotic variance-covariance matrix in mapping quantitative trait loci when using the EM algorithm. Biometrics 53: 653–665.

Kao CH, Zeng ZB, Teasdale R (1999) Multiple interval mapping for quantitative trait loci. Genetics 152: 1023–1216.

Lander ES, Botstein D (1989) Mapping Mendelian factors underlying quantitative traits using RFLP linkage maps. Genetics 121: 185–199.

Lander E, Green P, Abrahamson J, Barlow A, Daley M, Lincoln S, Newburg L (1987) MAPMAKER: An interactive computer package for constructing primary genetic linkage maps of experimental and natural populations. Genomics 1: 174–181.

Lin M, Lou XY, Chang M, Wu R (2003) A general statistical framework for mapping quantitative trait loci in nonmodel systems: issue for characterizing linkage phases. Genetics 165: 901–913.

Lincoln S, Daly M, Lander E (1992a) Constructing genetic maps with MAPMAKER/EXP 3.0. Whitehead Institute Technical Report, 3rd edn. Nine Cambridge Center, Cambridge, MA, USA.

Lincoln S, Daly M, Lander E (1992b) Mapping genes controlling quantitative traits with MAPMAKER/QTL 1.1. Whitehead Institute Technical Report, 2nd edn. Nine Cambridge Center, Cambridge, MA, USA.

Liu BH (1998) Statistical Genomics: Linkage, Mapping and QTL Analysis. CRC Press, Boca Raton, Florida, USA.

Lynch M, Walsh B (1997) Genetics and Analysis of Quantitative Traits. Sinauer Associates, Sunderland, Massachusetts, USA.

Mackay TFC (2001) The Genetic Architecture of Quantitative Traits. Annu Rev Genet 35: 303–339.

Maliepaard C, Jansen J, van Ooijen JW (1997) Linkage analysis in a full-sib family of an outbreeding plant species: Overview and consequences for applications. Genet Res 70: 237–250.

Malosetti M, Voltas J, Romagosa I, Ullrich SE, van Eeuwijk FA (2004) Mixed models including environmental variables for studying QTL by environment interaction. Euphytica 137: 139–145.

Malosetti M, Ribaut JM, Vargas M, Crossa J, van Eeuwijk F (2008) A multi-trait multi-environment QTL mixed model with an application to drought and nitrogen stress trials in maize (*Zea mays* L.). Euphytica 161: 241–257.

Manly KF, Cudmore Jr RH, Meer JM (2001) Map Manager QTX, cross-platform software for genetic mapping. Mam Genome 12: 930–932.

Margarido GRA, Souza AP, Garcia AAF (2007) OneMap: software for genetic mapping in outcrossing species. Hereditas 144: 78–79.

Mauricio R (2001) Mapping quantitative loci in plants: uses and caveats for evolutionary biology. Nat Rev Genet 2: 370–381.

McIntyre CL, Whan VA, Croft B, Magarey R, Smith GR (2005a) Identification and validation of molecular markers associated with Pachymetra Root Rot and brown rust resitance in sugarcane using map- and association-based approachs. Mol Breed 16: 151–161.

McIntyre CL, Casu RE, Drenth J, Knight D, Wham VA, Croft BJ, Jordan DR, Manners JM (2005b) Resistance gene analogues in sugarcane and sorghum and their association with quantitative trait loci for rust resistance. Genome 48: 391–400.

McIntyre CL, Jackson PA, Cordeiro GM, Amouyal O, Hermann S, Aitken KS, Eliott F, Henry RJ, Casu RE, Bonnett GD (2006) The identification and characterisation of alleles of sucrose phosphate synthase gene family III in sugarcane. Mol Breed 18: 39–50.

Meer JM, Manly KF, Cudmore RH (2002) Software for Genetic Mapping of Mendelian Markers and Quantitative Traits Loci. Roswell Cancer Park Institute, Buffalo, New York, USA.

Ming R, Liu SC, Lin YR, da Silva J, Wilson W, Braga D, Van Deynze A, Wenslaff TF, Wu KK, Moore PH, Burnquist W, Sorrells ME, Irvine JE, Paterson AH (1998) Detailed alignment of *Saccharum* and *Sorghum* chromosomes: Comparative organization of closely related diploid and polyploid genomes. Genetics 150: 1663–1682.

Ming R, Liu SC, Moore PH, Irvine JE, Paterson AH (2001) QTL analysis in a complex autopolyploid: genetic control of sugar content in sugarcane. Genome Res 11: 2075–2084.

Ming R, Liu SC, Bowers JE, Irvine JE, Paterson AH (2002a) Construction of a *Saccharum* consensus genetic map from two interspecific crosses. Crop Sci 42: 570–583.

Ming R, Wang YW, Draye X, Moore PH, Irvine JE, Paterson AH (2002b) Molecular dissection of complex traits in autopolyploids: mapping QTL's affecting sugar yield and related traits in sugarcane. Theor Appl Genet 105: 332–345.

Ming R, Del Monte TA, Hernandez E, Moore PH, Irvine JE, Paterson AH (2002c) Comparative analysis of QTLs affecting plant height and flowering among closely-related diploid and polyploid genomes. Genome 45: 794–803.

Mohan M, Nair S, Bhagwat A, Krishna, TG, Yano M, Bhatia CR, Sasaki T (1997) Genome mapping, molecular markers and markers-assisted selection in crop plants. Mol Breed 3: 87–103.

Morgante M, Salamini F (2003) From plant genomics to breeding practice. Curr Opin Plant Biotechnol 14: 214–219.

Mudge J, Andersen WR, Kehrer R, Fairbanks DJ (1996) A RAPD genetic map of *Saccharum officinarum*. Crop Sci 36: 1362–1366.

Oliveira KM, Pinto LR, Marconi TG, Margarido GRA, Pastina MM, Teixeira LHM, Figueira AM, Ulian EC, Garcia AAF, Souza AP (2007) Functional genetic linkage map on EST-markers for a sugarcane (*Saccharum* spp.) commercial cross. Mol Breed 20(3): 189–208.

Paterson A, Lander E, Lincoln S, Hewitt J, Peterson S, Tanksley S (1988) Resolution of quantitative traits into mendelian factors using a complete RFLP linkage map. Nature 225: 721–726.

Piperidis N, Jackson PA, D'Hont A, Besse P, Hoarau JY, Courtois B, Aitken KS, McIntyre CL (2008) Comparative genetics in sugarcane enables structured map enhancement and validation of marker-trait associations. Mol Breed 21: 233–247.

Porceddu A, Albertini E, Barcaccia G, Falistorco E, Falcinelli M (2002) Linkage mapping in apomictic and sexual kentucky blue grass (*Poa pratensis* L) genotypes using a two way pseudotestcross strategy based on AFLP and SAMPL markers. Theor Appl Genet 104: 273–280.

Pritchard JK, Stephens M, Donnelly P (2000) Inference of population structure using multilocus genotype data. Genetics 155: 945–959.

Raboin LM, Oliveira KM, Lecunff L, Telismart H, Roques D, Butterfield M, Hoarau JY, D'Hont A (2006) Genetic mapping in sugarcane, a high polyploid, using biparental progeny: identification of a gene controlling stalk colour and a new rust resistance gene. Theor Appl Genet 112: 1382–1391.

R Development Core Team (2009) R: A Language and Environment for Statistical Computing. R Foundation for Statistical Computing, Vienna, Austria. ISBN 3-900051-07-0:*http://www.R-project.org*.

Reffay N, Jackson PA, Aitken KS, Hoarau JY, D'Hont A, Besse P, McIntyre CL (2005) Characterisation of genome regions incorporated from an important wild relative into Australian sugarcane. Mol Breed 15: 367–381.

Ridout MS, Tong S, Vowden CJ, Tobutt KR (1998) Three-point linkage analysis in crosses of allogamous plant species. Genet Res 72: 111–121.

Ritter E, Salamini F (1996) The calculation of recombination frequencies in crosses of allogamous plant species with applications to linkage mapping. Genet Res 67: 55–65.

Rossi M, Araujo PG, Paulet F, Garsmeur O, Dias VM, Chen H, Van Sluys MA, D'Hont AD (2003) Genomic distribution and characterization of EST-derived resistance gene analogs (RGAs) in sugarcane. Mol Genet Genom 269: 406–419.

SAS Institute (1989a) SAS/STAT User's Guide. Version 6, 4th edn. SAS Institute, Cary, North Carolina, USA.

SAS Institute (1989b) JMP User's Guide. Version 2.0.2. SAS Institute, Cary, North Carolina, USA.

Satagopan JM, Yandell BS, Newton MA, Osborn TC (1996) A Bayesian approach to detect quantitative trait loci using Markov Chain Monte Carlo. Genetics 144: 805–816.

Schlötterer C (2004) The evolution of molecular markers - just a matter of fashion? Nat Rev Genet 5: 63–69.

Seaton G, Haley CS, Knott SA, Kearsey M, Visscher PM (2002) QTL Express: mapping quantitative trait loci in simple and complex pedigrees. Bioinformatics 18: 339–340.

Semagn K, Bjornstad A, Ndjiondjop MN (2006) Na overview of molecular marker methods for plants. Afr J Biotechnol 5: 2540–2568.

Shepherd M, Cross M, Dieters MJ, Henry R (2003) Genetic maps for *Pinus elliottii* var *hondurensis* using AFLP and microsatellite markers. Theor Appl Genet 106: 1409–1419.

Sillanpaa MJ, Arjas E (1998) Bayesian Mapping of Multiple Quantitative Trait Loci From Incomplete Inbred Line Cross Data. Genetics 148: 1373–1388.

Sills GR, Bridges WC, Al-Janabi SM, Sobral BWS (1995) Genetic analysis of agronomic traits in a cross between sugarcane (*Saccharum officinarum L.*) and its presumed progenitor (*S. robustum* Brandes and Jesw. Ex Grassl.). Mol Breed 1: 355–363.

da Silva JA, Bressiani JA (2005) Sucrose synthase molecular marker associated with sugar content in elite sugarcane progeny. Genet Mol Biol 28 (2): 294–298.

da Silva JAG, Sorrells ME, Burnquist W, Tanksley SD (1993) RFLP linkage map of *Saccharum spontaneum*. Genome 36: 782–791.

da Silva JAG, Honeycutt RJ, Burnquist W, Al-Janabi SM, Sorrells ME, Tanksley SD, Sobral WS (1995) *Saccharum spontaneum* L. 'SES 208' genetic linkage map combining RFLP and PCR based markers. Mol Breed 1: 165–179.

Soller M, Brody T, Genizi A (1976) On the power of experimental design for the detection of linkage between marker loci and quantitative loci in crosses between inbred lines. Theor Appl Genet 47: 35–39.

Sterky F, Lundeberg J (2000) Sequencing genes and genomes. J Bacteriol 76: 1–31.

Stuber CW, Edwards MD, Wendel JF (1987) Molecular-marker-facilitated investigations of quantitative trait loci in maize. II. Factors influencing yield and its component traits. Crop Sci 27: 639–648.

Syvänen AC (2001) Accessing genetic variation: genotyping single nucleotide polymorphisms. Nat Rev Genet 2: 930–942.

Takeda S, Matsuoka M (2008) Genetic approaches to crop improvement: responding to environmental and population changes. Nat Rev Genet 9: 444–457.

Tautz D (1989) Hypervariability of simple sequences as a general source of polymorphic DNA markers. Nucl Acids Res 17: 6463–6471.

Utz HF, Melchinger AE (2003) PLABQTL: a Program for Composite Interval Mapping of QTL. Institute of Plant Breeding, Seed Science, and Population Genetics, Univ of Hohenheim, Stuttgart, Germany: *www.uni-hohenheim.de/~ipspwww/soft.html.*

van Eeuwijk FA, Malossetti M, Yin X, Struik PC, Stam P (2005) Statistical models for genotype by environment data: From conventional ANOVA models to eco-physiological QTL models. Aust J Agri Res 56: 883–894.

van Eeuwijk FA, Malossetti M, Boer MP (2007) Modelling the genetic basis of response curves underlying genotype x environment interaction. In: JHJ Spiertz , PC Struik , HH van Laar (eds) Scale and Complexity in Plant Systems Research: Gene-Plant-Crop Relations. Springer, Dordrecht, The Netherlands, pp 115–126.

van Ooijen JW (2004) MapQTL 5, Software for the Mapping of Quantitative Trait Loci in Experimental Populations. Kyazma BV, Wageningen, The Netherlands.

van Ooijen JW, Voorrips RE (2001) JoinMap 3.0, Software for the Calculation of Genetic Linkage Maps. Plant Research International. Kyazma BV, Wageningen, The Netherlands.

Wang S, Basten CJ, Zeng ZB (2007) Windows QTL-Cartographer 2.5. Department of Statistics. North Carolina State Univ, Raleigh, North Carolina: *statgen.ncsu.edu/qtlcart/WQTLCart.htm.*

Wei X, Jackson PA, McIntyre CL (2006) Associations between DNA markers and resistance to diseases in sugarcane and effects of population substructure. Theor Appl Genet 114: 155–164.

Weller JI (1986) Maximum likelihood techniques for the mapping and analysis of quantitative trait loci with the aid of genetic markers. Biometrics 42: 627–640.

Williams JKF, Kubelik AR, Livak KG, Rafalki JA, Tingey SV (1990) DNA polymorphisms amplified by arbitrary primers are useful as genetic markers. Nucl Acids Res 18: 6531–6535.

Wu KK, Burnquist W, Sorrells ME, Tew TL, Moore PH, Tanksley SD (1992) The detection and estimation of linkage in polyploids using single-dose restriction fragments. Theor Appl Genet 83: 294–300.

Wu R, Ma CX, Painter I, Zeng ZB (2002) Simultaneous maximum likelihood estimation of linkage and linkage phases in outcrossing species. Theor Pop Biol 61: 349–363.

Yi N, Xu S (2001) Bayesian mapping of quantitative trait loci under complicated mating designs. Genetics 157: 1759–1771.

Yi N, Yandell B S, Churchill GA, Allison DB, Eisen E J, Pomp D (2005) Bayesian model selection for genome-wide epistatic QTL analysis. Genetics 170: 1333–1344.

Yi N, Banerjee S, Pomp D, Yandell BS (2007a) Bayesian mapping of genome-wide interacting quantitative trait loci for ordinal traits. Genetics 176: 1855–1864.

Yi N, Shriner D, Banerjee S, Mehta T, Pomp D, Yandell BS (2007b) An efficient Bayesian model selection approach for interacting quantitative trait loci models with many effects. Genetics 176: 1865–1877.

Zabeau M, Vos P (1993) Selective restriction fragment amplification: a general method for DNA fingerprinting. European Patent Application number: 92402629.7. Publication number 0534858A1.

Zeng ZB (1993) Theoretical basis of separation of multiple linked gene effects on mapping quantitative trait loci. Proc Natl Acad of Sci 90: 10972–10976.

Zeng ZB (1994) Precision mapping of quantitative trait loci. Genetics 136: 1457–1468.

Zeng ZB, Kao CH, Basten CJ (1999) Estimating the genetic architecture of quantitative traits. Genet Res 74: 279–289.

Structural Genomics and Genome Sequencing

Andrew H Paterson,[1] Glaucia Souza,[2] Marie-Anne Van Sluys,[3]
Ray Ming[4] and Angelique D'Hont[5]*

ABSTRACT

Sugarcane exemplifies many challenges associated with genetic and
genomic analysis of angiosperms, being a recently formed
autopolyploid with a large genome, and with its most economically
important forms being aneuploid interspecific hybrids. Despite nearly
two decades of vigorous activity, current genetic maps remain
"incomplete" (with some chromosomes/segments lacking informative
sequence-tagged polymorphism) and physical mapping tools are of
insufficient depth to cover each allele in any one genotype. Nonetheless,
considerable advances have been made by reduced-representation
sequencing of cDNA and repetitive DNA. The virtually-complete
sequencing of a close relative (sorghum) provides a valuable framework
for deducing the probable arrangement of much of the sugarcane
genome. Refined sequencing strategies and rapidly dropping costs
enhance the likelihood that one or more sugarcane genomes will be
sequenced in the near future, and the worldwide sugarcane community

[1]Plant Genome Mapping Laboratory, University of Georgia, 111 Riverbend Road, Athens,
GA 30602, USA.
[2]Instituto de Química—Departamento de Bioquímica, Universidade de São Paulo. Av.
Prof. Lineu Prestes, 748, B9S, sala 954. São Paulo, SP. BRAZIL 05508-900.
[3]GaTE lab, Departamento de Botânica-IB, USP, rua do Matão 277, 05508-090, São
Paulo, SP-Brazil.
[4]Department of Plant Biology, University of Illinois at Urbana—Champaign, Urbana, IL
61801, USA.
[5]CIRAD, UMR 1098 DAP, TAA96/03, Avenue Agropolis, 34398 Montpellier cedex 5,
France.
*Corresponding author: *paterson@uga.edu*

continues to discuss various possible sequencing strategies involving different cost levels, different genotypes, and with different expected outcomes. A major contribution is expected from sugarcane genomics regarding the understanding of allelic variation and expression profile in such a complex genomic context. A singular opportunity in the post-genomic era for sugarcane will be to reveal the early events in the adaptation of a genome to the duplicated (polyploid) state, and how this adaptation might relate to productivity of biomass and specific metabolites such as sucrose.

Keywords: autopolyploidy, comparative genomics, C4 photosynthesis, biomass, sucrose

8.1 Introduction

Tropical grasses are among the most efficient biomass accumulators known, thanks to "C4" photosynthesis, a complex combination of biochemical and morphological specializations discovered in sugarcane (Kortschak et al. 1965; Hatch and Slack 1966) that confer efficient carbon assimilation at high temperatures. The Saccharinae clade of tropical grasses is of singularly-large importance, including three leading candidate lignocellulosic biofuels crops, *Sorghum* (currently the #2 US biofuel crop), *Saccharum* (sugarcane and its relatives, currently the #1 biofuel crop worldwide), and *Miscanthus*, among the highest-yielding biomass crops with about twice the biomass of switchgrass in the Midwest in the US (Heaton et al. 2004).

Sugarcane is among the world's most important crops, and in particular is presently the leading biofuel crop worldwide. *Saccharum* genotypes are characterized by numerous (from 36 to more than 200) variably sized chromosomes. *S. officinarum* has been defined as having $2n = 80$, with clones having *S. officinarum* morphology but higher chromosome numbers being considered atypical or hybrids (reviewed by (Sreenivasan et al. 1987)). For *S. officinarum* and its probable wild progenitor, *S. robustum*, which exhibits from 60 to 200 chromosomes with major cytotypes of $2n = 60$ or 80, the most likely basic chromosome number is $x = 10$ based on quantitative karyotypping (Ha et al. 1999), fluorescence in situ hybridization(D'Hont et al. 1998)and periodicity among accessions for which chromosome numbers are known (Irvine 1999). A basic chromosome number of $x = 10$ appears likely to be ancestral to the Saccharinae, being consistent with *Sorghum* [and noting that $x = 5$ sorghums are thought to be recently derived from $x = 10$ types (Spangler et al. 1999)].

Both naturally occurring and human-mediated polyploidization have been central to sugarcane evolution and improvement. *Saccharum* and *Sorghum* are thought to have diverged from a common ancestor between 5 and 9 million years ago (Aljanabi et al. 1994; Jannoo et al. 2007) and some genotypes can still be crossed to one another (Dewet et al. 1976). *Saccharum* and *Sorghum* share more extensive genome-wide colinearity, and fewer chromosomal rearrangements (Dufour et al. 1997; Guimaraes et al. 1997; Ming et al. 1998), than either share with any other known grass. Many regions of the *Sorghum* genome correspond to eight or more homologous regions of *Saccharum*, showing that in the short period since their divergence from a common ancestor, *Saccharum* has been through at least two whole-genome duplications (Ming et al. 1998). These recent genome duplications are superimposed on at least one, and perhaps more, additional genome duplication(s) shared by most if not all cereals (Paterson et al. 2004).

Modern cultivars are both polyploid and aneuploid with 100–130 chromosomes (Simmonds 1976), 85–90% of which are from *S. officinarum* and 10–15% from a wild relative *S. spontaneum* (Dhont et al. 1996), due to a few interspecific crosses performed a century ago. These interspecific crosses were followed by backcrosses to *S. officinarum* to recover types adapted to cultivation (Arceneaux 1965; Price 1965). During this process, a high frequency of transmission of $2n$ chromosomes by the female (*S. officinarum*) parent was discovered (Bremer 1923, 1961), which facilitated the recovery of *S. officinarum* alleles for sugar production, while introgressing disease resistance, vigor, and adaptability from *S. spontaneum*. This "nobilization" process yielded interspecific poly-aneuploid genotypes of a complexity exceeding that of most if not all other crops.

Genetic analysis using a singularly-informative restriction fragment length polymorphism (RFLP) probe (BNL12.06) revealed a remarkable range of affinities among chromosomes within a single homologous group, from high levels of preferential pairing (although not true allelism among these particular restriction fragments), to frequent univalence for one chromosome that had a clear interspecific origin (noting that there was no evidence that interspecific hybridity was directly related to its univalence) (Jannoo et al. 2004).

8.2 Genetic Mapping

DNA marker-based genetic maps have been made in several *Saccharum* populations (reviewed by Ming et al. 2005), but none have a sufficient number of DNA markers to even link into the expected chromosome number. For the 80 chromosomes (each) of *S. officinarum* and *S. robustum*, one would need about 1,600 loci to have a minimally-complete map of even one parent (Paterson 1996), and 3,200 for both. Because only a subset of loci from such highly polymorphic reference maps are informative in any one elite cross, a much larger number of loci is justifiable to provide the informative subsets necessary for plant breeding applications. Moreover, most previously characterized *Saccharum* sequence-tagged markers are RFLPs—we are only aware of 221 non-redundant SSRs mapped to date in sugarcane (Cordeiro et al. 2000; da Silva 2001; Rossi et al. 2003; Edme et al. 2006; Pinto et al. 2006; Raboin et al. 2006; Aitken et al. 2007), although more than 2,005 have been located in sugarcane ESTs among which 342 have shown to detect an average of 7.55 alleles per locus with PIC content averaging 0.73 (Oliveira et al. 2009). SSRs from maize and other taxa will transfer to a modest degree (Cordeiro et al. 2001; Selvi et al. 2003). Ming and co-workers (unpublished) recently developed 3,199 pairs of SSR primers from the following sources: 2,644 from The Institute for Genomic Research (TIGR) sugarcane EST database; 53 from sugarcane bacterial artificial chromosome (BAC) end sequences generated in the Paterson laboratory; 458 from an SSR enriched genomic library reported recently (Parida et al. 2009), and 44 from 20 sequenced sugarcane BACs generated by Ming's group. Among these, 439 polymorphic markers have been mapped, generating 827 loci in a mapping population derived from LA Purple (*S. officinarum*, $2n = 80$) and Mol 5829 (*S. robustum*, $2n = 80$). The 827 SSR markers were added to the existing 442 RFLP and 2,011 AFLP markers mapped in this population. Additional SSR markers are being developed from genome survey sequences of LA Purple (*S. officinarum*) and Mol6081 (*S. robustum*) using Roche 454 Titanium.

8.3 Genome Sequencing

Genetic maps provide a first level of "bridging" between the collective wisdom and organismal experience of classical crop improvement, and the precision and potential of DNA-based analysis. What are the next steps to improve information about, and knowledge of, the sugarcane genome?

A truly comprehensive picture of the sugarcane genome including the entire suite of genes, their all-important regulatory elements, and their complete arrangement along the chromosomes will eventually require complete sequencing of the sugarcane genome. The time is right to develop strategies to resolve sugarcane-specific problems that will be required for eventual genome-wide sequencing, so that when sequencing costs fall far enough, the required sugarcane genomic infrastructure will be in place.

Some early steps in sequencing of the sugarcane genome are well advanced, and are described further below. A natural first step, the sequencing of the majority of abundantly expressed genes, is largely complete.

The autopolyploid and interspecific hybrid nature of the sugarcane genome poses a significant challenge that is shared by several other crops (particularly crops that allocate much of their photosynthate to biomass rather than seed, such as forage crops and other bioenergy crops or candidates). As has often been true in dealing with the complexities of sugarcane genomics, considerable guidance is offered by the closely related genome of sorghum. Its small genome (~730 Mb) and low level of gene duplication comparable to rice makes *Sorghum* an attractive model for functional genomics of C4 grasses in general and the Saccharinae in particular (the grass clade that includes sugarcane, *Miscanthus* and sorghum), and motivated its complete sequencing (Paterson et al. 2009).

8.3.1 Transcriptome Sequencing

A natural first step in exploring the information-encoding potential of a genome is the isolation and sequencing of cDNA synthesized from polyA-RNA that can be routinely isolated from plant tissues. Sugarcane enjoys an extensive public collection of such "expressed-sequence tags" (ESTs), largely derived from a Brazilian initiative, SUCEST (*http:// watson.fapesp.br/sucest.htm; http://sucest-fun.org*) that broadly sampled diverse genotypes, tissues, and physiological states (Vettore et al. 2003). Additional efforts in Australia, South Africa, and the USA added further depth and breadth to this large sampling.

The cDNA resources have provided the foundation for a host of investigations into pathways and processes active in specific sugarcane tissues, with a collection of studies published in Volume 24 of the Brazilian journal "Genetics and Molecular Biology". The sugarcane transcriptome is also discussed in further detail in Chapter 9 of this volume. Also relevant to sugarcane is similarly extensive EST resources for closely related sorghum (Pratt et al. 2005), largely derived from a US National Science Foundation initiative that emphasized the genotype that was subsequently used in genome sequencing (Paterson et al. 2009).

In sum, cDNA sequencing for the sugarcane-sorghum complex by traditional (Sanger) approaches has probably reached a point of diminishing returns, although the use of low-cost massively parallel approaches may reveal additional transcripts that have escaped prior detection.

8.3.2 Physical Mapping

Physical mapping refers to a number of technologies for studying DNA on a physical scale, using various metrics for determining the physical quantity of DNA between two or more reproducible reference points. In the context of genome sequencing, physical mapping often refers to the characterization of large-insert DNA clones such as BAC, by methods that attach sets of reproducible landmarks to individual BACs, then analysis of the landmarks using established software and approaches such as FPC (Soderlund et al. 2000) to identify BACs that share a sufficient number of landmarks to be deemed statistically likely to cover overlapping portions of the genome.

The most widely used BAC library for sugarcane (and indeed, the only one known to these authors) is for the cultivar R570 (Tomkins et al. 1999).

The R570 BAC library is thought to provide about 1.3 × total genome coverage, which is though to represent about 14 × coverage of the basic chromosome set (keeping in mind that sugarcane cultivars are high-dosage autopolyploids, with 10–12 somewhat different copies of each member of the set). For most taxa, this would be a very powerful resource adequate for virtually all purposes, including assembly of detailed fingerprint-based contigs. For sugarcane, however, this goal is confounded by two problems:

(1) An average of 15–25% of homologs in most cultivars are derived from the wild relative *S. spontaneum*, meaning that for each locus in the basic chromosome set, there may exist two diverse sub-populations of BACs (*S. officinarum*-derived and *S. spontaneum*-derived, respectively), which may each have additional variation within the subpopulation as a result of heterozygosity.

(2) Heterozygosity in homologous regions that are derived from *S. officinarum* will be indistinguishable from differences in the physical coverage of genomic DNA by different BACs, and will result in varying degrees of uncertainty in physical map assembly.

While heterozygosity has been manageable in assembly of BAC-based physical maps of several other organisms (human, for example), the level of heterozygosity in sugarcane is higher as a joint result of interspecific hybridity and recent polyploidy. In an exploratory study, Paterson and coworkers (unpublished) identified all BACs from the sugarcane R570 library corresponding to low-copy DNA probes in two targeted regions of the genome, and fingerprinted the BACs using then-standard approaches (Marra et al. 1997). If heterozygosity among alleles (BACs) were sufficiently low as to permit "contigging" (unambiguous identification of overlapping BACs), then "contigs" with average depth of 10–12 BACs would have been expected. However, what was found was contigs of average depth of 1–2 BACs, instead consistent with the per-nucleotide coverage of the library and indicating that heterozygosity caused sufficient divergence to separate most BACs into allelic groups. Thus, to build a robust physical map for this sugarcane cultivar would require production and analysis of about 10–12 fold coverage not just of the basic chromosome set, but of each nucleotide. This would be about 500,000–1,000,000 BACs, about 5–10x the number presently available. While technically feasible, such an undertaking is a large investment that has been hard to justify economically.

There is growing evidence of the degree of differentiation among alleles and homologs. Several hom(oe)ologous haplotypes (BAC clones) from the modern sugarcane cultivar R570 were sequenced and compared ((Jannoo et al. 2007 and unpublished data). These haplotypes belonged to two gene-rich regions with on average one gene every 9 kb. One of these regions that bear the *Adh*1 gene has been thoroughly studied within the Poaceae family

(Ilic et al. 2003). At the gene level, the sugarcane hom(oe)ologous haplotypes showed very high colinearity as well as very high conservation of gene structure and sequence, the average sequence identity for exons being about 95% between hom(oe)ologous sugarcane haplotypes. High homology was also observed along the non-transcribed regions, except for transposable elements (TEs). Conversely, the organization of TEs that represent on average 33% of the regions studied, was remarkably divergent between hom(oe)ologous haplotypes.

Compared to sorghum, the sugarcane haplotypes displayed high colinearity and homology at the exon level, and remarkable homology in most of the non-coding parts of the genome except TEs.

8.4 Approaches and Genotypes for Whole-Genome Sequencing

What is the best approach to sequence the sugarcane genome? This question is even more complex for sugarcane than for other crops, in that sugarcane is an interspecific autopolyploid hybrid, with uncertain chromosome number, high inter-and intragenomic heterozygosity, and inadequate genetic and physical maps.

Clone-by-clone sequencing of contiguous large-insert DNA clones simplifies a large genome into small pieces and delimits uncertainties to intervals of about 100 kb. However the costs of assembling large-insert libraries and ordering clones has motivated "whole-genome shotgun" approaches, which achieve contiguity based on overlaps among paired-end sequences from random clones (Paterson 2006). The respective merits of clone-by-clone and WGS sequencing approaches in general have previously been expertly reviewed (Green 2001). Herein, we focus on their merits in the context of the distinguishing features of sugarcane.

8.4.1 Sequencing Strategies

8.4.1.1 BAC-based Sequencing

BAC-based genome sequencing is based on the identification of a non-redundant "tiling path" of BAC clones that collectively cover an entire genome, with just enough overlap between consecutive BACs to be confident of their contiguity. In addition to the technical and logistical challenges detailed above for identifying such contiguous BACs for an autopolyploid interspecific hybrid such as sugarcane, there is considerable time and cost associated with production of subclone libraries for sequencing of each BAC clone. Moreover, low-complexity regions of a genome result in low-complexity BAC clones that may be refractory to clone-based physical mapping because of the presence of too few non-redundant features to unambiguously determine relationships among consecutive BACs (Bowers et al. 2005). It is important to consider that previous plant sequencing projects have concentrated mostly on homozygous diploid organisms, which clearly reduce problems resulting from allelic sequence variation. Computational tools need to be developed to assemble allelic versions of a locus. A way to circumvent this problem in sugarcane is to do a BAC by BAC approach even if it does not correspond to a true tiling path.

8.4.1.2 Whole-Genome Shotgun Sequencing

This approach, of sequencing paired ends of sufficient numbers of random DNA clones to cover an entire genome many times over, requires a minimum of *a priori* information, increasing speed and reducing cost. The whole-genome shotgun approach provides coverage of virtually all of a genome, including low-complexity regions that may be refractory to clone-based physical mapping (Bowers et al. 2005). Any genes in the regions are likely to be sequenced, even if it proves impossible to assemble their surroundings.

The WGS strategy also has disadvantages. Although it permits the distinction of alleles from errors by virtue of redundant sampling when high sequence coverage is obtained (Dehal et al. 2002), clone-based approaches allow assembly for one allele at a time, excluding the possibility of heterozygosity. In sugarcane, in which each of a dozen or so copies (alleles) of a chromosome may contain different haplotypes of otherwise identical genes, heterozygosity may be a considerable complication. A WGS sequence of highly heterozygous *Ciona intestinalis* with 1.2% rates of polymorphism (Dehal et al. 2002), resulted in some small sequence scaffolds that are in fact short divergent haplotypes. One could envision that the problem may be several-fold greater in sugarcane.

A recurring argument against WGS approaches in angiosperms focuses on the more recent origin of repetitive DNA than in animals or microbes. Consider a hypothetical element for which 1,000 identical copies are randomly dispersed throughout a genome. Using WGS, each time that one of these is sequenced, there are 999 equally likely choices (multiplied by the redundancy of coverage) for the next clone along the DNA strand. By contrast, in clone-by-clone sequencing, a given BAC clone (for example) may only contain one member of this family. This problem is especially serious in plants, in that individual members of repetitive DNA families are often recently derived and may have few distinguishing mutations. Long repetitive elements are especially problematic—for elements that are shorter than a sequencing read (often 500–1000 nt), flanking sequence might be locus-specific.

Sugarcane is no different from most plant species in that its genome harbors several classes of repetitive sequences among which transposable elements (TE) are largely represented. Comparative studies between sugarcane and sorghum genomes disclose significant amount of synteny within gene-rich regions although sugarcane analyzed regions display a large proportion of both class 1 and class 2 TEs (Jannoo et al. 2007; and GaTE lab unpublished). Previous work based on the SUCEST collection enabled the construction of a sugarcane TE database (Rossi et al. 2001; Araujo et al. 2005). Based on TE diversity as described by (Wicker et al. 2007), most lineages are identified. Each family is composed of a full-length autonomous representative and several non-autonomous copies spread along the chromosomes. These elements can be recognized within a range of sequence similarities and molecular structure and usually it is expected that these copies regardless of activity share about 80% nucleotide identity. It is clear that a WGS sequencing approach will require a high level of accuracy of reads (to distinguish different family members) and a strong bioinformatics strategy to help assembly. Knowledge of the insertion profile of the most abundant families could circumvent part of the problem and contribute to the assembly of this exceedingly complex genome.

As noted elsewhere (Green 2001), WGS and clone-based strategies continue to converge. WGS-based sequence assembly benefits from positional information, such as the genetic and physical maps and associated paired BAC-end sequences that are essential to clone-based strategies. Clone-based sequencing routinely employs random shotgun approaches to sequence each clone, and the availability of WGS sequence accelerates assembly and finishing. The optimal balance between these respective strategies is perhaps as complicated as the "equation" regarding the case for when to sequence a genome: the costs of clone production, genetic and physical mapping (and prior relevant data available), high-throughput sequencing, and directed sequence finishing all need to be considered. Features of genome organization also need to be taken into account, such as heterozygosity, polyploidy, and the abundance, distribution and homogeneity of repetitive DNA families.

One attractive option for autopolyploids such as sugarcane, is to "simplify" the genome by attempting to sequence only one representative for each of the dozen or so alleles that might be present for each locus in the basic chromosome set (Paterson 2006). While this would necessarily be a BAC-based approach, it would dramatically reduce the amount of sequencing needed relative to either a traditional BAC-based approach or a whole-genome shotgun, and also would not require enormous BAC libraries (see above). Essentially, the basic gene set might be revealed by sequencing a tiling path of clones, setting aside the task of detecting allelic variation until sequencing is cheaper or alternative approaches emerge.

This "modified BAC-based approach" would still require the identification of a tiling path of BACs, albeit covering the basic chromosome set rather than each allele. With the sorghum sequence as a guide, and with the *a priori* knowledge that sorghum and sugarcane have very similar gene content (but not dosage!) and order, the identification of such BACs might be accomplished by probe hybridization approaches, BAC end-sequencing, or preferably both.

Moreover, it may not be necessary to tile and sequence the entire genome. The vast majority of sorghum genes are found in about one-third of the genome (Paterson et al. 2009), that also is relatively repeat-poor, and shows a high degree of colinearity with rice (Bowers et al. 2005). Anticipating that this putatively euchromatic fraction of the genome is as well conserved in sugarcane as it was over the much greater evolutionary distance between sorghum and rice (Bowers et al. 2005), then about 2,500–3,000 BACs may be sufficient to sequence representative members of most of the basic gene set for sugarcane in its native context.

8.4.2 Genotypes that are Early Priorities for Sugarcane Genome Sequencing

In the near future sequencing a genome may cost sufficiently little, so that decisions about which genotype to sequence will become of minor importance—however, today, for practical reasons a community still must prioritize one or a few genotypes over others. The decision as to which sugarcane genotype(s) to sequence is somewhat different than for other taxa, in that one can envision each sugarcane genotype as a "collection" of alleles, many of which may also be found in other genotypes. Nonetheless, a range of economic, scientific and historical considerations favor some genotypes over others as early priorities.

Ideally, one wishes to sequence a hybrid cultivar, as well as the ancestral genotypes from which it and other commercial hybrids originated. Due to their lower complexity, it will be important to obtain sequence data for ancestral species *S. officinarum* (such as LA Purple or Black Cheribon), *S. spontaneum* (such as SES208 or Coimbatore) and *S. robustum* (such as Mol6081). These ancestors represent relatively pure autopolyploids that can aid in the assembly of sequences for inter-species hybrids. R570 and Q165 are the modern cultivars with the most dense genetic maps (with RFLP, SSR, DArT, AFLP and SNP markers). SP80-3280 has the largest EST collection (over 43,000 transcripts), and R570 as noted above has BAC library widely used by the community. However, no single cultivar stands out as the ideal target as of writing of this chapter. In practice, one can envisage the initial phases of sugarcane genome sequencing as a series of surveys of several genotypes and cultivars to assess technologies and bioinformatic methods, using the sorghum genome as the reference species. As a first step the community decided to conduct surveys of all cultivars and genotypes mentioned above (LA Purple, Black Cheribon, SES208, Coimbatore, Mol6081, R570, Q165, SP80–3280). As sequencing data increases and new technologies are implemented it will be desirable that gene-rich regions be covered to a large extent, allowing for alleles to be identified. Knowledge of repetitive DNA structure may also help in assembly as noted above.

8.5 Translational Benefits to Sugarcane from the Sequences of Other Plants

Its small genome (~730 Mb) and low level of gene duplication comparable to rice makes *Sorghum* an attractive model for functional genomics of C4 grasses in general and the Saccharinae in particular, and motivated its complete sequencing under the US Department of Energy Joint Genome Institute (JGI) "Community Sequencing Program" (Paterson et al., in preparation). In contrast, the large and complex polyploid genomes of *Saccharum* species lack genomics tools required for comparative and bioinformatic approaches to enhance fundamental knowledge of genome structure, function, and organization. The relatively small *Sorghum* genome (~740 Mb) is representative of tropical grasses in that it has "C4" photosynthesis, using complex biochemical and morphological specializations to improve carbon assimilation at high temperatures. By contrast, rice is more representative of temperate grasses, using "C3" photosynthesis. Sorghum and maize (the leading US crop) diverged from a common ancestor ~12 my ago (Gaut et al. 1997; Swigonova et al. 2004) versus ~50 my ago for rice and the maize/sorghum lineage (Paterson et al. 2009). Sugarcane may have shared ancestry with sorghum as little as 8–9 my ago (Sobral et al. 1994; Jannoo et al. 2007), retains similar gene order (Ming et al. 1998), and even produces viable progeny in some intergeneric crosses (deWet et al. 1976, Morrell et al. unpublished data). As noted above, DNA marker-based genetic maps have been made in several *Saccharum* populations (reviewed by Ming et al. 2005), but none have a sufficient number of DNA markers to even link into the expected chromosome number.

Strong synteny, their relatively low level of sequence duplication, and extensive colinearity facilitates the use of cereal models such as sorghum and rice to develop DNA markers to support crop improvement and other applications (Lohithaswa et al. 2007). We identified about 71,000 SSRs in the sorghum genome, somewhat enriched in the genic regions and with many in gene-poor regions associated with repeated DNA. Further, conserved-intron scanning primers for 6,760 genes provide DNA markers useful across many Poaceae and even non-Poaceae monocots (Feltus et al. 2006), particularly useful for "orphan cereals" that lack maps.

8.6 Toward the Post-Genomic Era for Sugarcane

The importance of sugarcane as the world's leading sugar and biomass crop offers a host of applied questions, challenges, and opportunities that will benefit from its eventual sequencing. The unusual genome structure of sugarcane, having experienced two whole-genome duplications in a relatively short time and also preserving a high level of intragenomic heterozygosity by autopolyploidy, also makes it suitable for addressing fundamental questions about the nature of crop productivity.

The availability of high throughput resequencing technologies will allow for an in-depth analysis of allelic variation in sugarcane and empower gene discovery projects. Once gene promoters are identified for a great majority of the genes, even if a whole-genome assembly is not yet accomplished, one can start to envisage studies on regulatory networks and development of tools for a systems biology approach in this complex grass. This will be a starting point to characterize complex traits such as yield, combining studies on the transcriptome with the development of robust computational tools to integrate several levels of information. ChIP-Seq technology is being implemented in sugarcane to identify transcription factor (TF) targets and gene promoters. The results will have multiple direct consequences on breeding programs that frequently select for cis-regulatory elements and TF changes in search of genotypes better adapted to the environment and with increased agronomic performance. Whole-genome sequencing will also lead to a catalog of complete ORFeomes, which are valuable tools for the development of transgenics. With the identification of new genes and their functional evaluation by generating transgenics we will need to integrate the immense amount of data in a robust computational infrastructure with a database holding sequences, promoters, CREs, expression data, agronomic, physiological and biochemical characterization of each of a diverse panel of sugarcane cultivars and their progenitors. This will provide a foundation for further grass comparative studies toward establishment and comparison of sugarcane, rice, maize and sorghum conserved regulatory networks.

A singular opportunity in the post-genomic era for sugarcane will be to reveal the early events in the adaptation of a genome to the duplicated state, and how this adaptation might relate to productivity of biomass and specific metabolites such as sucrose. Its high polyploidy and relative intolerance of inbreeding provided the first evidence of the importance of the combination of polyploidy and heterozygosity to sugarcane productivity (Ming et al. 2005). Early studies using genetic markers have shown complex relationships between productivity and QTL allele dosage (Ming et al. 2001, 2002a, b), suggesting that for alleles at many loci, "one copy is good but two is not better".

Investigation into the phenotypic consequences of polyploidy is being invigorated by the discovery that most if not all angiosperms are paleopolyploids (Bowers et al. 2003), and we are only at the beginning of understanding the processes by which a genome "adapts" to the duplicated state (Chapman et al. 2006; Paterson et al. 2006). The comparison of sugarcane and sorghum promises new insights into adapation to the duplicated state. The sorghum genome has not undergone duplication in ~ 70 million years (Paterson et al. 2004), making it an ideal outgroup for unraveling the consequences of more recent genome duplications in related grasses. Sugarcane has undergone at least two genome duplications since its divergence from sorghum 8–9 MYA (Jannoo et al. 2007). In comparison of sorghum to another polyploidy being sequenced, maize, individual sorghum regions are distributed over two distinct regions resulting from maize-specific genome doubling (Wei et al. 2007)—gene fractionation is evident, and subfunctionalization is probable (Paterson et al. 2009). *Saccharum* BACs show substantially conserved gene order with sorghum (Jannoo et al. 2007; Paterson et al. 2009), and sorghum is likely to prove even more valuable for deducing the consequences of *Saccharum* genome duplications. In the long term, comparative sequencing of *S. spontaneum*, *S. officinarum*, and elite sugarcane cultivars will provide insight into the roles of polyploidy in evolution of the ancestral *Saccharum* lineage, its speciation, and the consequences of partly-reuniting the respective species (genomes) in a common nucleus.

Acknowledgements

We thank FAPESP, the NSF Plant Genome Research Program, the International Consortium for Sugarcane Biotechnology (ICSB), Genoscope, and the B3I program (University of GA) for funding relevant aspects of our research. GMS is recipient of a CNPq Productivity fellowship. We are indebted to the participants of the BIOEN Workshop on Sugarcane Genome Sequencing in August 2008 and PAG Workshop in January 2009 for valuable discussions on this topic.

References

Aitken KS, Jackson PA, McIntyre CL (2007) Construction of a genetic linkage map for *Saccharum officinarum* incorporating both simplex and duplex markers to increase genome coverage. Genome 50: 742–756.

Aljanabi SM, McClelland M, Petersen C, Sobral BWS (1994) Phylogenetic analysis of organellar DNA-sequences in the Andropogoneae, Saccharinae. Theor Appl Genet 88: 933–944.

Araujo P, Rossi M, de Jesus E, Saccaro NJ, Kajihara D, Massa R, de Felix J, Drummond R, Falco M, Chabregas S, Ulian E, Menossi M, Van Sluys M (2005) Transcriptionally active transposable elements in recent hybrid sugarcane. Plant J 44: 707–717.

Arceneaux G (1965) Cultivated sugarcanes of the world and their botanical derivation. Proc Int Soc Sugar Cane Technol 12: 844–854.

Bowers JE, Chapman BA, Rong J, Paterson AH (2003) Unravelling angiosperm genome evolution by phylogenetic analysis of chromosomal duplication events. Nature 422: 433–438.

Bowers JE, Arias MA, Asher R, Avise JA, Ball RT, Brewer GA, Buss RW, Chen AH, Edwards TM, Estill JC, Exum HE, Goff VH, Herrick KL, Steele CLJ, Karunakaran S, Lafayette GK, Lemke C, Marler BS, Masters SL, McMillan JM, Nelson LK, Newsome GA, Nwakanma CC, Odeh RN, Phelps CA, Rarick EA, Rogers CJ, Ryan SP, Slaughter KA, Soderlund CA, Tang HB, Wing RA, Paterson AH (2005) Comparative physical mapping links conservation of microsynteny to chromosome structure and recombination in grasses. Proc Natl Acad Sci USA, 102: 13206–13211.

Bremer G (1923) A cytological investigation of some species and species-hybrids of the genus *Saccharum*. Genetica 5: 273–326.

Bremer G (1961) Problems in breeding and cytology of sugar cane. 4. Origin of increase of chromosome number in species hybrids of *Saccharum*. Euphytica 10: 325–342.

Chapman BA, Bowers JE, Feltus FA, Paterson AH (2006) Buffering crucial functions by paleologous duplicated genes may impart cyclicality to angiosperm genome duplication. Proc Natl Acad Sci USA 103: 2730–2735.

Cordeiro GM, Taylor GO, Henry RJ (2000) Characterisation of microsatellite markers from sugarcane (*Saccharum* sp.), a highly polyploid species. Plant Sci 155: 161–168.

Cordeiro GM, Casu R, McIntyre CL, Manners JM, Henry RJ (2001) Microsatellite markers from sugarcane (*Saccharum* spp.) ESTs cross transferable to erianthus and sorghum. Plant Sci 160: 1115–1123.

D'Hont A, Ison D, Alix K, Roux C, Glaszmann JC (1998) Determination of basic chromosome numbers in the genus *Saccharum* by physical mapping of ribosomal RNA genes. Genome 41: 221–225.

Dehal P, Satou Y, Campbell RK, Chapman J, Degnan B, De Tomaso A, Davidson B, Di Gregorio A, Gelpke M, Goodstein DM, Harafuji N, Hastings KEM, Ho I, Hotta K, Huang W, Kawashima T, Lemaire P, Martinez D, Meinertzhagen IA, Necula S, Nonaka M, Putnam N, Rash S, Saiga H, Satake M, Terry A, Yamada L, Wang HG, Awazu S, Azumi K, Boore J, Branno M, Chin-bow S, DeSantis R, Doyle S, Francino P, Keys DN, Haga S, Hayashi H, Hino K, Imai KS, Inaba K, Kano S, Kobayashi K, Kobayashi M, Lee BI, Makabe KW, Manohar C, Matassi G, Medina M, Mochizuki Y, Mount S, Morishita T, Miura S, Nakayama A, Nishizaka S, Nomoto H, Ohta F, Oishi K, Rigoutsos I, Sano M, Sasaki A, Sasakura Y, Shoguchi E, Shin-i T, Spagnuolo A, Stainier D, Suzuki MM, Tassy O, Takatori N, Tokuoka M, Yagi K, Yoshizaki F, Wada S, Zhang C, Hyatt PD, Larimer F, Detter C, Doggett N, Glavina T, Hawkins T, Richardson P, Lucas S, Kohara Y, Levine M, Satoh N, Rokhsar DS (2002) The draft genome of Ciona intestinalis: Insights into chordate and vertebrate origins. Science 298: 2157–2167.

deWet JM J, Gupta SC, Harlan JR, Grassl CO (1976) Cytogenetics of introgression from *Saccharum* into *Sorghum*. Crop Sci 16: 568–572.

Dhont A, Grivet L, Feldmann P, Rao S, Berding N, Glaszmann JC (1996) Characterisation of the double genome structure of modern sugarcane cultivars (*Saccharum* spp.) by molecular cytogenetics. Mol Gen Genet 250: 405–413.

Dufour P, Deu M, Grivet L, Dhont A, Paulet F, Bouet A, Lanaud C, Glaszmann JC, Hamon P (1997) Construction of a composite sorghum genome map and comparison with sugarcane, a related complex polyploid. Theor Appl Genet 94: 409–418.

Edme SJ, Glynn NG, Comstock JC (2006) Genetic segregation of microsatellite markers in *Saccharum officinarum* and *S. spontaneum*. Heredity 97: 366–375.

Feltus FA, Singh HP, Lohithaswa HC, Schulze SR, Silva T, Paterson AH (2006) Conserved intron scanning primers: Targeted sampling of orthologous DNA sequence diversity in orphan crops. Plant Physiol 140: 1183–1191.

Gaut BS, Clark LG, Wendel JF, Muse SV (1997) Comparisons of the molecular evolutionary process at rbcL and ndhF in the grass family (Poaceae). Mol Biol Evol 14: 769–777.

Green ED (2001) Strategies for the systematic sequencing of complex genomes. Nat Rev Genet 2: 573–583.

Guimaraes CT, Sills GR, Sobral BWS (1997) Comparative mapping of Andropogoneae: *Saccharum* L. (sugarcane) and its relation to sorghum and maize. Proc Natl Acad Sci USA, 94: 14261–14266.

Ha S, Moore P, Heinz D, Kato S, Ohmido N, Fukui K (1999) Quantitative chromosome map of the polyploid *Saccharum spontaneum* by multifluorescence in situ hybridization and imaging methods. Plant Mol Biol 39: 1165–1173.

Hatch MD, Slack CR (1966) Photosynthesis by sugarcane leaves—a new carboxylation reaction and pathway of sugar formation. Biochem J 101: 103.

Heaton E, Voigt T, Long SP (2004) A quantitative review comparing the yields of two candidate C-4 perennial biomass crops in relation to nitrogen, temperature and water. Biomass Bioenerg 27: 21–30.

Ilic K, SanMiguel P, Bennetzen J (2003) A complex history of rearrangement in an orthologous region of the maize, sorghum, and rice genomes. Proc Natl Acad Sci USA, 100: 12265–12270.

Irvine JE (1999) *Saccharum* species as horticultural classes. Theor Appl Genet 98: 186–194.

Jannoo N, Grivet L, David J, D'Hont A, Glaszmann JC (2004) Differential chromosome pairing affinities at meiosis in polyploid sugarcane revealed by molecular markers. Heredity 93: 460–467.

Jannoo N, Grivet L, Chantret N, Garsmeur O, Glaszmann JC, Arruda P, D'Hont A (2007) Orthologous comparison in a gene-rich region among grasses reveals stability in the sugarcane polyploid genome. Plant J 50: 574–585.

Kortschak HP, Hartt CE, Burr GO (1965) Carbon dioxide fixation in sugarcane leaves. Plant Physiol 40: 209–213.

Lohithaswa HC, Feltus FA, Singh HP, Bacon CD, Bailey CD, Paterson AH (2007) Leveraging the rice genome sequence for comparative genomics in monocots. Theor Appl Genet 115: 237–243.

Marra M, Kucaba T, Dietrich N, Green E, Brownstein B, Wilson R, McDonald K, Hillier L, McPherson J, Waterston R (1997) High-throughput fingerprint analysis of large-insert clones. Genome Res 7: 1072–1084.

Ming R, Liu SC, Lin YR, da Silva J, Wilson W, Braga D, van Deynze A, Wenslaff TF, Wu KK, Moore PH, Burnquist W, Sorrells ME, Irvine JE, Paterson AH (1998) Detailed alignment of *Saccharum* and *Sorghum* chromosomes: Comparative organization of closely related diploid and polyploid genomes. Genetics 150: 1663–1682.

Ming R, Liu S-C, Irvine JE, Paterson AH (2001) Comparative QTL analysis in a complex autopolyploid: Candidate genes for determinants of sugar content in sugarcane. Genome Res 11: 2075–2084.

Ming R, Del Monte TA, Hernandez E, Moore PH, Irvine JE, Paterson AH (2002a) Comparative analysis of QTLs affecting plant height and flowering among closely-related diploid and polypiold genomes. Genome 45: 794–803.

Ming R, Wang YW, Draye X, Moore PH, Irvine JE, Paterson AH (2002b) Molecular dissection of complex traits in autopolyploids: mapping QTLs affecting sugar yield and related traits in sugarcane. Theor Appl Genet 105: 332–345.

Ming R, Moore PH, Wu KK, D'Hont A, Tew TL, Mirkov TE, Da Silva J, Schnell RJ, Brumbley SM, Lakshmanan P, Jifon J, Rai M, Comstock JC, Glaszmann JC, Paterson AH (2005) Sugarcane improvement through breeding and biotechnology. Plant Breed Rev 27: 15–118.

Oliveira K, Pinto LR, Marconi TG, Mollinari M, Ulian EC, Chabregas SM, Falco MC, Burnquist W, Garcia AA F, Souza AP (2009) Characterization of new polymorphic functional markers for sugarcane. Genome 52: in press.

Parida SK, Kalia SK, Kaul S, Dalal V, Hemaprabha G, Selvi A, Pandit A, Singh A, Gaikwad K, Sharma TR, Srivastava PS, Singh NK, Mohapatra T (2009) Informative genomic microsatellite markers for efficient genotyping applications in sugarcane. Theor Appl Genet 118: 327–338.

Paterson AH (ed) (1996) Genome Mapping in Plants. Academic Press/Landes Bioscience, Austin, USA.

Paterson AH (2006) Leafing through the genomes of our major crop plants: strategies for capturing unique information. Nat Rev Genet 7: 174–184.

Paterson AH, Bowers JE, Chapman BA (2004) Ancient polyploidization predating divergence of the cereals, and its consequences for comparative genomics. Proc Natl Acad Sci USA, 101: 9903–9908.

Paterson AH, Chapman BA, Kissinger J, Bowers JE, Feltus FA, Estill J, Marler BS (2006) Convergent retention or loss of gene/domain families following independent whole-genome duplication events in *Arabidopsis*, *Oryza*, *Saccharomyces*, and *Tetraodon*. Trends Genet 22: 597–602.

Paterson AH, Bowers JE, Bruggmann R, Dubchak I, Grimwood J, Gundlach H, Haberer G, Hellsten U, Mitros T, Poliakov A, Schmutz J, Spannagl M, Tang H, Wang X, Wicker T, Bharti AK, Chapman J, Feltus FA, Gowik U, Lyons E, Maher C, Narechania A, Penning B, Zhang L, Carpita NC, Freeling M, Gingle AR, Hash CT, Keller B, Klein PE, Kresovich S, McCann MC, Ming R, Peterson DG, Ware D, Westhoff P, Mayer KFX, Messing J, Rokhsar DS (2009) The *Sorghum bicolor* genome and the diversification of grasses. Nature 457: 551–556.

Pinto LR, Oliveira KM, Marconi T, Garcia AAF, Ulian EC, de Souza AP (2006) Characterization of novel sugarcane expressed sequence tag microsatellites and their comparison with genomic SSRs. Plant Breed 125: 378–384.

Pratt LH, Liang C, Shah M, Sun F, Wang HM, Reid SP, Gingle AR, Paterson AH, Wing R, Dean R, Klein R, Nguyen HT, Ma HM, Zhao X, Morishige DT, Mullet JE, Cordonnier-Pratt MM (2005) Sorghum expressed sequence tags identify signature genes for drought, pathogenesis, and skotomorphogenesis from a milestone set of 16,801 unique transcripts. Plant Physiol 139: 869–884.

Price S (1965) Interspecific hybridization in sugarcane breeding. Proc Int Soc Sugar Cane Technol 12: 1021–1026.

Raboin L, Oliveira K, Lecunff L, Telismart H, Roques D, Butterfield M, Hoarau J, D'Hont A (2006) Genetic mapping in sugarcane, a high polyploid, using bi-parental progeny: identification of a gene controlling stalk colour and a new rust resistance gene. Theor Appl Genet 112: 1382–1391.

Rossi M, Araújo PG, Van Sluys MA (2001) Survey of transposable elements in sugarcane expressed sequence tags (ESTs). Genet Mol Biol 24: 147–154.

Rossi M, Araujo P, Paulet F, Garsmeur O, Dias V, Hui C, Van Sluys MA, D'Hont A (2003) Genome distribution and characterization of EST derived sugarcane resistance gene analogs. Mol Genet Genom 269: 406–419.

Selvi A, Nair NV, Balasundaram N, Mohapatra T (2003) Evaluation of maize microsatellite markers for genetic diversity analysis and fingerprinting in sugarcane. Genome da Silva JAG (2001) Preliminary analysis of microsatellite markers derived from sugarcane expressed sequence tags (ESTs). Genet Mol Biol 24: 155–159.

Simmonds NW (1976b) Sugarcanes. In Simmons NW (ed) Evolution of Crop Plants. Longman Scientific & Technical, Essex.

Sobral BWS, Braga DPV, Lahood ES, Keim P (1994) Phylogenetic analysis of chloroplast restriction enzyme site mutations in the Saccharinae Griseb Subtribe of the Andropogoneae Dumort tribe. Theor Appl Genet 87: 843–853.

Soderlund C, Humphray S, Dunham A, French L (2000) Contigs built with fingerprints, markers, and FPCV4.7. Genome Res 10: 1772–1787.

Spangler R, Zaitchik B, Russo E, Kellogg E (1999) Andropogoneae evolution and generic limits in Sorghum (Poaceae) using ndhF sequences. Syst Bot 24: 267–281.

Sreenivasan T, Ahloowalia B, Heinz D (1987) Cytogenetics. In: D Heinz (ed) Sugarcane Improvement through Breeding. Elsevier, New York, USA.

Swigonova Z, Lai J, Ma J, Ramakrishna W, Llaca V, Bennetzen JL, Messing J (2004) Close split of sorghum and maize genome progenitors. Genome Res 14: 1916–1923.

Tomkins JP, Yu Y, Miller-Smith H, Frisch DA, Woo SS, Wing RA (1999) A bacterial artificial chromosome library for sugarcane. Theor Appl Genet 99: 419–424.

Vettore AL, da Silva FR, Kemper EL, Souza GM, da Silva AM, Ferro MIT, Henrique-Silva F, Giglioti EA, Lemos MVF, Coutinho LL, Nobrega MP, Carrer H, Franca SC, Bacci M, Goldman MHS, Gomes SL, Nunes LR, Camargo LEA, Siqueira WJ, Van Sluys MA, Thiemann OH, Kuramae EE, Santelli RV, Marino CL, Targon M, Ferro JA, Silveira HCS, Marini DC, Lemos EGM, Monteiro-Vitorello CB, Tambor JHM, Carraro DM, Roberto PG, Martins VG, Goldman GH, de Oliveira RC, Truffi D, Colombo CA, Rossi M, de Araujo PG, Sculaccio SA, Angella A, Lima MMA, de Rosa VE, Siviero F, Coscrato VE, Machado MA, Grivet L, Di Mauro SMZ, Nobrega FG, Menck CFM, Braga MDV, Telles GP, Cara FAA, Pedrosa G, Meidanis J, Arruda P (2003) Analysis and functional annotation of an expressed sequence tag collection for tropical crop sugarcane. Genome Res 13: 2725–2735.

Wei F, Coe E, Nelson W, Bharti AK, Engler F, Butler E, Kim H, Goicoechea JL, Chen M, Lee S, Fuks G, Sanchez-Villeda H, Schroeder S, Fang Z, McMullen M, Davis G, Bowers JE, Paterson AH, Schaeffer M, Gardiner J, Cone K, Messing J, Soderlund C, Wing RA (2007) Physical and genetic structure of the maize genome reflects its complex evolutionary history. PLoS Genet 3: e123.

Wicker T, Sabot F, Hua-Van A, Bennetzen J, Capy P, Chalhoub B, Flavell A, Leroy P, Morgante M, Panaud O, Paux E, SanMiguel P, Schulman A (2007) A unified classification system for eukaryotic transposable elements. Nat Rev Genet 8: 973–982.

Functional Genomics: Transcriptomics of Sugarcane —Current Status and Future Prospects

Rosanne E. Casu,[1] Carlos Takeshi Hotta[2] and Glaucia Mendes Souza[3]

ABSTRACT

Systems biology is the science of relating genes and metabolic pathways across multiple levels, which may range from an individual to an entire crop. Transcriptomics, the global analysis of expression of RNA is an integral component of this and is widely applied in plant biology in both model systems and crop plants, such as sugarcane. This chapter reviews sugarcane transcriptomics, including the tools developed, the use of the transcriptome to improve marker discovery and the relevance of resources developed for related species. The use of the new high-throughput sequencing technologies is discussed as is the possible impact of sugarcane transcriptomics in genomics-assisted breeding. All of these techniques will inform efforts to expedite

[1]CSIRO Plant Industry, Queensland Bioscience Precinct, 306 Carmody Rd, St. Lucia, QLD, 4067, Australia and CRC Sugar Industry Innovation through Biotechnology, Level 5, John Hines Building, The University of Queensland, St. Lucia, QLD, 4072, Australia.
[2]Instituto de Química, Departamento de Bioquímica, Universidade de São Paulo. Av. Prof. Lineu Prestes 748, B9S, sala 954, São Paulo, SP, 05508-900, Brazil.
[3]Instituto de Química, Departamento de Bioquímica, Universidade de São Paulo. Av. Prof. Lineu Prestes 748, B9S, sala 954, São Paulo, SP, 05508-900, Brazil.
*Corresponding author: *Rosanne.Casu@csiro.au*

enhanced breeding of sugarcane either by traditional or precision breeding strategies.

Keywords: sugarcane, transcriptome, expressed sequence tag, transcript profile, array, de novo sequencing

9.1 Introduction

Sugarcane is an important crop throughout the tropical regions of the world. A mature sugarcane plant contains a very large amount of biomass in addition to the sucrose accumulated in its stems. Total dry matter yields, i.e. sugar and non sugar components, are normally approximately 20–60 t/ha in commercial environments, but can go as high as 70 t/ha (Muchow et al. 1994). Sucrose can comprise 12–16% of the fresh weight of stems and approximately 50% of its dry weight (Bull and Glasziou 1963), translating through to approximately 0.7 M sucrose (Moore 1995). Ethanol derived from the fermentation of sucrose is now an increasingly important product that can be used as a fuel, either directly or mixed with refined petroleum. It has been estimated that sugarcane ethanol may replace up to 10% of the world's refined petroleum consumption in the next 15 to 20 years (Goldemberg 2007).

In order to increase sugarcane ethanol production to supply an increasing worldwide demand as well as ensure that sucrose production remains viable, issues associated with the expansion of sugarcane cultivation must also be addressed. For example, in Brazil, sugarcane cultivation has expanded over pasturelands in Brazil, most of which are nutrient deprived and also into regions experiencing seasonal drought. In Australia, many of the sugarcane growing regions rely on irrigation for optimal yields. These situations may be amenable to crop management but the breeding or engineering of new sugarcane varieties that can give superior yield under these circumstances may be a longer-term solution.

The traditional method for identifying a gene responsible for a particular trait is to first demonstrate that the trait is heritable, followed by isolation of a candidate gene that is postulated to be responsible for the trait. This "single-gene" approach is fundamentally flawed for many traits since it assumes that every trait is governed by a single gene. Genomics and all of the related "omics" techniques, e.g. transcriptomics, break this formula and rely on the in-depth assembly of large amounts of data followed by data-mining to determine connections between a particular trait and any number of associated genes. Its highest form is "systems biology", which seeks to relate genes and metabolic pathways across levels ranging from the individual molecule through the crop (Moore 2005).

Transcriptomics is ideally the global analysis of expression of the total complement of RNA in a given cell or tissue. It has many inherent advantages, particularly since it targets the genes active in the cells or tissue at the time of sampling, thereby reducing the complexity of the data generated. It also allows for robust, relatively simple comparisons between e.g. tissues within the same organism, the same tissue at different developmental stages and tissues exposed to various stresses or treatments (Schnable et al. 2004; Brady et al. 2006; Galbraith 2006). It directly complements gene marker analysis, assaying the complete genome for changes associated with a particular trait and proteomics, the assay of the total protein complement present in a given cell or tissue (Alwala and Kimbeng 2009; Watt et al 2009). This technique has been widely applied in plant biology, both in model systems such as *Arabidopsis thaliana* and also in crop plants, e.g. rice, wheat, maize, barley, and, recently, in sugarcane.

This chapter will concentrate on the current status of sugarcane transcriptomics, the tools developed, the use of the transcriptome to identify markers for quantitative trait loci (QTL) mapping, assessing the impact of resources derived from related species, the impact of new high-throughput sequencing technologies and the possible impact of sugarcane transcriptomics in genomics-assisted breeding.

9.2 Tools for Sugarcane Transcriptomics

Numerous techniques have been developed for large-scale gene expression profiling, mostly falling into one of two groups. Techniques in the first group assay RNA samples by hybridization to nucleotide probes attached to various supports while those in the second require the de novo generation of sequence tags from individual RNA samples. Array technologies include the nylon-based cDNA macroarrays (Lennon and Lehrach 1991; Piétu et al. 1996), cDNA microarrays (Schena et al. 1995), long oligonucleotide microarrays (e.g. Agilent Technologies) and short oligonucleotide microarrays (e.g. Affymetrix). The latter three technologies rely on either attachment of the DNA to glass by cross-linking or by in situ oligonucleotide synthesis. Serial analysis of gene expression (SAGE), reported by Velculescu et al. (1995), and Massively Parallel Signature Sequencing (MPSS), from Brenner et al. (2000a, b), both involve de novo generation of sequence tags. However, all techniques currently available, except for anonymous cDNA microarrays, require access to either a genome sequence or a deeply sequenced transcriptome in order to provide the source sequence for either probe construction or identification of sequenced tags.

Transcriptome studies in sugarcane were first undertaken in South Africa (Carson and Botha 2000, 2002). A small collection of expressed sequence tags (ESTs) were generated from both leaf roll and stem. This collection was sufficient to infer diverse physiological functions and provided an impetus for other groups to invest in the area. The largest collection of ESTs was generated by SUCEST, a large consortium of Brazilian researchers who sequenced approximately 238,000 ESTs from 26 diverse cDNA libraries (Vettore et al. 2001, 2003). Collections of approximately 10,000 ESTs each were also generated in Australia (Casu et al. 2003, 2004; Bower et al. 2005) and in the US (Ma et al. 2004), each group sequencing from three different cDNA libraries each. All of the collections generated were clustered and annotated by the originating research groups, but the entire EST set was clustered by the Center for Genomic Research (TIGR) in 2004 as the Sugarcane Gene Index 2.1, and on 29th July, 2008 by the Computational Biology and Functional Genomics Laboratory at the Dana-Farber Cancer Institute as the Sugarcane Gene Index 2.2. The current Sugarcane Gene Index clustered 255,635 ESTs and 499 expressed transcripts (fully sequenced cDNAs), resulting in 40,016 theoretical contigs, 76,529 singleton ESTs and 43 singleton expressed transcripts. Additional sugarcane transcript clustering activities were also undertaken by TIGR (now the J. Craig Venter Institute) and by PlantGDB (Casu 2009).

9.2.1 Array Technologies

The first transcriptomics tools developed for sugarcane were macroarrays, prepared using plasmid DNA or DNA fragments spotted on to nylon membranes followed by cross-linking. They have been used to investigate differences between immature and mature leaf as well as immature and mature internodal tissue (Carson et al. 2002a, b), the effect of exposure of young sugarcane plantlets to cold (Nogueira et al. 2003), tissue profiling of transcriptionally active transposable elements (De Araujo et al. 2005), expression profiling of genes with established roles in sucrose accumulation through stem development and across various genotypes (Watt et al. 2005), identification of methyl jasmonate-responsive genes in sugarcane leaves (De Rosa Jr. et al. 2005), identification of genes responsive to the application of ethanol on sugarcane leaves (Camargo et al. 2007), and, most recently, expression profiling of sink activities after source perturbation by shading (McCormick et al. 2008a) and identification of new ABA- and MEJA-activated sugarcane *bZIP* transcription factors (Schlögl et al. 2008). A feature of their continued use, despite the advent of new technologies, is their convenience, especially if assaying a smaller number of genes, low cost and adaptability, particularly since all the groups using macroarrays also have access to EST clone collections. Carson et al. (2002b) were able to identify 132 sugarcane cDNA clones differentially expressed in immature and maturing internodal tissue, using a 400 clone nylon macroarray derived from reciprocal subtractive hybridization libraries. Most of the genes with putative identities were involved in stress responses, regulatory processes, carbohydrate metabolism and cell wall metabolism but not specifically in sucrose metabolism. A slightly later study profiling immature leaf, mature leaf, immature stem and mature stem using a 1,000 clone macroarray detected 61 transcripts that accumulated more highly in leaf tissue and 25 transcripts that had higher expression in the stem. Sucrose synthase 2 was the only differentially expressed sucrose metabolism-related transcript and, as reported previously, it was preferentially expressed in immature leaf (Carson et al. 2002a). A pilot study assessing the use of suppression-subtractive hybridization technology followed by macroarray hybridization of 288 putatively up-regulated targets successfully determined that it was possible to identify conclusively genes that were up-regulated in roots that had been exposed to Al^{3+}, and that commonalities exist between the response to Al^{3+} and oxidative stress (Watt 2003). The use of larger arrays (2x 768 random EST targets), combined with a clearly defined and short cold treatment, resulted in the identification of 14 of 34 identified cold-inducible ESTs that were homologous to previously described cold- or drought-inducible genes (Nogueira et al. 2003).

The advent of microarrays represented a quantum leap in expression profiling. The ability to deposit larger numbers of probes robotically on to a glass slide resulted in the first truly large-scale expression profiling analyses and also offered the first opportunity to simultaneously assay two samples using the one microarray. These were first used in sugarcane by Casu et al. (2003, 2004) to investigate gene expression differences between immature and maturing stem of sugarcane, using glass microarrays assaying up to 4,715 non-redundant random ESTs derived from immature stem, maturing stem and roots in duplicate. The first study focused on carbohydrate metabolism-related transcripts and found that putative sugar transporters were highly expressed in maturing stem and that coordinated expression of enzymes involved in sucrose synthesis and cleavage was also evident. The second study concentrated on transcripts associated with stem maturation, finding that transcripts associated with fiber metabolism, defense and stress mechanisms, especially putative dirigent proteins, were the most highly expressed transcripts in maturing stem. Defense signaling was further explored in roots by Bower et al. (2005) who used these arrays to profile the transcriptional response of sugarcane roots to methyl jasmonate. They identified several transcripts induced by exogenous application of methyl jasmonate, including PR-10, lipoxygenase and dirigent as well as enzymes involved the metabolism of phenolics and oxidative stress. The microarray format also lends itself to smaller custom arrays such as those used by Papini-Terzi et al. (2005) which contained 1,632 elements (1,280 analyzed) derived mainly from ESTs annotated as members of signal transduction gene families. They used this to profile individual variation of plants cultivated in the field and transcript abundance in six plant organs (flowers, roots, leaves, lateral buds, and 1st and 4th internodes). The data indicated that 153 genes were ubiquitously expressed and 217 whose expression was enriched in at least one of the tissues analyzed. This microarray was further customized with additional clones, mainly derived from sugarcane kinase genes and used to study signal transduction-related responses to phytohormones and environmental challenges in sugarcane (Rocha et al. 2007) including methyl jasmonate, abscisic acid, insect (*Diatraea saccharalis*), and endophytic bacteria (*Gluconacetobacter* and *Herbaspirillum*) elicited responses. Adopting an outliers searching method, 179 genes with strikingly different expression levels were identified as differentially expressed in at least one of the treatments analyzed.

The most recent report uses a custom 3,598 clone microarray to profile the effect of elevated CO_2 on sugarcane leaves (de Souza et al. 2008). The data indicated that increased CO_2 concentration resulted in a 50% increase of biomass, altered photosynthesis and an increase in cellulose content. A total of 33 genes were found responsive to high CO_2 treatment including those involved in photosynthesis and carbohydrate metabolism genes. This may be directly related to the increased biomass phenotype observed.

Microarrays are very powerful tools for the dissection of differential transcript expression but two-color microarray experiments nearly always use different clone sets attached to the substrate and, in any case, are designed as either direct comparison between pairs of samples, closed loops or use a common reference sample (Churchill 2002). This makes individual microarrays difficult to incorporate into any later experiments, particularly those generated in other laboratories. In addition, standard microarrays prepared using DNA fragments exhibit problems with hybridization kinetics and often show great variability in spot fluorescence intensity for the same transcript represented by DNA fragments of varying lengths. Discrimination of transcript alleles or closely related members of multi-gene families is also diminished since the DNA fragments tend to be quite long and mainly consist of coding regions which show less variability between members of multi-gene families than the 3' untranslated regions. The variability of hybridization kinetics can be lessened by the use of oligonucleotide arrays since all of the probes attached to the substrate are of equal length. Oligonucleotide array design can also assist in the ability to discriminate closely related members of multi-gene families and the use of a one-color system allows for the building of an array bank, either within a laboratory or in a public database e.g. Gene Expression Omnibus (GEO—*http:// www.ncbi.nlm.nih.gov/geo/*) or ArrayExpress (*http://www.ebi.ac.uk/microarray-as/ae/*) (Barrett et al. 2008; Parkinson et al. 2008).

Oligonucleotide array technology consists of either multiple short oligonucleotides representing one transcript or several long oligonucleotides per transcript synthesized in situ on a glass or quartz substrate. These arrays offer the opportunity to use all transcript sequence data that is publicly available since there is no requirement to physically possess the DNA fragments. The first company to offer short oligonucleotide technology was Affymetrix, Inc. Their expression arrays routinely represent each transcript using at least one probe set of 11 probe pairs each, with each probe being 25 nucleotides long. Each probe pair contains one perfect match oligonucleotide and one with a central mismatch, the latter acting as a control for non-specific binding and the probe set is preferentially complementary to regions of 3′ untranslated region sequence. Affymetrix offers a standard suite of arrays designed to assay each of five monocot crops (rice, barley, wheat, maize and sugarcane) but the number of sugarcane genes represented are relatively low (around 7,000). Alternatively, custom arrays can be designed by Affymetrix or other companies such as Agilent Technologies or Roche NimbleGen to represent larger numbers of transcripts.

The GeneChip® Sugar Cane Genome Array produced by Affymetrix was used by Casu et al. (2007) to profile the expression differences between immature stem (meristem and internodes 1–3), maturing stem (internode 8) and mature stem (internode 20) of sugarcane. They were able to distinguish transcript expression differences clearly between various members of the cellulose synthase subunit family and the allied cellulose synthase-like gene family and also identified many co-expressed transcripts with involvement in cell wall synthesis and degradation as well as lignification. These arrays were also used by McCormick et al. (2008b) to profile leaves whose sugar content had been manipulated by cold-girdling. Numerous differentially expressed transcripts involved in photosynthesis, assimilate partitioning, cell wall synthesis, phosphate metabolism and stress were identified. In addition, what appears to be antagonistic regulation of trehalose 6-phosphate phosphatase and trehalose 6-phosphate synthase was observed. The expression profiles generated in these two experiments can be considered to be foundation sets since, due to the consistency of array design and the use of one-color technology, these expression profiles could easily be reanalyzed in the context of other expression profiles generated using these arrays.

No matter which generic array platform is used, as the number of array datasets generated increases, it will be possible to analyze these datasets together in order to generate new information. For example, it would be possible to infer transcription networks by analyzing batches of array datasets (Hashimoto et al. 2004). These networks may suggest key elements that were previously unknown and nodes of regulation that may be conserved across grasses. For example, it is already known through analyses conducted by Vincentz et al. (2004) that 27.5% of sugarcane genes as defined by EST clusters appear to be specific to monocots and 13.5% are likely to be specific to this species. These genes could represent important trends in the evolution of grasses and, more specifically, sugarcane. The transcription factors included in these genes could well be important in transcription networks governing processes that are dominant in sugarcane, such as sucrose accumulation. A survey of two promoter elements, ABRE and CE3, that are known to be ABA-responsive, has shown that these elements are not equally common in *A. thaliana* and rice (Gomez-Porras et al. 2007). Comparison of these types of results with array datasets and transcription networks may result in the identification of important targets for manipulation that are specific to monocots. Analysis of cross-species microarray data is particularly well-implemented at Genevestigator (*https:/ /www.genevestigator.ethz.ch*) where it is possible to interrogate curated microarray datasets from both animals e.g. human, mouse and rat, and plants e.g. *Arabidopsis*, rice and barley in order to identify underlying metabolic pathways for traits that are important across species (Zimmermann et al. 2004).

Even though oligonucleotide array technologies have solved many of the well-documented problems encountered with microarrays, there is still the problem of a limited dynamic range, usually only detecting more abundant transcripts to the detriment of low-copy transcripts. More importantly, array technologies only report the transcription levels of the transcripts that are included in the array, possibly missing out on key information for identifying critical pathways that underpin cellular function.

9.2.2 De novo Generation of Sequence Tags

Sequence tag generation technologies such as serial analysis of gene expression (SAGE) and massively parallel signature sequencing (MPSS) can be used both qualitatively and quantitatively to identify transcriptionally active regions of a genome or to profile differences in gene expression between different tissues and treatments (Vega-Sánchez et al. 2007). Both of these techniques are most useful when a well-annotated genome or a large clustered EST collection is available for interrogation. SAGE tags were originally designed to be only 14 bp in length (Velculescu et al. 1995), but further refinements of the method have lead to the use of 21 bp tags in the LongSAGE method (Saha et al. 2002) and 26 bp tags for SuperSAGE (Matsumura et al. 2003). Numerous variations of these methods are now available e.g. 3' LongSAGE and 5' LongSAGE (Wei et al. 2004), Robust-LongSAGE (Gowda et al. 2004) and those relying on new sequencing technologies such as DeepSAGE (Nielsen et al. 2006), 5'RATE (Gowda et al. 2006) and MS-PET (Ng et al. 2006). In all cases, the tags are sequenced and then matched to an existing genome or transcriptome. The other de novo tag technology, MPSS, originally started as a combination technique involving microbead arrays, hybridization and sequencing (Brenner et al. 2000a, b), but is also used to refer to the Solexa "sequencing by synthesis" technique offered by Illumina, Inc.

SAGE and its derivatives have been used extensively in plant biology (Vega-Sánchez et al. 2007). Several databases are now available organizing MPSS data generated for *A. thaliana*, rice, *Medicago truncatula*, *Magnoporthe grisea* and grapevines (Nakano et al. 2006). However, there is only one report so far of the use of de novo tag technology for expression profiling of sugarcane. Calsa and Figueira (2007) used standard 14 bp SAGE to characterize the sugarcane mature leaf transcriptome. They analyzed 5,227 unique tags from mature leaf tissue, finding only 173 that did not match any sugarcane EST. Photosynthesis, carbon fixation and chlorophyll biosynthesis were the most abundant processes represented. In addition, it was noted that of the three C4 photosynthetic cycles present, based on tag numbers and confirmatory RT-qPCR, it was most likely that phosphoenolpyruvate carboxykinase-type decarboxylation predominated over NADP+ -malic enzyme in mature leaves, an exciting development since it is the opposite to that previously assumed (Calsa and Figueira 2007). However, this hypothesis still requires further confirmatory biochemical analysis. Expression profiling of sugarcane using MPSS has not yet been reported, however, given the recent advances in next generation sequencing technologies (reviewed below), the window of opportunity for the use of MPSS for expression profiling in sugarcane has passed.

9.2.3 *Other Expression Profiling*

Once large, diverse EST collections have been established, it is possible to mine these for candidate genes for further study. In addition, it becomes feasible to use these databases as a basis for "digital expression profiling". This requires access to large numbers of EST sequences derived from non-normalized libraries across diverse tissues and conditions. However, this can even be performed on smaller collections to identify candidate genes for further validation. Expression profiling based on in silico analysis of ESTs, followed up by microarray analysis and Northern analysis resulted in the identification of a putative hexose transporter (Casu et al. 2003) and dirigent, a glycoprotein involved in both lignin and lignan biosynthesis (Casu et al. 2004), both of which are highly differentially expressed and extremely abundant in the maturing stem of sugarcane. More recently, a similar approach using semi-quantitative PCR and digital expression profiling was employed to identify and profile the alternative oxidases and uncoupling proteins of sugarcane, determining that these gene families were differentially expressed across tissue types and also displayed different profiles when subjected to chilling stress (Borecký et al. 2006).

9.3 Transcriptome QTL Mapping

The advent of large EST collections provided an immediate source of sequences with the potential to be developed into useful DNA markers. These have the advantage over anonymous markers of potential attribution of a putative function to the protein products of these transcripts. This offers the possibility of finding a "perfect" marker that can be closely associated with a trait or even causative.

The sugarcane EST collections have been extensively surveyed by various groups to identify several different marker types. ESTs can be tested without modification as restriction fragment length polymorphism (RFLP) markers as long as one has access to the cDNA clones. Cross-species use of sugarcane ESTs as RFLP probes was established by their use as anonymous markers on a *Sorghum bicolor* genetic map (Tao et al. 1998; Tao et al. 2000). The first report for sugarcane was a preliminary study that examined a small number of ESTs from the SUCEST collection for utility as RFLP markers across individuals from a mapping population segregating for various traits as well as a number of commercial varieties displaying resistance or susceptibility to leaf scald (da Silva et al. 2001). The first ESTs to be added to an existing sugarcane map were a set of resistance gene analogs (RGAs) that had been identified from the SUCEST collection, used as RFLP markers, which were then localized on to an existing AFLP map together with some additional SSR markers (Rossi et al. 2003). The clear association between an EST-RFLP marker and a trait was established with the identification of an EST-derived RFLP corresponding to sucrose synthase which was associated with sugar content, making this marker useful for marker-assisted selection in sugarcane breeding (da Silva and Bressiani 2005). A combination approach of direct identification of resistance gene analogues from EST cluster data and de novo PCR from RNA from sugarcane tissues resulted in the determination of map location in sugarcane for 31 RGAs. Comparative mapping between sugarcane and sorghum indicated that there was syntenic localization of several RGA clusters. Three brown rust-associated RGAs mapped to the same linkage group in sorghum and the third to a region previously shown to contain a major rust resistance-QTL in sorghum, indicating the value of RGAs for identification of disease resistance markers and simultaneous comparative mapping between sugarcane and sorghum (McIntyre et al. 2005).

Other marker technologies e.g. simple sequence repeats (SSRs) and single nucleotide polymorphisms (SNPs) only require access to sequence data to commence marker discovery. SSR discovery was initiated from both the Australian (Cordeiro et al. 2001) and SUCEST (da Silva 2001) EST collections. These collections were both rich sources of dinucleotide and trinucleotide SSRs, with a number of SSRs identified as being polymorphic across diverse genotypes of sugarcane and even genera e.g. *Erianthus* and *Sorghum* species (Cordeiro et al. 2001), unlike SSRs derived from intergenic regions which were more polymorphic within sugarcane. This points to EST-SSRs being ideally suited for diversity studies. Further studies have identified additional SSRs (Pinto et al. 2004, 2006; Oliveira et al. 2009). Their use in large-scale mapping was a logical next step. Examples from other species such as barley (Thiel et al. 2003) and cotton (Park et al. 2005) proved the value of EST-SSRs for genetic map creation. For example, Park et al. (2005) used SSRs and complex sequence repeats (CSRs) developed from a fiber transcriptome database to create a genetic map for cotton. Preliminary QTL analysis for fiber-related traits identified one QTL for fiber length that was linked to a marker homologous to endo-β-1,4-glucanase, which has an important role in cell extension during rapid polar elongation of developing fibers. In sugarcane, the major genetic maps available have been produced using RFLPs, RAPDs, AFLPs and genomic SSRs (da Silva et al. 1995; Grivet et al. 1996; Guimarães et al. 1999; Hoarau et al. 2001; Aitken et al. 2005, 2007, 2008; Reffay et al. 2005; Garcia et al. 2006; Raboin et al. 2006). EST-derived markers (515 SSRs and 37 RFLPs) were added to the map of Garcia et al. (2006) which allowed for the mapping of several genes involved in carbohydrate metabolism: sugar transporter, fructose-1,6-bisphosphate aldolase, fructose-1,6-bisphosphatase, diphosphatefructose-6-phosphate 1-phosphotransferase, glyceraldehyde-3-phosphate dehydrogenase, sucrose synthase, triosephosphate isomerase 1 and cellulose synthase-4 (Oliveira et al. 2007).

The search for sugarcane SNPs commenced when the sugarcane EST collections were surveyed electronically for sequence polymorphisms that could be associated with genuine sequence differences rather than those associated with sequencing errors. An analysis of sequence polymorphism of ESTs from the SUCEST database corresponding to 6-phosphogluconate dehydrogenase led to the identification of two gene clusters, with one reliable SNP discovered for one set and 39 reliable SNPs for the other (Grivet et al. 2001). A later study analyzed 219 sequences encoding alcohol dehydrogenase derived from SUCEST, identifying reliably three paralogous genes (one *Adh2* and two *Adh1*) for this enzyme in sugarcane. In addition, 37 highly reliable SNPs plus eight insertion-deletions across the three genes were also identified (Grivet et al. 2003). These "digital SNP" discovery exercises determined conclusively the utility of trawling the EST databases for potentially useful SNPs. This approach was subsequently used to examine the sucrose phosphate synthase gene family (McIntyre et al. 2006). All sugarcane ESTs homologous to sucrose-phosphate synthase were retrieved and aligned, with regions being identified that differentiated the five known SPS gene families. One of these families, gene family III, was selected for further study with de novo sequences for this family being generated and analyzed from both parents of a mapping population. The results suggested that two SPS family III genes were present per genome, each with different numbers of alleles. In addition, two haplotypes were mapped near to QTL for increased sucrose content but were not themselves associated with increased sucrose accumulation. SNP identification across multiple genes was used by Cordeiro et al. (2006) who identified SNPs in a set of EST clusters that had previously been described as being differentially expressed between immature stem and maturing stem of the sugarcane variety Q117 by microarray analysis by Casu et al. (2003, 2004). The 47 identified polymorphic SNP loci were tested across nine genotypes, with a five marker set being able to uniquely identify each genotype, therefore, being potentially useful for the fingerprinting of varieties to maintain integrity. However, the identification of single dose SNP markers using EST sequence data in sugarcane remains difficult due to the possibility of developing assays for SNPs that either are absent or not segregating in a particular population. Recently, an approach using 454 sequencing (Margulies et al. 2005) to identify SNPs in over 200 genes in the parents of an Australian mapping population has resulted in a more cost-effective way to identify usable SNPs (Bundock et al. 2009).

De novo technologies such as amplified fragment length polymorphism (AFLP) can also be used with cDNA for transcriptional profiling and, potentially, mapping. AFLP was initially developed as a technique for DNA fingerprinting (Vos et al. 1995) but was rapidly adapted for use with cDNA in transcriptional profiling or RNA fingerprinting (Bachem et al. 1996). In sugarcane, cDNA-AFLP has been used several times in profiling transcripts involved in infection of sugarcane by *Ustilago scitaminea* (Thokoane and Rutherford 2001; Butterfield et al. 2004; Borrás-Hidalgo et al. 2005; Lao et al. 2008), *Bipolaris sacchari* (Borrás-Hidalgo et al. 2005) and *Puccinia melanocephala* (Carmona et al. 2004), with all studies identifying differentially expressed transcripts involved in signal transduction and disease resistance. Butterfield et al. (2004) also used their identified transcripts as RFLP probes across a population of 78 genotypes with known ratings for sugarcane smut, eldana borer and SCMV and also scored a subset of these markers across additional 53 genotypes used in their breeding program, informing their crossing strategies to obtain coordinated assembly of resistance factors in progeny.

The use of cDNA-AFLP for transcriptome mapping was first applied in *A. thaliana* by Brugmans et al. (2002) to prepare a simple map. Recently, it was used by Ritter et al. (2008) to construct a much larger transcriptome linkage map in potato by using non-differential cDNA-AFLP. This map is about 800 centi-Morgans (cM) in length and contains nearly 700 transcriptome derived fragments (TDFs). These TDFs generally have biological meaning unlike most markers generated by AFLP using genomic

DNA. Ritter et al. (2008) analyzed four TDFs co-localized to known QTL for resistance and found that two of the four TDFs were homologous to resistance genes. However, to date, transcriptome mapping has not been performed in sugarcane, possibly due to the enormous number of markers required to generate a meaningful map in sugarcane where even a map containing over 1,000 markers is still estimated to have only approximately 50% genome coverage (Aitken et al. 2005). However, it remains an approach that could be exploited in the future in order to add value to existing maps derived using other marker systems.

The detection of expression QTL (eQTL) that are responsible for differences in gene expression by the profiling of many individuals from a population segregating for a particular trait is the ultimate combination of mapping and transcriptome analysis. This technique, "genetical genomics" (Jansen and Nap 2001), requires that gene expression as measured by intensity values from a microarray or GeneChip be assessed as a quantitative trait (Brem et al. 2002; Eaves et al. 2002; Schadt et al. 2003; Grattapaglia and Kirst 2008). These studies have been the most successful in either simple organisms such as yeast (Brem et al. 2002), or in laboratory mice (Eaves et al. 2002; Schadt et al. 2003), plants derived from inbred lines (Schadt et al. 2003; Shi et al. 2007) and human reference families (Schadt et al. 2003). Larger-scale genetical genomics of less-related individuals in plants have been mainly confined to trees such as eucalyptus hybrids (Kirst et al. 2004) and *Populus* (Street et al. 2006). In the latter study, genes differentially expressed between genotypes exhibiting extreme sensitivity and insensitivity to drought on the basis of leaf abscission were identified, followed by the observation that 13 differentially expressed genes co-located to genomic regions identified by QTL mapping. A number of these genes have known functions that could impact on drought tolerance or response to drought stress. In sugarcane, a small-scale microarray study profiled 12 individuals selected from a population generated from a cross between sugarcane varieties Q117 and MQ77-340, a population that segregates for sugar accumulation (Casu et al. 2005). This resulted in the identification of 62 ESTs that were differentially expressed between individuals with low or high sucrose accumulation phenotypes. Only two of the identified genes had known functions in carbohydrate metabolism, implying that numerous genes with presently unknown functions may have a role in sucrose accumulation. The current difficulties encountered in standard gene mapping in sugarcane are also likely to be encountered in genetical genomics. Population size and structure, trait complexity (Pérez-Enciso et al. 2003) and the ability to detect the expression of specific alleles (much improved now with the advent of the Affymetrix GeneChip® Sugar Cane Genome Array—Casu et al. 2007) as well as the necessity to factor ploidy into mapping analysis makes this approach challenging. As with *Eucalyptus* (Grattapaglia and Kirst 2008), the possession of a genetic map containing significant numbers of mapped genes and, ideally, at least a base genome sequence (currently under discussion—G. Souza, pers. comm.) will render this approach more tractable.

9.4 Use of Resources of Related Species

Comparative transcriptomics can be a useful strategy to identify conserved and divergent pathways in plants as well as to establish evolutionary trends that may be important to understand plant physiology. Furthermore, when important functional information about certain pathways is missing for the plant species being studied, comparative transcriptomics may allow the generation of more specific hypotheses for further testing.

Sugarcane is taxonomically classified as a member of the Poaceae. Therefore, transcriptomic resources generated for other members of the Poaceae are likely to have some relevance, even as background for dissecting sugarcane's transcriptome. Large clustered EST collections curated at The J. Craig Venter Institute (*http://plantta.tigr.org/index.shtml*) and the Computational Biology and Functional Genomics Laboratory at the Dana-Farber Cancer Institute and Harvard School of Public Health (*http://compbio.dfci.harvard.edu/tgi/plant.html*) exist for economically important crop grasses e.g. rice, sorghum, wheat, maize and barley and sugarcane (Table 9-1). However, of this group, sorghum and maize are taxonomically the closest to sugarcane, making resources generated for these species the most relevant.

The first major draft of the *S. bicolor* genome has been released recently at Phytozome (*http://www.phytozome.org*) with the Maize Genome Project

Table 9-1 Comparison of the number of input and output sequences available for the Plant Transcript Assemblies and Plant Gene Indices (surveyed on 31st August, 2009).

Species	Plant Transcript Assemblies			Plant Gene Indices		
	ESTs + cDNAs	Clusters	Singletons	ESTs + cDNAs	Clusters	Singletons
Hordeum vulgare	457,719	30,171	93,180	483,840	41,206	39,517
Oryza sativa	1,205,038	49,870	197,646	1,278,120	77,158	104,638
Sugarcane[#]	–	–	–	256,134	40,016	76,572
Saccharum hybrid cultivar[#]	9,367	831	7035	–	–	–
Saccharum officinarum[#]	246,460	26,894	130,296	–	–	–
Sorghum bicolor	203,700	20,714	28,218	209,434	23,442	22,583
Triticum	843,257	61,121	257,828	1,038,118	91,464	124,988
Zea mays	1,048,825	64,601	232,592	1,698,645	112,156	202,931

[#]Sugarcane, *Saccharum* hybrid cultivar and *Saccharum officinarum* have been treated as different taxonomic entries by the two databases (an artificial distinction in this case).

also underway (Casu 2009). The considerable transcriptomic resources in the form of EST clusters (Table 9-1) will aid in the annotation of these genomes and inform the annotation of sugarcane transcript clusters. However, the major sink tissue for both of these species is not the stem as it is in sugarcane, making these resources slightly less useful for specifically understanding the mode of sucrose accumulation in the sugarcane stem.

The other grass of recent interest is *Miscanthus x giganteus*. This grass is another member of the PACCAD clade of the Poaceae and is even more closely related to sugarcane than *S. bicolor* (Pan et al. 2000). This temperate grass accumulates a high biomass, making it highly suitable for development as a crop for biofuel production based on ligno-cellulosic digestion (Christian et al. 2008; Hastings et al. 2008). However, little gene or EST information is currently available for this grass, which should be remedied in a project currently funded by the Energy Biosciences Institute (*http:// www.energybiosciencesinstitute.org*), a collaboration between the University of California, Berkeley; the Lawrence Berkeley National Laboratory; the University of Illinois and BP. The aim of this project is to generate ESTs, a draft genome and expression analysis platforms to assist in the exploitation of this grass as a feedstock for bioenergy production. These resources will add value to existing sugarcane data, particularly informing research on biomass traits.

Besides EST comparative analysis, it is also possible to use microarray information generated in one species to add value to data generated in another. For example, it is possible to compare array datasets of the same species, available in the GEO database, using the sequence accession number, or among different species, using sequence similarity searching via the BLAST algorithm. For example, two low temperature induced transcripts (LTI; CA186860 and CA096029) that were induced by drought in sugarcane (Rocha et al. 2007), had significant identity with two rice genes. Os05g0138300, annotated as "hydrophobic protein RCI2A (Low temperature and salt responsive protein LTI6A)", and Os08g0408500, a pathogenesis-related transcriptional factor and ERF domain containing protein, were both induced when rice varieties grown under salt stress, suggesting that these genes are involved in stress signaling in both species. However, provision of high quality transcriptomic data specifically derived from sugarcane will still be necessary even with the depth of resources available now and in the near future for related species. Some traits of sugarcane, specifically those affecting sucrose accumulation in the stem, cannot be adequately addressed only using resources developed in other species.

9.5 Impact of Advances in Sequencing Technology

Next-generation sequencing (NGS) technologies are currently revolutionizing whole-genome sequencing and re-sequencing projects (Margulies et al. 2005). These technologies are much more cost- and time-effective than traditional Sanger sequencing, and are likely to provide the next great leap forward for sugarcane transcriptomics since they make large-scale transcriptome analysis using de novo sequencing viable (Asmann et al. 2008). In contrast to the SAGE technique reviewed above, NGS does not require the use of bacterial cloning, which significantly increases the throughput and avoids possible biases introduced by the cloning steps. The main platforms currently available for NGS are 454/Roche, Illumina/Solexa and ABI/SOLiD. In *Arabidopsis*, the use of massively parallel pyrosequencing via the 454 Genome Sequencer GS20 System has shown a dynamic range that approached six orders of magnitude, detecting transcripts that correspond to less than 0.001% of the total transcripts (Weber et al. 2007). In addition, the expression of previously undetected transcripts or those that had been annotated incorrectly was also detected. Since NGS has proved to be more quantitative and more sensitive than array technologies, it is entirely possible that NGS will allow the conclusive identification of different alleles of the same gene, a feat currently extremely difficult in sugarcane, due to its aneuploid, polyploid genome. It has already been noted that many sugarcane transcript sequence clusters do not contain a full-length coding region and it would be expected that NGS could provide additional sequence to link current sequence clusters into full-length transcripts.

Even though the cost per sequenced base-pair of NGS is cheaper than other sequencing technologies, it is still more costly than current array techniques and is, therefore, unlikely to completely replace them, at least in the short term. There is also the limitation on the number and time taken to process multiple samples. Furthermore, the bioinformatics involved in the sequence analysis are still not fully developed, requiring new methods to store, process and retrieve massive volumes of data (Asmann et al. 2008). For example, none of the current assembly algorithms were able to assemble full contiguous sequences from the ESTs generated by pyrosequencing in *A. thaliana* (Weber et al. 2007). The advent of the longer reads that are now available is expected to improve this. The current sequence analysis limitations could be a problem in a genome with highly repetitive sequences as the sugarcane genome but, once the bioinformatics challenges are solved, NGS will be also useful to rapidly identify new and relevant SNPs. Ultimately, it will be an additional valuable tool to facilitate the sequencing of the sugarcane genome.

9.6 Concluding Remarks: Application of Functional Genomics to the Improvement of Sugarcane

Functional genomics can be used to improve sugarcane in two ways: (1) to provide new genetic markers to accelerate variety development in a classical breeding program and (2) to identify candidate genes for use in precision breeding of transgenic varieties with novel traits.

The generation of a new sugarcane variety through classical breeding may take 12 to 15 years from the first crosses to commercial release (Roach 1989). Genetic markers derived from functional genomics may speed this process by aiding the selection of individuals to be used in the crosses, as parents or as progeny, and by reducing the time and resources needed for agronomic evaluation. However, it is still necessary to effectively translate new gene findings into scalable tools to aid the generation of new varieties. Most genetic marker development is still centered on markers derived from genomic DNA, however, the identification of EST-SSRs (Garcia et al. 2006) and SNPs (Cordeiro et al. 2006) from transcriptome data do represent early attempts to integrate functional genomics with genetic diversity studies.

An alternative to traditional breeding is the generation of transgenic varieties by precision breeding. Genes can be down-regulated or over-expressed in order to study their function and also to produce novel phenotypes not possible through conventional breeding. The engineering of a novel trait into an otherwise elite background would still take a few years due to the time requirements for plant generation and field evaluation of trait stability, however, this time is likely to be many years shorter than that required for classical breeding. In the next couple of years we may see a surge in such studies now that many groups have candidate genes to test for improvement of sucrose content and other characteristics of interest. In this context it will be also desirable to expedite the development of tools such as tissue-enriched promoters that increase efficiency of transgene expression in a controlled fashion, making it possible to restrict gene expression to the tissues of interest and, possibly, aid in the generation of improved varieties.

Many novel gene functions have been unravelled by the transcriptome studies of classically bred varieties of sugarcane but few of these have been functionally tested by the generation of new transgenic varieties. Transformation of sugarcane is a proven technology and has been applied in many laboratories, mainly to introduce known herbicide and insect-resistance traits (Bower and Birch 1992; Gallo-Meagher and Irvine 1996; Falco et al. 2000; Butterfield et al. 2002; Leibbrandt and Snyman 2003; Manickavasagam et al. 2004). However, the engineering of sugarcane for novel sugar traits has recently commenced, with attempts having been made to create new pathways altering the behavior of the stem sink with the purpose of increasing sugar content (Chong et al. 2007; Wu and Birch 2007).

Given the importance of sugar and various fiber components of sugarcane as feedstocks for biofuel and other alternative products, increased attention will be paid to knowledge gained from studying the physiology and functional genomics of sugarcane. Transcriptomics can offer important insights into traits of interest in sugarcane, e.g. the identification of non-obvious genes associated with traits of interest, the provision of new markers with functional annotations to improve current genetic maps and provide functional links between genes and traits and as a source of potential transgenes to be used in the genetic modification of sugarcane but it will need to be complemented by at least a basic genome in order to increase its relevance. There will also need to be additional research in dissecting the roles of the non-obvious transcripts identified as being associated with traits of interest. Ideally, this will involve transgenic testing and/or fine mapping, both of which are resource-hungry endeavors. What is clear is that transcriptomics cannot be used alone for the improvement of sugarcane but is an essential component of a system biology approach. Ultimately, this will yield the best results and is a likely route to the timely production of novel sugarcane varieties that are optimized for a range of traits relating to sucrose, alternative product and biofuel production.

Acknowledgments

The authors would like to thank Dr. John M. Manners and Dr. Graham D. Bonnett for critical reading of the manuscript.

References

Aitken KS, Jackson PA, McIntyre CL (2005) A combination of AFLP and SSR markers provides extensive map coverage and identification of homo(eo)logous linkage groups in a sugarcane cultivar. Theor Appl Genet 110: 789–801.

Aitken KS, Jackson PA, McIntyre CL (2007) Construction of a genetic linkage map for *Saccharum officinarum* incorporating both simplex and duplex markers to increase genome coverage. Genome 50: 742–756.

Aitken KS, Hermann S, Karno K, Bonnett GD, McIntyre CL, Jackson PA (2008) Genetic control of yield related stalk traits in sugarcane. Theor Appl Genet 117: 1191–1203.

Alwala S, Kimbeng CA (2010) Molecular genetic linkage mapping in *Saccharum*: strategies, resources and achievements. In: RJ Henry, C Kole (eds) Genetics, Genomics and Breeding of Sugarcane. Science Publ, New Hampshire, Plymouth, Jersey, USA, pp 69–96.

de Araujo PG, Rossi M, de Jesus EM, Saccaro Jr NL, Kajihara D, Massa R, de Felix JM, Drummond RD, Falco MC, Chabregas SM, Ulian EC, Menossi M, Van Sluys M-A (2005) Transcriptionally active transposable elements in recent hybrid sugarcane. Plant J 44: 707–717.

Asmann YW, Wallace MB, Thompson E (2008) Transcriptome profiling using next-generation sequencing. Gastroenterology 135: 1466–1468.

Bachem CWB, van der Hoeven RS, de Bruijn SM, Vreugdenhil D, Zabeau M., Visser RGF (1996) Visualization of differential gene expression using a novel method of RNA fingerprinting based on AFLP: Analysis of gene expression during potato tuber development. Plant J 9: 745–753.

Barrett T, Troup DB, Wilhite SE, Ledoux P, Rudnev D, Evangelista C, Kim IF, Soboleva A, Tomashevsky M, Marshall KA, Phillippy KH, Sherman PM, Muertter RN, Edgar R (2008) NCBI GEO: archive for high-throughput functional genomic data. Nucl Acids Res 37: D885–D890.

Borecký J, Nogueira FTS, de Oliveira KAP, Maia IG, Vercesi AE, Arruda P (2006) The plant energy-dissipating mitochondrial systems: depicting the genomic structure and the expression profiles of the gene families of uncoupling protein and alternative oxidase in monocots and dicots. J Exp Bot 57: 849–864.

Borrás-Hidalgo O, Thomma BPHJ, Carmona E, Borroto CJ, Pujol M, Arencibia A, Lopez J (2005) Identification of sugarcane genes induced in disease-resistant somaclones upon inoculation with *Ustilago scitaminea* or *Bipolaris sacchari*. Plant Physiol Biochem 43: 1115–1121.

Bower NI, Casu RE, Maclean DJ, Reverter A, Chapman SC, Manners JM (2005) Transcriptional response of sugarcane roots to methyl jasmonate. Plant Sci 168: 761–772.

Bower R, Birch RG (1992) Transgenic sugarcane plants via microprojectile bombardment. Plant J 2: 409–416.

Brady SM, Long TA, Benfey PN (2006) Unravelling the dynamic transcriptome. Plant Cell 18: 2101–2111.

Brem RB, Yvert G, Clinton R., Kruglyak L (2002) Genetic dissection of transcriptional regulation in budding yeast. Science 296: 752–755.

Brenner S, Williams S, Vermaas E, Storck T, Moon K, McCollum C, Mao J, Luo SJ, Kirchner J, Eletr S, DuBridge R, Burcham T, Albrecht G (2000a) *In vitro* cloning of complex mixtures of DNA on microbeads: physical separation of differentially expressed cDNAs. Proc Natl Acad Sci USA 97: 1665–1670.

Brenner S, Johnson M, Bridgham J, Golda G, Lloyd DH, Johnson D, Luo S, McCurdy S, Foy M, Ewan M, Roth R, George D, Eletr S, Albrecht G, Vermaas E, Williams SR, Moon K, Burcham T, Pallas M, DuBridge RB, Kirchner J, Fearon K, Mao J-I, Corcoran K (2000b) Gene expression analysis by massively parallel signature sequencing (MPSS) on microbead arrays. Nat Biotechnol 18: 630–634.

Brugmans B, del Carmen AF, Bachem CWB, van Os H, van Eck HJ, Visser RGF (2002) A novel method for the construction of genome wide transcriptome maps. Plant J 31: 211–222.

Bull TA, Glasziou KT (1963) The evolutionary significance of sugar accumulation in *Saccharum*. Aust J Biol Sci 16: 737–742.

Bundock PC, Eliott FG, Ablett G, Benson AD, Casu RE, Aitken KS, Henry RJ (2009) Targeted single nucleotide polymorphism (SNP) discovery in a highly polyploid plant species using 454 sequencing. Plant Biotechnol J 7: 347–354.

Butterfield MK, Irvine JE, Valdez Garza M, Mirkov TE (2002) Inheritance and segregation of virus and herbicide resistance transgenes in sugarcane. Theor Appl Genet 104: 797–803.

Butterfield MK, Rutherford RS, Carson DL, Huckett BI (2004) Application of gene discovery to varietal improvement in sugarcane. S Afr J Bot 70: 167–172.

Calsa T, Figueira A (2007) Serial analysis of gene expression in sugarcane (*Saccharum* spp.) leaves revealed alternative C4 metabolism and putative antisense transcripts. Plant Mol Biol 63: 745–762.

Camargo SR, Cançado GM, Ulian EC, Menossi M (2007) Identification of genes responsive to the application of ethanol on sugarcane leaves. Plant Cell Rep 26: 2119–2128.

Carmona E, Vargas D, Borroto CJ, Lopez J, Fernández AI, Arencibia A, Borrás-Hidalgo O (2004) cDNA-AFLP analysis of differential gene expression during the interaction between sugarcane and *Puccinia melanocephala*. Plant Breed 123: 499–501.

Carson DL, Botha FC (2000) Preliminary analysis of expressed sequence tags for sugarcane. Crop Sci 40: 1769–1779.

Carson DL, Botha FC (2002) Genes expressed in sugarcane maturing internodal tissue. Plant Cell Rep 20: 1075–1081.

Carson DL, Huckett BI, Botha FC (2002a) Differential gene expression in sugarcane leaf and internodal tissues of varying maturity. S Afr J Bot 68: 434–442.

Carson DL, Huckett BI, Botha FC (2002b) Sugarcane ESTs differentially expressed in immature and maturing internodal tissue. Plant Sci 162: 289–300.

Casu RE (2010) Role of bioinformatics as a tool for sugarcane research. In: RJ Henry , C Kole (eds) Genetics, Genomics and Breeding of Sugarcane. Science Publ, New Hampshire, Plymouth, Jersey, USA, pp 229–248.

Casu RE, Grof CPL, Rae AL, McIntyre CL, Dimmock CM, Manners JM (2003) Identification of a novel sugar transporter homologue strongly expressed in maturing stem vascular tissues of sugarcane by expressed sequence tag and microarray analysis. Plant Mol Biol 52: 371–386.

Casu RE, Dimmock CM, Chapman SC, Grof CPL, McIntyre CL, Bonnett GD, Manners JM (2004) Identification of differentially expressed transcripts from maturing stem of sugarcane by *in silico* analysis of stem expressed sequence tags and gene expression profiling. Plant Mol Biol 54: 503–517.

Casu RE, Manners JM, Bonnett GD, Jackson PA, McIntyre CL, Dunne R, Chapman SC, Rae AL, Grof CPL (2005) Genomics approaches for the identification of genes determining important traits in sugarcane. Field Crops Res 92: 137–147.

Casu RE, Jarmey JM, Bonnett GD, Manners JM (2007) Identification of transcripts associated with cell wall metabolism and development in the stem of sugarcane by Affymetrix GeneChip Sugarcane Genome Array expression profiling. Funct Integr Genom 7: 153–167.

Chong BF, Bonnett GD, Glassop D, O'Shea MG, Brumbley SM (2007) Growth and metabolism in sugarcane are altered by the creation of a new hexose-phosphate sink. Plant Biotechnol J 5: 240–253.

Christian DG, Riche AB, Yates NE (2008) Growth, yield and mineral content of *Miscanthus × giganteus* grown as a biofuel for 14 successive harvests. Indust Crops Prod 28: 320–327.

Churchill GA (2002) Fundamentals of experimental design for cDNA microarrays. Nat Genet 30: 490–495.

Cordeiro GM, Casu R, McIntyre CL, Manners JM, Henry RJ (2001) Microsatellite markers from sugarcane (*Saccharum* spp.) ESTs cross transferable to *Erianthus* and sorghum. Plant Sci 160: 1115–1123.

Cordeiro GM, Eliott F, McIntyre CL, Casu RE, Henry RJ (2006) Characterisation of single nucleotide polymorphisms in sugarcane ESTs. Theor Appl Genet 113: 331–343.

Eaves IA, Wicker LS, Ghandour G, Lyons PA, Peterson LB, Todd JA, Glynne RJ (2002) Combining mouse congenic strains and microarray gene expression analyses to study a complex trait: the NOD model of Type 1 Diabetes. Genome Res 12: 232–243.

Falco MC, Tulmann Neto A, Ulian EC (2000) Transformation and expression of a gene for herbicide resistance in a Brazilian sugarcane. Plant Cell Rep 19: 1188–1194.

Galbraith DW (2006) DNA microarray analyses in higher plants. OMICS: J Integr Biol 10: 455–473.

Gallo-Meagher M, Irvine JE (1996) Herbicide resistant transgenic sugarcane plants containing the bar gene. Crop Sci 36: 1367–1374.

Garcia AAF, Kido EA, Meza AN, Souza HMB, Pinto LR, Pastina MM, Leite CS, Silva J, Ulian EC, Figueira A, Souza AP (2006) Development of an integrated genetic map of a sugarcane (*Saccharum* spp.) commercial cross, based on a maximum-likelihood approach for estimation of linkage and linkage phases. Theor Appl Genet 112: 298–314.

Goldemberg J (2007) Ethanol for a sustainable energy future. Science 315: 808–810.

Gomez-Porras J, Riano-Pachon D, Dreyer I, Mayer J, Mueller-Roeber B (2007) Genome-wide analysis of ABA-responsive elements ABRE and CE3 reveals divergent patterns in *Arabidopsis* and rice. BMC Genom 8: 260.

Gowda M, Jantasuriyarat C, Dean RA, Wang G-L (2004) Robust-LongSAGE (RL-SAGE): A substantially improved LongSAGE method for gene discovery and transcriptome analysis. Plant Physiol 134: 890–897.

Gowda M, Li H, Alessi J, Chen F, Pratt R, Wang G-L (2006) Robust analysis of 5'-transcript ends (5'-RATE): a novel technique for transcriptome analysis and genome annotation. Nucl Acids Res 34: e126.

Grattapaglia D, Kirst M (2008) *Eucalyptus* applied genomics: from gene sequences to breeding tools. New Phytol 179: 911–929.

Grivet L, D'Hont A, Roques D, Feldmann P, Lanaud C, Glaszmann J (1996) RFLP mapping in cultivated sugarcane (*Saccharum* spp.): genome organization in a highly polyploid and aneuploid interspecific hybrid. Genetics 142: 987–1000.

Grivet L, Glaszmann JC, Arruda P (2001) Sequence polymorphism from EST data in sugarcane: a fine analysis of 6-phosphogluconate dehydrogenase genes. Genet Mol Biol 24: 161–167.

Grivet L, Glaszmann J-C, Vincentz M, da Silva F, Arruda P (2003) ESTs as a source for sequence polymorphism discovery in sugarcane: example of the *Adh* genes. Theor Appl Genet 106: 190–197.

Guimarães C, Honeycutt R, Sills G, Sobral B (1999) Genetic maps of *Saccharum officinarum* L. and *Saccharum robustum* Brandes & Jew. ex grassl. Genet Mol Biol 22: 125–132.

Hashimoto RF, Kim S, Shmulevich I, Zhang W, Bittner ML, Dougherty ER (2004) Growing genetic regulatory networks from seed genes. Bioinformatics 20: 1241–1247.

Hastings A, Clifton-Brown J, Wattenbach M, Stampfl P, Mitchell CP, Smith P (2008) Potential of *Miscanthus* grasses to provide energy and hence reduce greenhouse gas emissions. Agron Sustain Dev 28: 465–472.

Hoarau J-Y, Offmann B, D'Hont A, Risterucci A-M, Roques D, Glaszmann J-C, Grivet L (2001) Genetic dissection of a modern sugarcane cultivar (*Saccharum* spp.). I. Genome mapping with AFLP markers. Theor Appl Genet 103: 84–97.

Jansen RC, Nap JP (2001) Genetical genomics: the added value from segregation. Trends Genet 17: 388–391.

Kirst M, Myburg AA, de León JPG, Kirst ME, Scott J, Sederoff R (2004) Coordinated genetic regulation of growth and lignin revealed by quantitative trait locus analysis of cDNA microarray data in an interspecific backcross of *Eucalyptus*. Plant Physiol 135: 2368–2378.

Kriventseva EV, Rahman N, Espinosa O, Zdobnov EM (2008) OrthoDB: the hierarchical catalog of eukaryotic orthologs. Nucl Acids Res 36: D271–275.

Lao M, Arencibia AD, Carmona ER, Acevedo R, Rodríguez E, Leín O, Santana I (2008) Differential expression analysis by cDNA-AFLP of *Saccharum* spp. after inoculation with the host pathogen *Sporisorium scitamineum*. Plant Cell Rep 27: 1103–1111.

Leibbrandt NB, Snyman SJ (2003) Stability of gene expression and agronomic performance of a transgenic herbicide-resistant sugarcane line in South Africa. Crop Sci 43: 671–677.

Lennon GG, Lehrach H (1991) Hybridization analyses of arrayed cDNA libraries. Trends Genet 7: 314–317.

Ma HM, Schulze S, Lee S, Yang M, Mirkov E, Irvine J, Moore P, Paterson A (2004) An EST survey of the sugarcane transcriptome. Theor Appl Genet 108: 851–863.

Manickavasagam M, Ganapathi A, Anbazhagan VR, Sudhakar B, Selvaraj N, Vasudevan A, Kasthurirengan S (2004) *Agrobacterium*-mediated genetic transformation and development of herbicide-resistant sugarcane (*Saccharum* species hybrids) using axillary buds. Plant Cell Rep 23: 134–143.

Margulies M, Egholm M, Altman WE, Attiya S, Bader JS, Bemben LA, Berka J, Braverman MS, Chen Y-J, Chen Z, Dewell SB, Du L, Fierro JM, Gomes XV, Godwin BC, He W, Helgesen S, Ho CH, Irzyk GP, Jando SC, Alenquer MLI, Jarvie TP, Jirage KB, Kim J-B, Knight JR, Lanza JR, Leamon JH, Lefkowitz SM, Lei M, Li J, Lohman KL, Lu H, Makhijani VB, McDade KE, McKenna MP, Myers EW, Nickerson E, Nobile JR, Plant R, Puc BP, Ronan MT, Roth GT, Sarkis GJ, Simons JF, Simpson JW, Srinivasan M, Tartaro KR, Tomasz A, Vogt KA, Volkmer GA, Wang SH, Wang Y, Weiner MP, Yu P, Begley RF, Rothberg JM (2005) Genome sequencing in microfabricated high-density picolitre reactors. Nature 437: 376–380.

Matsumura H, Reich S, Ito A, Saitoh H, Kamoun S, Winter P, Kahl G, Reuter M, Krüger DH, Terauchi R (2003) Gene expression analysis of plant host–pathogen interactions by SuperSAGE. Proc Natl Acad Sci USA, 100: 15718–15723.

McCormick AJ, Cramer MD, Watt DA (2008a) Changes in photosynthetic rates and gene expression of leaves during a source sink perturbation in sugarcane. Ann Bot 101: 89–102.

McCormick AJ, Cramer MD, Watt DA (2008b) Differential expression of genes in the leaves of sugarcane in response to sugar accumulation. Trop Plant Biol 1: 142–158.

McIntyre CL, Casu RE, Drenth J, Knight D, Whan VA, Croft BJ, Jordan DR, Manners JM (2005) Resistance gene analogues in sugarcane and sorghum and their association with quantitative trait loci for rust resistance. Genome 48: 391–400.

McIntyre CL, Jackson M, Cordeiro GM, Amouyal O, Hermann S, Aitken KS, Eliott F, Henry RJ, Casu RE, Bonnett GD (2006) The identification and characterisation of alleles of sucrose phosphate synthase gene family III in sugarcane. Mol Breed 18: 39–50.

Moore PH (1995) Temporal and spatial regulation of sucrose accumulation in the sugarcane stem. Aust J Plant Physiol 22: 661–679.

Moore PH (2005) Integration of sucrose accumulation processes across hierarchical scales: towards developing an understanding of the gene-to-crop continuum. Field Crops Res 92: 119–135.

Muchow RC, Spillman MF, Wood AW, Thomas MR (1994) Radiation interception and biomass accumulation in a sugarcane crop grown under irrigated tropical conditions. Aust J Agri Res 45: 37–49.

Nakano M, Nobuta K, Vemaraju K, Tej SS, Skogen JW, Meyers BC (2006) Plant MPSS databases: signature-based transcriptional resources for analyses of mRNA and small RNA. Nucl Acids Res 34: D731–735.

Ng P, Tan JJS, Ooi HS, Lee YL, Chiu KP, Fullwood MJ, Srinivasan KG, Perbost C, Du L, Sung W-K, Wei C-L, Ruan Y (2006) Multiplex sequencing of paired-end ditags (MS-PET): a strategy for the ultra-high-throughput analysis of transcriptomes and genomes. Nucl Acids Res 34: e84.

Nielsen KL, Hogh AL, Emmersen J (2006) DeepSAGE—digital transcriptomics with high sensitivity, simple experimental protocol and multiplexing of samples. Nucl Acids Res 34: e133.

Nogueira FTS, de Rosa Jr VE, Menossi M, Ulian EC, Arruda P (2003) RNA expression profiles and data mining of sugarcane response to low temperature. Plant Physiol 132: 1811–1824.

Oliveira KM, Pinto LR, Marconi TG, Margarido GRA, Pastina MM, Teixeira LHM, Figueira AV, Ulian EC, Garcia AAF, Souza AP (2007) Functional integrated genetic linkage map based on EST markers for a sugarcane (*Saccharum* spp.) commercial cross. Mol Breed 20: 189–208.

Oliveira KM, Pinto LR, Marconi TG, Mollinari M, Ulian EC, Chabregas SM, Falco MC, Burnquist W, Garcia AAF, Souza AP (2009) Characterization of new polymorphic functional markers for sugarcane. Genome 52: 191–209.

Pan YB, Burner DM, Legendre BL (2000) An assessment of the phylogenetic relationship among sugarcane and related taxa based on the nucleotide sequence of 5S rRNA intergenic spacers. Genetica 108: 285–295.

Papini-Terzi FS, Rocha FR, Vêncio RZN, Oliveira KC, Felix JM, Vicentini R, de Souza Rocha C, Simões ACQ, Ulian EC, di Mauro SMZ, da Silva AM, Pereira CAB, Menossi M, Souza GM (2005) Transcription profiling of signal transduction-related genes in sugarcane tissues. DNA Res 12: 27–38.

Park Y-H, Alabady MS, Ulloa M, Sickler B, Wilkins TA, Yu J, Stelly DM, Kohel RJ, El-Shihy OM, Cantrell RG (2005) Genetic mapping of new cotton fiber loci using EST-derived microsatellites in an interspecific recombinant inbred line cotton population. Mol Genet Genom 274: 428–441.

Parkinson H, Kapushesky M, Kolesnikov N, Rustici G, Shojatalab M, Abeygunawardena N, Berube H, Dylag M, Emam I, Farne A, Holloway E, Lukk M, Malone J, Mani R, Pilicheva E, Rayner TF, Rezwan F, Sharma A, Williams E, Bradley XZ, Adamusiak T, Brandizi M, Burdett T, Coulson R, Krestyaninova M, Kurnosov P, Maguire E, Neogi SG, Rocca-Serra P, Sansone S-A, Sklyar N, Zhao M, Sarkans U, Brazma A (2008) ArrayExpress update—from an archive of functional genomics experiments to the atlas of gene expression. Nucl Acids Res 37: D868–D872.

Pérez-Enciso M, Toro MA, Tenenhaus M, Gianola D (2003) Combining gene expression and molecular marker information for mapping complex trait genes: a simulation study. Genetics 164: 1597–1606.

Piétu G, Alibert O, Guichard V, Lamy B, Bois F, Leroy E, Mariage-Sampson R, Houlgatte R, Soularue P, Auffray C (1996) Novel gene transcripts preferentially expressed in human muscles revealed by quantitative hybridization of a high density cDNA array. Genome Res 6: 492–503.

Pinto LR, Oliveira KM, Ulian EC, Garcia AAF, de Souza AP (2004) Survey in the sugarcane expressed sequence tag database (SUCEST) for simple sequence repeats. Genome 47: 795–804.

Pinto LR, Oliveira KM, Marconi T, Garcia AAF, Ulian EC, de Souza AP (2006) Characterization of novel sugarcane expressed sequence tag microsatellites and their comparison with genomic SSRs. Plant Breed 125: 378–384.

Raboin L-M, Oliveira KM, Lecunff L, Telismart H, Roques D, Butterfield M, Hoarau J-Y, D'Hont A (2006) Genetic mapping in sugarcane, a high polyploid, using bi-parental progeny: identification of a gene controlling stalk colour and a new rust resistance gene. Theor Appl Genet 112: 1382–1391.

Reffay N, Jackson PA, Aitken KS, Hoarau J-Y, D'Hont A, Besse P, McIntyre CL (2005) Characterisation of genome regions incorporated from an important wild relative into Australian sugarcane. Mol Breed 15: 367–381.

Ritter E, Ruiz de Galarreta J, Eck H, Sánchez I (2008) Construction of a potato transcriptome map based on the cDNA–AFLP technique. Theor Appl Genet 116: 1003–1013.

Roach BT (1989) Origin and improvement of the genetic base of sugarcane. Proc Aust Soc Sugar Cane Technol 12: 34–47.

Rocha F, Papini-Terzi F, Nishiyama M, Vêncio R, Vicentini R, Duarte R, de Rosa V, Vinagre F, Barsalobres C, Medeiros A, Rodrigues F, Ulian E, Zingaretti S, Galbiatti J, Almeida R, Figueira A, Hemerly A, Silva-Filho M, Menossi M, Souza G (2007) Signal transduction-related responses to phytohormones and environmental challenges in sugarcane. BMC Genom 8: 71.

de Rosa Jr VE, Nogueira FTS, Menossi M, Ulian EC, Arruda P (2005) Identification of methyl jasmonate-responsive genes in sugarcane using cDNA arrays. Braz J Plant Physiol 17: 173–180.

Rossi M, Araujo PG, Paulet F, Garsmeur O, Dias VM, Chen H, Van Sluys M-A, D'Hont A (2003) Genomic distribution and characterization of EST-derived resistance gene analogs (RGAs) in sugarcane. Mol Genet Genom 269: 406–419.

Saha S, Sparks AB, Rago C, Akmaev V, Wang CJ, Vogelstein B, Kinzler KW, Velculescu VE (2002) Using the transcriptome to annotate the genome. Nat Biotechnol 20: 508–512.

Schadt EE, Monks SA, Drake TA, Lusis AJ, Che N, Colinayo V, Ruff TG, Milligan SB, Lamb JR, Cavet G, Linsley PS, Mao M, Stoughton RB, Friend SH (2003) Genetics of gene expression surveyed in maize, mouse and man. Nature 422: 297–302.

Schena M, Shalon D, Davis R, Brown P (1995) Quantitative monitoring of gene expression patterns with a complementary DNA microarray. Science 270: 467–470.

Schlögl PS, Nogueira FTS, Drummond R, Felix JM, Rosa VE, Vicentini R, Leite A, Ulian EC, Menossi M (2008) Identification of new ABA- and MEJA-activated sugarcane bZIP genes by data mining in the SUCEST database. Plant Cell Rep 27: 335–345.

Schnable PS, Hochholdinger F, Nakazono M (2004) Global expression profiling applied to plant development. Curr Opin Plant Biol 7: 50–56.

Shi C, Uzarowska A, Ouzunova M, Landbeck M, Wenzel G, Lubberstedt T (2007) Identification of candidate genes associated with cell wall digestibility and eQTL (expression quantitative trait loci) analysis in a Flint x Flint maize recombinant inbred line population. BMC Genom 8: 22.

da Silva JAG (2001) Preliminary analysis of microsatellite markers derived from sugarcane expressed sequence tags (ESTs). Genet Mol Biol 24: 155–159.

da Silva JA, Bressiani JA (2005) Sucrose synthase molecular marker associated with sugar content in elite sugarcane progeny. Genet Mol Biol 28: 294–298.

da Silva J, Honeycutt RJ, Burnquist W, Al-Janabi SM, Sorrells ME, Tanksley SD, Sobral, BWS (1995) *Saccharum spontaneum* L. 'SES 208' genetic linkage map combining RFLP- and PCR-based markers. Mol Breed 1: 165–179.

da Silva JAG, Ulian EC, Barsalobres CF (2001) Development of EST-derived RFLP markers for sugarcane breeding. Proc Int Soc Sugar Cane Technol 24: 318–322.

de Souza AP, Gaspar M, da Silva EA, Ulian EC, Waclawovsky AJ, Nishiyama Jr MY, Dos Santos RV, Teixeira MM, Souza GM, Buckeridge MS (2008) Elevated CO_2 increases photosynthesis, biomass and productivity, and modifies gene expression in sugarcane. Plant Cell Environ 31: 1116–1127.

Street NR, Skogström O, Sjödin A, Tucker J, Rodríguez-Acosta M, Nilsson P, Jansson S, Taylor G (2006) The genetics and genomics of the drought response in *Populus*. Plant J 48: 321–341.

Tao YZ, Jordan DR, Henzell RG, McIntyre CL (1998) Construction of a genetic map in a sorghum recombinant inbred line using probes from different sources and its comparison with other sorghum maps. Aust J Agri Res 49: 729–736.

Tao YZ, Henzell RG, Jordan DR, Butler DG, Kelly AM, McIntyre CL (2000) Identification of genomic regions associated with stay green in sorghum by testing RILs in multiple environments. Theor Appl Genet 100: 1225–1232.

Thiel T, Michalek W, Varshney RK, Graner A (2003) Exploiting EST databases for the development and characterization of gene-derived SSR-markers in barley (*Hordeum vulgare* L.). Theor Appl Genet 106: 411–422.

Thokoane LN, Rutherford RS (2001) cDNA-AFLP differential display of sugarcane (*Saccharum* spp. hybrids) genes induced by challenge with the fungal pathogen *Ustilago scitaminea* (sugarcane smut). Proc Ann Congr—S Afr Sugar Technol Assoc 75: 104–107.

Vega-Sánchez ME, Gowda M, Wang GL (2007) Tag-based approaches for deep transcriptome analysis in plants. Plant Sci 173: 371–380.

Velculescu VE, Zhang L, Vogelstein B, Kinzler KW (1995) Serial analysis of gene expression. Science 270: 484–487.

Vettore AL, da Silva FR, Kemper EL, Arruda P (2001) The libraries that made SUCEST. Genet Mol Biol 24: 1–7.

Vettore AL, da Silva FR, Kemper EL, Souza GM, da Silva AM, Ferro MIT, Henrique-Silva F, Giglioti EA, Lemos MVF, Coutinho LL, Nobrega MP, Carrer H, Franca SC, Bacci Jr M, Goldman MHS, Gomes SL, Nunes LR, Camargo LEA, Siqueira WJ, Van Sluys M-A, Thiemann OH, Kuramae EE, Santelli RV, Marino CL, Targon MLPN, Ferro JA, Silveira HCS, Marini DC, Lemos EGM, Monteiro-Vitorello CB, Tambor JHM, Carraro DM, Roberto PG, Martins VG, Goldman GH, de Oliveira RC, Truffi D, Colombo CA, Rossi M, de Araujo PG, Sculaccio SA, Angella A, Lima MMA, de Rosa Jr VE, Siviero F, Coscrato VE, Machado MA, Grivet L, Di Mauro SMZ, Nobrega FG, Menck CFM, Braga MDV, Telles GP, Cara FAA, Pedrosa G, Meidanis J, Arruda P (2003) Analysis and functional annotation of an expressed sequence tag collection for tropical crop sugarcane. Genome Res 13: 2725–2735.

Vincentz M, Cara FAA, Okura VK, da Silva FR, Pedrosa GL, Hemerly AS, Capella AN, Marins M, Ferreira PC, Franca SC, Grivet L, Vettore AL, Kemper EL, Burnquist WL, Targon MLP, Siqueira WJ, Kuramae EE, Marino CL, Camargo LEA, Carrer H, Coutinho LL, Furlan LR, Lemos MVF, Nunes LR, Gomes SL, Santelli RV, Goldman MH, Bacci Jr M, Giglioti EA, Thiemann OH, Silva FH, Van Sluys M-A, Nobrega FG, Arruda P, Menck CFM (2004) Evaluation of monocot and eudicot divergence using the sugarcane transcriptome. Plant Physiol 134: 951–959.

Vos P, Hogers R, Bleeker M, Reijans M, van de Lee T, Hornes M, Frijters A, Pot J, Peleman JJ, Kuiper M (1995) AFLP: a new technique for DNA fingerprinting. Nucl Acids Res 23: 4407–4414.

Watt DA (2003) Aluminium-responsive genes in sugarcane: identification and analysis of expression under oxidative stress. J Exp Bot 54: 1163–1174.

Watt DA, McCormick AJ, Govender C, Carson DL, Cramer MD, Huckett BI, Botha FC (2005) Increasing the utility of genomics in unravelling sucrose accumulation. Field Crops Res 92: 149–158.

Watt D, Butterfield M, Huckett B (2010) Proteomics and metabolomics. In: RJ Henry, C Kole (eds) Genetics, Genomics and Breeding of Sugarcane. Science Publ, New Hampshire, Plymouth, Jersey, USA, pp 193–228.

Weber APM, Weber KL, Carr K, Wilkerson C, Ohlrogge JB (2007) Sampling the *Arabidopsis* transcriptome with massively parallel pyrosequencing. Plant Physiol 144: 32–42.

Wei CL, Ng P, Chiu KP, Wong CH, Ang CC, Lipovich L, Liu ET, Ruan Y (2004) 5' Long serial analysis of gene expression (LongSAGE) and 3' LongSAGE for transcriptome characterization and genome annotation. Proc Natl Acad Sci USA, 101: 11701–11706.

Wu L, Birch RG (2007) Doubled sugar content in sugarcane plants modified to produce a sucrose isomer. Plant Biotechnol J. 5: 109–117.

Zimmermann P, Hirsch-Hoffmann M, Hennig L, Gruissem W (2004) GENEVESTIGATOR. *Arabidopsis* microarray database and analysis toolbox. Plant Physiol 136: 2621–2632.

Proteomics and Metabolomics

Derek Watt,[1,2] Mike Butterfield[1,3] and Barbara Huckett[1,2,4]*

ABSTRACT

Sugarcane improvement, whether by conventional or molecular means, relies on an understanding of the biology of the plant, particularly interactions occurring across hierarchical scales of organization. In this regard, the application of metabolomics and proteomics to the study of sugarcane is poised to deliver large volumes of data on protein and metabolite fluctuations associated with developmental and environmental cues, and in response to genetic perturbations. The challenge for the sugarcane research community will be to ensure that such data are generated, interpreted and integrated in a manner that will contribute meaningfully to crop improvement. This challenge is discussed in terms of the limitations to currently available technologies, existing knowledge of sugarcane metabolite and protein composition and lessons learned from metabolomic and proteomic studies of other plant species. Perspectives are presented on potential applications of these technologies to molecular breeding, metabolic engineering, elucidation of stress-response pathways and regulation of sucrose accumulation.

Keywords: sugarcane improvement, systems biology, high-dimensional biology, predictive metabolic engineering, metabolic modeling

[1]SA Sugarcane Research Institute, Private Bag X02, Mount Edgecombe, 4300, South Africa.
[2]School of Biological and Conservation Sciences, University of KwaZulu-Natal, Private Bag X54001, Durban, 4000, South Africa.
[3]Current Address: ICRISAT—Patancheru, Patancheru 502324, Andhra Pradesh, India.
[4]Current Address: Chase Place, Westville 3630, South Africa.
*Corresponding author: *derek.watt@sugar.org.za*

10.1 Terms of Reference

The concepts of "proteome" and "metabolome" have been part of scientific thinking for several decades and these terms emerged in the literature quite early , particularly in prokaryotic studies. Association of the concepts with systematic analysis of eukaryotes became more prominent in the late 1990s, for example in functional analysis of the yeast genome (Oliver et al. 1998). However, "proteomics" and "metabolomics" are more recent linguistic developments. Nikolau and Wurtele (2007) state that the first use of the word "metabolomics" in the title of a research paper was in 2001. Its first use in the title of a publication based on plant studies followed shortly thereafter (Fiehn 2002; Hall et al. 2002), coinciding with the first International Congress on Plant Metabolomics (Wageningen, The Netherlands, April 2002). A metabolomic study in sugarcane came a year later (Bosch et al. 2003).

Although the term "metabolomics" is no longer new, its meaning has been the subject of some uncertainty. It has been contrasted with and seen as distinct from "metabolite profiling", which had been used in medical, diagnostic and other contexts prior to the development of functional genomics (Villas-Bôas et al. 2005). In this chapter, we take a broad view and use the terms "proteomic" and "metabolomic" to refer to both targeted and comprehensive screening of proteins and metabolites designed to identify and measure components contributing to function and gene-phenotype relationships in sugarcane.

10.2 Introduction

10.2.1 Omic Philosophy and its Relevance to Sugarcane

In the suite of omic technologies used to describe biological systems, proteomics and metabolomics follow on from genomics and transcriptomics. Together, and with the addition of networks or arrays of physiological processes, these basic omics aim to provide insight in the "phenome"—the complete phenotype for all traits of individuals within populations. The integration of these fields of study has been labeled "systems biology" (Hood 1998). More recently, the term "high-dimensional biology" or HDB has been coined to reflect the specific challenges of analyzing, interpreting and extracting sense from gigabytes of data from high-throughput sequencing, gene expression and metabolite analysis studies (e.g., Gadbury et al. 2004; Mehta et al. 2006; Quackenbush 2007).

Drawing meaningful inferences from HDB data requires both valid statistical methods and a sound epistemological foundation in terms of appropriate questions that can be posed to the data (Mehta et al. 2006).

Many individual metabolites are common to different metabolic processes or pathways, and may also be produced differentially in different cellular and subcellular compartments (Perera and Nikolau 2007). Measuring the level of an individual metabolite per se may thus provide limited information on its role in a particular metabolic pathway or in a specific cellular compartment or organ. This is one of the limitations of metabolomic studies on a whole-plant basis, and must be accounted for by either correct experimental design or by only framing meaningful experimental hypotheses (Mehta et al. 2006).

An additional complication in analyzing and interpreting HDB and omic data is that there is often not a direct correlation between one omic paradigm and the next. For example, studies have found low correlations between mRNA levels and protein levels (Osuna et al. 2007; Carpentier et al. 2008). Post-transcriptional regulation of gene expression by trans-acting microRNAs adds levels of complexity to integrating omic data, as it has been shown that single miRNAs can influence the expression of hundreds, and even thousands of proteins (Baek et al. 2008; Selbach et al. 2008). Although these intricate regulatory networks complicate the interpretative synthesis of phenome from genome, proteomics can be used as a tool to address questions of gene regulation (Mourelatos 2008).

Sugarcane is grown as a commercial crop as it has the ability to accumulate high levels of sucrose: up to 60% of dry matter weight. Sucrose content of sugarcane is viewed by plant breeders as a quantitative trait, and in the context of breeding new varieties has been studied using the tools of quantitative genetics. Sucrose, however, is also a metabolite involved in many plant processes, and sucrose levels in different tissues and organs will vary over time, dependent on rates of synthesis, transport, breakdown, respiration and other processes. For example, Whittaker and Botha (1997) reported that cycles of sucrose synthesis and degradation occur in all internodes of a single sugarcane genotype, characterized by changing fluxes of partitioning of carbon between sucrose, the hexose pool and respiration along the culm. They went on to demonstrate that the flux of carbon from sucrose/glucose to respiration could also vary between different genotypes (Whittaker and Botha 1999). Information based solely on absolute sucrose levels in mature sugarcane internodes thus tells us little about cytosolic sucrose synthesis, vacuolar storage mechanisms or the involvement of sucrose in other metabolic processes. In an omics context, however, the quantitative trait of whole-plant sucrose content can be broken down into its constituents of metabolomics (levels in different tissues in different biochemical pathways, and levels of related substrate sugars and breakdown products, respiration, etc.), proteomics (enzyme activity in all the relevant pathways), transcriptomics (see Casu et al., Chapter 9 of this volume) and genomics (high density genetic markers across the whole genome) (see

Pastina et al., Chapter 7 of this volume). An appropriate HDB dataset containing all the relevant hierarchies of omic data, along with valid bioinformatic tools, would allow biologically meaningful hypotheses to be interrogated (see Casu, Chapter 11 of this volume).

10.2.2 A Brief History of Sugarcane Metabolic Biology

Early research into sugarcane sucrose metabolism in the leaf and culm provided insights that were not only of agricultural significance but also of major importance in the general fields of plant biochemistry and physiology. During the 1950s and 1960s, complementary research programs at the Hawaiian Sugar Planters' Association Experiment Station in Aiea and the Colonial Sugar Refining Company in Brisbane elucidated the C_4 dicarboxylic acid (C_4; Hatch-Slack) pathway of photosynthesis (Kortschak et al. 1965; Hatch and Slack 1966). Identification of the metabolites involved in this carboxylation pathway (Hatch and Slack 1966) facilitated the later discovery and characterization of unique plant enzymes catalyzing these reactions, including pyruvate orthophosphate dikinase (EC 2.7.9.1) and NADP-malate dehydrogenase (EC 1.1.1.82) (Hatch and Slack 1969). During those early years, additional parallel investigations by the Australian and Hawaiian groups of sugarcane metabolite translocation and culm metabolism yielded an enduring conceptual framework of the mechanisms by which plants accumulate sugars to high concentration (e.g., Glasziou 1960, 1961; Hartt et al. 1963, 1964; Hatch and Glasziou 1963, 1964; Sacher et al. 1963; Glasziou and Gayler 1972).

Subsequent to the early seminal investigations of the Australian and Hawaiian research groups, evidence emerged which suggested that sugarcane is not source-limited in terms of yield and productivity and that sucrose storage potential may reside in processes within the culm (Irvine 1975). This observation, together with a widespread perception of sugarcane as being a good model system for the study of sucrose accumulation, provided considerable impetus for the detailed analysis of culm metabolism. Hence, by the 1990s, sucrose metabolism of sugarcane had been the most intensively studied amongst all crops (Hawker 1985; Moore 1995), yielding insights into the metabolite composition (e.g., Welbaum and Meinzer 1990) and enzyme complement (e.g., Goldner et al. 1991; Lingle and Smith 1991) of the storage parenchyma tissue, as well as the mechanisms for long-distance, intracellular and intercellular sugar transport (e.g., Komor et al. 1981; Lingle 1989; Hawker et al. 1991; Priesser and Komor 1991; Jacobsen et al. 1992).

Progress in recombinant DNA technology during the 1970s and 1980s and the development in the 1990s of an effective genetic transformation protocol for sugarcane (Gallo-Meagher and Irvine 1996) heralded a new era for sugarcane metabolic research. The means to improve sugarcane through transgenesis arrived at a time when the relatively small gains in sugarcane yield achieved by conventional breeding were becoming a cause for concern in a number of sugar industries (subsequently reviewed by Cock 2003 and Moore 2005). Although controversial, it was argued at the time that the ceiling to culm sucrose content, in particular, was partly due to the narrow gene pool used in sugarcane breeding programs (Roach 1989). Hence, much of sugarcane metabolic research began to focus on the identification of regulatory points in metabolism that could be targeted for transgenic manipulation towards increasing sucrose yield (Grof and Campbell 2001).

Research developments around the turn of the millennium, in both sugarcane and general plant molecular physiology, began to reveal the true complexity of metabolic regulation and the plasticity of metabolic responses to genetic perturbations (Botha et al. 2001; Halpin et al. 2001). In particular, the observation that variation in the expression of a gene does not always correlate to a change in the concentration of the corresponding protein or, indeed, the metabolic phenotype (Sumner et al. 2003) emphasized the need for a holistic approach to the identification of regulatory points in metabolism. It is likely that recent and ongoing advances in high-throughput technologies, when used in concert with modeling approaches, will enable the unraveling of sugarcane metabolism on a scale and to a level of resolution previously unattainable with standard technologies. Information on parallel fluctuations in gene expression and protein and metabolite levels in response to developmental and environmental cues will provide an invaluable basis for the design of strategies to not only enhance sucrose accumulation, but also to confer quantitative resistance to abiotic and biotic stress.

10.2.3 Perspectives on QTL Mapping of the Sugarcane Proteome/Metabolome

Most organisms have the ability to buffer both genetic and environmental variation to produce a stable, "optimum" phenotype through processes referred to as canalization, plasticity, genetic homeostasis or developmental stability (Debat and David 2001; Siegal and Bergman 2002; Hall et al. 2007). Metabolic networks consisting of enzymes and substrates/products are examples of canalized networks, in that the same amount (level) of a particular metabolite can be produced through different biochemical pathways, with differing levels of enzyme activity and differing amounts of intermediate metabolites along each section of the pathways involved. This poses a problem for genetic mapping, as it means there is not a one to one relationship between individual genes involved in the metabolic network and the phenotype of the end product metabolite. In order to definitively map quantitative trait loci (QTL) for sucrose content and other metabolites, an HDB dataset containing information on all metabolites and enzymes involved in the relevant pathways would be required. We are not yet in a position to do that in sugarcane.

QTL mapping studies in sugarcane have, however, identified markers ascribing 3% to 9% of the phenotypic variation in sucrose phenotype (Aitken et al. 2006) and 4% to 13% of the variation in sugar content (Ming et al. 2001; see also Pastina et al. in Chapter 7 of this volume). These results may seem to contradict the position stated above regarding the difficulties on mapping metabolites involved in canalized networks. The resolution of the apparent contradiction is likely to lie in the fundamental nature of networks. In the language of graph theory, metabolic networks have been shown to exhibit scale-free topology (Diaz-Mejia et al. 2007) in which numbers of connections between individual nodes (enzymes/metabolites) follow a power-law distribution. This results in networks where many nodes have few connections while a small number of nodes are highly connected and have a larger influence on the behavior of the network as a whole. Genetic variation at the loci coding for these "super-nodes" or their transcription factors may cause sufficient perturbation of the network to result in a measurable, heritable phenotype, hence allowing the identification of trait-associated QTL.

10.3 Tools and Applications

10.3.1 *Proteomics*

Since correlation between mRNA and protein levels has been shown to be remarkably low (Gygi et al. 1999; Sumner et al. 2003; Carpentier et al. 2008), proteins are important indicators of gene expression. Hence quantification of proteins and their post-translational derivatives is one of the prerequisites for the study of biological systems. Two-dimensional gel electrophoresis, mass spectrometry and bioinformatics are key components of proteomics technology (Beranova-Giorgiani 2003; Hart and Gaskell 2005). The technologies per se have been reviewed elsewhere (Dunn and Görg 2000; Aerbersold and Mann 2003; Lin et al. 2003). Two-dimensional gel electrophoresis (2-DGE) is sufficiently powerful that small changes in translational gene expression resulting from cues within the organism or external environmental shifts can be discriminated and scored. Several hundreds of proteins can be revealed on a single 2-D gel and differences in spot intensity can be measured using computer software packages. Because of this, the 2-DGE approach has been used on its own to measure the extent of variation between sugarcane genotypes (Ramagopal 1990). In the simpler genetic context of maize, more extensive characterization of proteome variations and mapping of the loci controlling them has been undertaken in a "molecular quantitative genetics" approach (de Vienne et al. 1999) and quantified proteins used as genetic descriptors to measure diversity among inbred lines (Burstin et al. 1994). In all of these studies, data are provided to demonstrate the reproducibility and significance of the proteomic variations measured.

Proteins separated by two-dimensional gel electrophoresis are usually highly purified and this facilitates identification by mass spectrometry (MS). The MS platform consists of a source to generate ions from the sample and an analyzer to separate and detect the ions according to mass. The various sources used to generate ions from the samples and the available range of analyzers for MS in protein analysis have been usefully reviewed in Dubey and Grover (2001) (Table 10-1).

The combined application of 2-D gel electrophoresis technology and MALDI-TOF-MS has led to the development of a proteome reference map for the diazotrophic bacterial associate of sugarcane, *Herbaspirillum seropedicae* (Chaves et al. 2007). Proteomes of prokaryotes, such as *Herbaspirillum*, make suitable targets for cataloging and characterization of most abundant protein species as they are relatively simple organisms and provide useful insights into endophytic function within the crop. In a similar fashion, plant organelle proteomes are of manageable size and have the potential to provide spatial resolution of protein expression within cells (Lilley and Dupree 2007). To date, comparative proteomic analyses of sugarcane itself have proven difficult due to the incompleteness of databases for identifying peptide fragments (G.D. Bonnett, pers. comm.).

Despite the rapid rate of expansion of the field of proteomics and the technological power of the tools that are available, it has been an increasing point of concern that the level of rigor required for proteomic data generation and analysis has been underestimated, resulting in published findings that are questionable and in need of validation (Wilkins et al. 2006). The key issues these 16 authors from 14 laboratories have identified are: experimental design, differential display leading to biomarker discovery, protein identification and analytical incompleteness. A set of guidelines for proteomics research is described (Wilkins et al. 2006). The fact that these now serve as criteria for editors of the journal *Proteomics* to use in reviewing manuscript submissions is a measure of the seriousness of the predicament and the critical need for stringent thinking in future proteomics endeavors. Similar problems in the omics in general have been identified and categorized as: bias, statistics, methodology and fitness of use (Lay et al. 2006).

Table 10-1 Systems for Mass Spectrometry in Proteomics (information from Dubey and Grover 2001).

Components and Systems	Acronym
Sources	
Matrix-assisted laser desorption-ionization	MALDI
Electrospray ionization	ESI
Analyzers	
Simple: Time of Flight	TOF
Complex: Fourier transform infra red	FTIR
Complex: Fourier transform ion cyclotron resonance	FTICR
Systems	
Mass measurement only	MALDI-TOF
Mass determination and amino acid sequencing	Tandem MS

10.3.2 Metabolomics

Metabolites are the products of regulatory processes in cells and their levels often reflect a biological response to a genetic or environmental perturbation (Fiehn 2002). In addition, it has been argued that since mRNA and proteins are mostly identified indirectly through sequence similarity, the metabolome may provide the most "functional" information of all the omics technologies (Sumner et al. 2003). Since plants are highly variable physiologically and metabolically, with many aspects of secondary metabolism still unexplored, they represent a major technical challenge in this field. It is estimated that up to 200,000 primary and secondary metabolites exist in the plant kingdom (Fiehn 2002; Oksman-Caldentey and Inzé 2004). At the same time, in any one species, the number of metabolites may be lower than the number of genes and proteins, one of several advantages of analyzing the metabolome rather than the genome or proteome (Dunn and Ellis 2005). Other advantages include the likelihood of changes in the metabolome being amplified compared to proteome or transcriptome, the high level of responsiveness of the metabolome to stress factors and other perturbations, and the considerably lower costs of metabolomic experiments (Dunn and Ellis 2005). Among the difficulties inherent in the approach are the dynamic behavior of metabolites in time and space and their fundamental chemistry (Stitt and Fernie 2003) as well as the in vivo concentration variation, which extends over 7–9 magnitudes, from pmole to mmole (Dunn and Ellis 2005). An excellent review of metabolic diversity in plants is included in the larger state-of-the-art analysis of metabolic phytochemistry by Fernie (2007).

The most robust and widely used methods for metabolome analysis combine chromatographic separation of extracts with detection and validation by mass spectrometry. Most common in the basic repertoire are gas chromatography mass spectrometry (GC-MS) and liquid chromatography mass spectrometry (LC-MS). The former is the principal technique for the analysis of metabolites that are volatile at temperatures up to 250°C or so (e.g., alcohols, esters and monoterpenes) as well as non-volatile polar metabolites such as amino acids, sugars and organic acids as long as they are converted to volatile and thermostable compounds by chemical derivatization. It is therefore a method of choice for analyzing primary metabolites. However, thermolabile compounds fail to be detected in GC-MS. The latter technique, LC-MS, adds versatility to the methodological range as it allows analysis of many of the families of secondary metabolites present in plants. Capillary electrophoresis-MS (CE-MS) is an alternative platform which offers speed, sensitivity and significant coverage of the whole metabolome in a single analysis, even in plants (Sato et al. 2004). The recently introduced Fourier transform mass spectrometric analyzers (Table 10-1) have the advantage of higher powers of resolution and allow more accurate estimations of molecular mass, thus reducing the need for refined chromatographic separation. In direct injection methods, extracts are delivered to a mass analyzer without any prior separation at all and a single mass spectrum is produced. This approach can overcome the bottlenecks experienced when rigorous identification and quantification of compounds is required, and lends itself to rapid comparative screening applications. A quite different overall approach to investigating the metabolome is the analysis of extracts by nuclear magnetic resonance (NMR) spectroscopy (Krishnan et al. 2005), which can identify abundant metabolites unambiguously but is not useful for analyzing a broad range of compounds due to the low level of resolution (Stitt and Fernie 2003; Hall 2006). MS technology has been reviewed by Aerbersold and Mann (2003) and the system components summarized in Table 10-1. Broad accounts of analytical methodologies for metabolite profiling and platforms for metabolomics are provided in various reviews (e.g., Glassbrook and Ryals 2001; Dunn and Ellis 2005; Hall 2006; Allwood et al. 2008). Glossaries of relevant technical terms are to be found in Nielsen and Oliver (2005) and Breitling et al. (2006). No one system of extraction, separation and detection can identify and quantify metabolites comprehensively (Fiehn and Weckwerth 2003). However, the throughput of metabolite profiling systems is high and the output information rich. A typical run on a conventionally designed GC-TOF-MS platform can reveal 500-1,000 peaks in approximately 30 minutes (Hall et al. 2002). However, these do not necessarily correspond to actual metabolites: when comparisons are made against library or authentic standards, the typical number of metabolites per run is in the 100–200 range.

In a useful overview and breakdown of technologies and applications, Fiehn and Weckwerth (2003) have split metabolic characterization into four classes: (1) metabolic fingerprinting for rapid genotype discrimination, (2) metabolic target analysis i.e. target-driven hypothesis testing of 1, 2 or a few single compounds, (3) quantitation of analytes in a set of related compounds or biochemical pathway and (4) analysis of all/many metabolites simultaneously, including unknowns. This last category represents true metabolomics, and requires ultra-high resolution techniques such as FT-ICR-MS or CE-MS, but fewer experiments have been carried out in this class compared to the others. In the plant kingdom as a whole there is a limited number of case studies of this kind (Breitling et al. 2006) and none exists for sugarcane at present. However, many authors interpret metabolomics broadly, as we have in this chapter, and in that context the diversity of applications is striking, contributing significantly to a greater understanding of biological mechanism and function in plants (Schauer and Fernie 2006).

10.3.3 Integration of Biochemical Pathways

The need to develop new ways of thinking about the complexity of living systems was recognized philosophically decades ago (Waddington 1977), long before the emergence of "systems biology" and approaches such as the omics which helped feed its development. Metabolism is the *sine qua non* of biochemistry. However, as inferred previously, straightforward hierarchical regulation of genes to produce linear reaction sequences along predefined pathways is an old paradigm allowing limited insight. There is widespread recognition that dynamic metabolic networks must be a new focus of attention (Fiehn and Weckwerth 2003; Weckworth 2003; Breitling et al. 2006) and that biology is about interactions rather than constituents (Schuster et al. 2006). The beauty of proteomics and metabolomics in biochemical investigation is the way they can help pioneer understanding of systems through the generation of comprehensive data sets based on accurate measurements of complex mixtures (Stitt and Fernie 2003; Kell 2004; Morgenthal et al. 2006). Metabolomics in particular has been viewed as the key to integrated systems biology as it is the most direct gauge of phenotype (Fiehn 2002; Edwards and Batley 2004; Fernie et al. 2004).

Reaching meaningful interpretation of biological processes from large sets of nontargeted protein or metabolite profile data requires the application of mathematical and statistical methods. For discriminating genotypic differences based on the variance of known individual compounds, principal component analysis (PCA) and other "black box" methods have been found to be useful (Fiehn and Weckwerth 2003; Morgenthal et al. 2006). However, more attention has been paid to the construction of biochemical reaction networks using mathematical concepts based on convex analysis. The 28-year history of this field of thinking is summarized by Papin et al. (2004). Two approaches have become prominent in recent years, "elementary flux modes" (or "basic reaction modes") (Schuster et al. 1999) and extreme pathways (Schilling et al. 2000). These have been clearly expounded and reviewed (Klamt and Stelling 2003; Papin et al. 2003, 2004). In a theoretical comparison of the two approaches, Klamt and Stelling (2003) conclude that unification of both approaches into one common framework is necessary and can be achieved. However, in a comparison illustrated using published metabolic reconstruction data from the human red blood cell and the human pathogen *Helicobacter pylori*, Papin and coworkers (2004) caution that careful interpretation is needed in describing network properties using these approaches if realistic outcomes are to be reached.

New information generated through metabolomics in a systems biology context still relies on comparison to background knowledge; this in turn is supported by detailed studies of reaction components and pathways conducted in the classical step-by-step manner. Various authors (e.g., Fiehn and Weckwerth 2003) have suggested that network analyses do not contest or supersede traditional biochemical methods for elucidating metabolic pathways; rather they are likely to act in a complementary way to achieve greater understanding of plant function. A seminal example and step forward in this regard is the investigation by Schwender and coworkers (2004) which combined classical substrate feeding experiments with elementary flux mode analysis to great effect, unraveling the role of Rubisco in previously undescribed metabolic interactions that increase the efficiency of carbon use in oil formation in developing seeds of *Brassica napus*. Another example of profitably coordinated approaches is work combining proteomics with traditional enzymatic biochemistry, flux analysis and cell biology (Giege et al. 2003), which showed that the glycolytic pathway is located on the outer membrane of mitochondria in *Arabidopsis*. The importance of such integrative studies to the future of functional genomics is discussed usefully and at length by Fernie et al. (2005) in the context of flux as an important aspect of the whole.

10.4 Insights into Regulation

10.4.1 Stress Responses

To understand stress biology, changes in the activities of individual enzymes or transporters in response to abiotic and biotic perturbations is of key importance. In theory, protein profiling presents a major opportunity to investigate these interactions in a comprehensive way. In reality, though, large-scale protein studies are limited (Peck 2005). Stress induced changes in protein quantity measured by 2-DGE have been few, mainly because the critical adjustments taking place are protein modifications, undetectable with core technology in broadly based analyses. This has led to more emphasis being put on investigating specific focus areas of the proteome using optimized systems. For example, in *Arabidopsis*, a high-throughput protein microarray-based proteomics technique was developed to study MAP kinase substrates (Feilner et al. 2005) and nano liquid chromatography used in an investigation of S-nitrosylated proteins (Lindermayr et al. 2005). The development of protein kinase chips for large-scale evaluation holds some promise (Williams and Cole 2001) as does shotgun proteomic analysis of complex protein samples (Wolters et al. 2001). The latter, if developed with capacity for high mass accuracy and unbiased statistical treatment of data, has an elevated capacity for detecting unique proteins (Weckwerth 2008).

In sugarcane, the 2-DGE proteomics approach has been used to characterize the heat shock response in cultured cells (Moisyadi and Harrington 1989) and to show that a low molecular mass auto-phosphorylating protein found among the heat shock products is a nucleoside diphosphate kinase (Moisyadi et al. 1994). In addition, in a preliminary comparison of stressed and unstressed sugarcane leaf material, it has allowed drought-induced proteomic expression patterns to be differentiated (Jangpromma et al. 2007).

Many categories of plant metabolites are involved in stress responses and their levels shift more markedly than do those of proteins. Not surprisingly, then, characterizing metabolic reactions to abiotic and biotic stress factors has been a more prominent focus of attention. Schauer and Fernie (2006) argue that vigorous research interest in this area is driven partly by the new technological tools that allow meaningful measurements to be made and partly by increasing crop losses resulting from stress. Climate change, expansion of agricultural activities into marginal lands and the resultant prospect of heightened stresses in the future add to these factors. The use of metabolomics in investigating stress responses of plants has been reviewed by Shulaev et al. (2008).

Cold is one of the most critical stresses experienced by plants (Last and Willmitzer 2001) and one that has been best characterized by metabolite profiling. The metabolomics of temperature stress was the subject of a recent review (Guy et al. 2008a). The most thorough studies in cold acclimation have been in *Arabidopsis* (e.g., Cook et al. 2004; Kaplan et al. 2004; Gray and Heath 2005; Goulas et al. 2006; Kaplan et al. 2007). However, studies of specific enzymatic and metabolic responses to cold stress have been reported for rice (Morsy et al. 2007) and sugarcane (Du et al. 1998, 1999). In a comparative study on the effects of chilling on photosynthesis in three species of sugarcane differing in origin and cold sensitivity, target enzymes and metabolites were measured using classical methods (Du et al. 1999). At 10°C, the tropical species *Saccharum officinarum* showed significantly decreased activities of pyruvate orthophosphate dikinase and NADP-malate dehydrogenase compared to the subtropical *S. sinense* and an intermediate hybrid (*S. officinarum* x *S. spontaneum* x *S. barberi*). In addition, there was substantial accumulation of aspartate and an increase in the level of alanine in the leaves of *S. officinarum* in response to chilling.

Water deficit is another key stress occurring during plant growth and development. In a review of metabolic networks regulating drought stress responses in plants (Seki et al. 2007), the importance of the phytohormone abscisic acid (ABA) as a primary stress trigger is highlighted, although it is pointed out that ABA-independent regulatory systems for dealing with drought stress also exist. ABA causes stomatal closure and induction of signaling cascades that lead to biochemical and physiological effects favorable to the plant in the absence of water, including the production and accumulation of metabolites with osmolytic function such as sugars, sugar alcohols, and amino acids. A specific drought-inducible gene, *SoDip22*, identified in the leaves of *Saccharum officinarum*, has been shown to encode a hydrophilic protein with a molecular mass of 15.9 kD (Sugiharto et al. 2002). The amino acid sequence is similar to proteins in the ABA-, stress- and ripening-inducible (Asr) family isolated from various other plant species, suggesting that the signaling pathway is mediated by ABA in this case. Expression of *SoDip22* appears to be localized in bundle sheath cells. In prior work on sugarcane, the suppression and interactions of leaf stomatal conductance, carbon exchange rate and photosynthetic enzyme activity in response to gradually developing water stress were measured and evaluated (Du et al. 1996). Until relatively recently, the compatible (non-toxic) osmolyte 3-dimethylsulphoniopropionate (DMSP), which occurs in significant amounts in marine algae, had been reported in only a few higher plants, for example the grass *Spartina* (Larher et al. 1977) and the Pacific strand plant *Wollastonia* in the family *Compositae* (Hanson et al. 1994). DMSP is thought to play a role in stabilizing the cytoplasm under saline or dry conditions. A GC-MS and NMR-based study of DMSP across 23 genera in the family Gramineae by the Hanson group showed that high levels also occur in mature leaves of sugarcane (Paquet et al. 1994). The *Saccharum* species *S. robustum*, *S. officinarum*, *S. sinese*, *S. edule*, *S. barberi* and commercial interspecific hybrids tested accumulated up to 6 µmol g^{-1} fresh weight while genera such as *Miscanthus* and *Erianthus* contained no more than 0.3 µmol g^{-1} fresh weight. Glycine betaine levels were ten-fold lower in these *Saccharum* species suggesting that DMSP may have replaced glycine betaine as an osmolyte in sugarcane. *S. spontaneum* showed variable DMSP levels according to genotype and location (the USA or Australia). The authors note this as surprising in view of *S. spontaneum* being a source of drought and salt stress tolerance in breeding programs (Paquet et al. 1994). However, results are for growth under favorable conditions and do not show the responsiveness of DMSP to stress in any of the sugarcane species tested, including *S. spontaneum*.

Plants, being sessile and nutrient-rich, suffer constant pressure and sometimes considerable damage from interacting organisms. These include parasites, symbionts, various categories of pathogen, and grazers such as herbivorous insects, mammals or members of other taxa. There is an overview of metabolomic approaches to all these kinds of interaction in the review by Allwood et al. (2008). There are several more specific discussions of the potential of proteomics in advancing understanding of plant-pathogen interactions (e.g., Xing et al. 2002; Thurston et al. 2005). Sugarcane research in the area of biotic stress is limited. In early "target analysis" work focusing on specific metabolites, Rutherford (1998) showed that chlorogenates and flavonoids in nodal budscales of cane, separated from crude extracts by HPLC, are differentially associated with resistance and susceptibility to the African stalk borer *Eldana saccharina*. An Australian study tested the hypothesis that the production of soluble and/or cell wall phenolic compounds in roots in response to whitegrub grazing damage is greater in sugarcane cultivars with more prominent antibiosis phenotypes (Nutt et al. 2004). Results did not support the hypothesis; they showed that while types and amounts of phenolic compounds changed significantly in a range of sugarcane clones tested under feeding stress, there are no characteristic profiles that can be used as markers of resistance. In other experiments driven by the need to investigate subsoil stress caused by pests like whitegrubs and pathogens such as the oomycete *Pachymetra chaunorhiza*, the transcriptional effect of methyl jasmonate (MJ) on root tissue was investigated (Bower et al. 2005). That work led to the identification of lipoxygenase and PR-10 protein homologs in the early stages of response to MJ (up to 24 hour) and other proteins in later stages. The lipoxygenase homolog was induced in untreated stem and leaf tissues following MJ application to the roots. Proteomic markers such as these provide lead subjects for further study of induced resistance pathways in sugarcane roots and cross-talk between root and leaf tissues.

The most intensive research to date has dwelt on unravelling the effects of specific stresses imposed in isolation under laboratory conditions. In reality, plants are subjected to a variety of stresses in combination and some studies have attempted to tackle the complexity of this situation (e.g., Rizhsky et al. 2004). Mittler (2006) reports a dual stress case study that draws on several sources of data based on drought and heat effects in grass species. Venn diagrams are used to show overlaps between transcripts and metabolites enhanced or suppressed during individual and combined stresses. A "stress matrix" is presented that demonstrate potential abiotic and biotic interactions that have important implications for agriculture. Overall, the review emphasizes the severity of stress combinations. The author suggests that these should be seen as new states of stress in plants and not just the sum of more than one stress effect.

10.4.2 Carbon Partitioning

Underlying the final sequestration of sucrose into the vacuoles of the culm storage parenchyma is a multitude of metabolic reactions and transport processes, located throughout the entire plant body. The aim of most sugarcane metabolic research has been to identify which of these reactions and processes ultimately regulate the level to which sucrose accumulates in the culm, with a view of applying the knowledge within sugarcane improvement strategies. Recent decades have seen some success in the use of reductionist strategies in the study of sugarcane metabolic biology; such approaches being adopted out of necessity due to the daunting complexity of the numerous networks involved in the regulation of sucrose accumulation. The majority of those studies relied on the use of classical biochemical and genetic tools to examine the response of sugarcane sucrose metabolism to developmental and environmental cues and to genetic perturbations induced through transgenesis. A snapshot of the complexity of metabolic regulation and the plasticity of metabolic responses has been revealed by those studies, reinforcing the need for global analysis of the metabolic networks. A trend towards greater access to the technologies for metabolome and proteome analysis and interpretation will permit the sugarcane research community to embrace the challenge of unraveling the complexity of sucrose metabolism regulatory networks and, in so doing, to build upon the information previously gained through classical approaches.

10.4.2.1 Prevalent Systems for Metabolic Analysis

In sugarcane metabolic research, considerable focus has been placed on the identification of the processes that underlie the dramatic increases in dry matter accumulation and sucrose concentration that occur during internode maturation (Moore 1995). Over the years, this transition in the culm from a growth to a storage function has been frequently used as a system for the discovery of regulatory points in sucrose accumulation (e.g., Whittaker and Botha 1997; Lingle 1999; Bindon and Botha 2002; Carson and Botha 2002; Bosch et al. 2003; Rae et al. 2005; Casu et al. 2007; Glassop et al. 2007; Lingle and Tew 2008). Studies conducted during this developmental shift have cataloged changes in carbon partitioning (Whittaker and Botha 1997; Bindon and Botha 2002), enzyme activity (Lingle 1997; Whittaker and Botha 1997; Lingle 1999; Terauchi et al. 2000; Grof et al. 2006; Lingle and Tew 2008), gene expression (Carson and Botha 2002; Casu et al. 2007), activity of sucrose transporters (Rae et al. 2005) and metabolite concentrations (Bosch et al. 2003; Glassop et al. 2007), either in single genotypes (Bindon and Botha 2002; Carson and Botha 2002; Rae et al. 2005; Casu et al. 2007; Glassop et al. 2007) or in a set of two or more genotypes that accumulate sucrose to different concentrations (Lingle 1997; Whittaker and Botha 1997; Lingle 1999; Terauchi et al. 2000; Bosch et al. 2003; Grof et al. 2006; Lingle and Tew 2008). A further interesting and potentially powerful approach has been the investigation of morphological, biochemical and gene expression correlates to sucrose accumulation amongst parents and progeny of families segregating for differences (Zhu et al. 1997; Casu et al 2005; Grof et al. 2007).

10.4.2.2 Analysis of Proteins and Metabolites: Current State-of-Play

To date, the analysis of sugarcane proteins has focused strongly on the determination of variations in the activity of specific enzymes during internode maturation and amongst genotypes; an approach that has provided an essential overview of enzymes that could potentially serve as control points of sucrose accumulation (Table 10-2). Many of these enzymes catalyze metabolic reactions central to the synthesis and breakdown of sucrose in the storage parenchyma, while others are involved in reactions that compete with sucrose metabolism for fixed carbon and were identified by tracing the fate of radioisotope-labeled metabolites (Whittaker and Botha 1997; Bindon and Botha 2002). The targeted functional analysis of these proteins by transgenesis has also proved invaluable in the further interrogation of their role in sucrose metabolism (e.g., Botha et al. 2001; Bekker 2007; Groenewald and Botha 2007a; Rossouw et al. 2007). While these approaches have yielded useful information regarding the involvement of specific enzyme proteins in the regulation of sucrose accumulation, much less is known about other proteins. The role of sugar transporter proteins remains obscure, having been examined primarily at the transcription level (Casu et al. 2003; McCormick et al. 2008a). However, recent work has investigated the in vitro activity of the sucrose transporter, ShSUT1 (Reinders et al. 2006), which is reported to play a role in sucrose export within metabolic sinks (Rae et al. 2005) (Table 10-2).

Pioneering functional analysis of sugarcane leaf proteins was instrumental in the characterization of the C_4 photosynthetic pathway (reviewed by Hatch 2002). Much of subsequent research has been targeted to the characterization of changes in leaf enzyme activity, particularly related to photosynthetic carbon fixation, in response to either environmental stimuli or climate change. Elevation of carbon dioxide to a level double that of ambient was shown to result in the up-regulation of the activity of the enzymes of photosynthesis and leaf sucrose metabolism, particularly during the early stage of leaf development (Vu et al. 2006). The associated improvement in leaf water-use efficiency reported in that study could ultimately lead to enhanced biomass and sucrose accumulation under elevated carbon dioxide. Of particular relevance to the functional analysis of sugarcane leaf proteins was the documentation of strong diurnal rhythms

Table 10-2 Proteins identified by classical metabolic analysis as playing a key role in sucrose accumulation in sugarcane. Proteins listed have been subjected to kinetic (enzymes) or in vitro (transporter) characterization (Characterization) and assessment of activity during development or in response to environmental cues (Localization).

Purpose of Analysis	Enzyme	Reference
Characterization	Fructokinase (EC 2.7.1.4)	Hoepfner and Botha (2004)
	Neutral invertase (EC 3.2.1.26)	Vorster and Botha (1998)
	Pyrophosphate:fructose 6-phosphate 1-phosphotransferase (PFP) (EC 2.7.1.90)	Groenewald and Botha (2007b)
	Sucrose synthase (EC 2.4.1.13)	Schäfer et al. (2004a, 2005)
	UPD-glucose dehydrogenase (EC 1.1.1.22)	Turner and Botha (2002)
	Sucrose transporter (ShSUT1)	Reinders et al. (2006)
Localization	Fructokinase	Hoepfner and Botha (2003)
	Neutral invertase	Vorster and Botha (1999); Rose and Botha (2000); Bosch et al. (2004); Lingle and Tew (2008)
	Soluble acid invertase (EC 3.2.1.26)	Zhu et al. (1997); Lingle (1999); Lingle and Tew (2008)
	Sucrose phosphate synthase (EC 2.4.1.14)	Zhu et al. (1997); Lingle (1999); Botha and Black (2000); Teruachi et al. (2000); Grof et al. (2007); Lingle and Tew (2008)
	Sucrose synthase	Buczynski et al. (1993); Lingle (1999); Botha and Black (2000); Schäfer et al. (2004b)
	Sucrose transporter (ShSUT1)	Rae et al. (2005)

in the activity of key enzymes of sucrose and starch synthesis (Du et al. 2000). In contrast, the activity of sucrose-phosphate synthase has been shown to correlate to leaf sucrose content diurnally, under varying plant nutrient regimes and across genotypes (Grof et al. 1998), suggesting that the activity of this protein could potentially serve as a biochemical marker for sucrose content.

Sugarcane metabolite analysis has been largely restricted to the determination of the levels of the most abundant sugars in internodes at different stages of development, across genotypes or in response to environmental factors (e.g., Zhu et al. 1997; Lingle 1999; Hoepfner and Botha 2003; Bonnett et al. 2006). In most instances, such metabolite analyses were conducted in parallel to the determination of the activity of allied enzymes of central sucrose metabolism (Table 10-2). The first definitive investigation of small suite of metabolites other than sucrose, glucose and fructose aimed at establishing the temporal relationship between sucrose accumulation and carbon partitioning during internode maturation (Whittaker and Botha

1997). Although the scope for metabolite analysis was limited at the time by available technology, the profound influence of that study on the understanding of sugarcane sucrose metabolism gave the first inkling of the potential power of global metabolite analysis, especially when conducted in concert with enzyme activity and metabolite flux analyses.

Reports on the application of GC-MS or allied mass spectrometric-based technologies to the analysis of sugarcane metabolite composition are currently quite limited. As was the case for studies using classical tools for biochemical analysis, these more recent studies have also focused on identifying changes to the primary metabolite profile of internodes during the internodal growth-to-storage transition (Bosch et al. 2003; Bosch 2005; van der Merwe 2005; Glassop et al. 2007). In addition, the resolution and discernment of the technology were examined through the comparison of internode metabolite profiles between genotypes (Bosch et al. 2003; Bosch 2005) and amongst transgenic and wild type sugarcane lines (van der Merwe 2005). Many of the data obtained by those GC-MS analyses (Bosch et al. 2003; Glassop et al. 2007) were confirmatory of earlier reports (Welbaum and Meizner 1990; Whittaker and Botha 1997; Asis et al. 2003), for example the higher levels of amino acids and organic acids in immature as compared to mature internodes. However, such metabolite profiling strategies are already beginning to illuminate the complexity of the metabolic context in which sucrose accumulation occurs within the sugarcane culm. For example, variations in concentration of the intermediates of trehalose metabolism during sucrose accumulation are of particular interest due to the proposed role of trehalose or trehalose-phosphate as a central metabolic regulator (Paul et al. 2008). Interesting similarities and dissimilarities in the relationship between sucrose and trehalose concentrations during internode development in different sugarcane genotypes have emerged (Bosch 2005; Glassop et al. 2007). Further extended metabolite profiling strategies within the context of the recently proposed involvement of trehalose metabolism in photosynthetic modulation (McCormick et al. 2008b) may provide an important insight into at least one of the mechanisms governing sucrose accumulation in sugarcane.

10.4.2.3 *Essential Complementary Power of Modeling*

Using currently available in vitro kinetic information on the enzymes and transporters of sucrose metabolism from sugarcane (Table 10-2) and other plants, kinetic modeling has been used to simulate the metabolic behavior of storage parenchyma cells during sugarcane internode maturation (Rohwer and Botha 2001; Uys et al. 2007). Although subject to limitations (reviewed by Poolman et al. 2004), this approach revealed that the inward transport of fructose and sucrose across the plasmalemma and tonoplast, respectively, exerted negative control over the futile cycling of sucrose within the cytosol, while having positive control on sucrose accumulation (Uys et al. 2007). Given the proposed central role of neutral invertase in mediating the futile cycling of sucrose (Vorster and Botha 1999), it is not surprising that the control profile of this enzyme was the reverse. The expression of a heterologous vacuolar sucrose transporter gene and down-regulation of neutral invertase (Rossouw et al. 2007), either alone or in concert, represents a potentially viable strategy to enhance sucrose accumulation. Hence, metabolic modeling of this nature represents a valuable tool for the interpretation of biochemical data towards the identification of suitable targets for the genetic engineering of sugarcane.

While the relatively small data sets derived by means of classical biochemical tools may be interrogated adequately by kinetic modeling or similar approaches, the vast data that are being generated during the increasingly widespread application of the multiple omic technologies to sugarcane biology will require the development of substantial capacity in bioinformatics and integrated modeling. However, once enabled, such HDB, coupled to conventional and high-throughput reductionist approaches, will have enormous potential to illuminate sucrose accumulation in sugarcane (Moore 2005).

10.4.3 Engineering Specific Traits

10.4.3.1 Enabling Knowledge and Technologies

Metabolic engineering aims to alter the metabolic pathways of an organism to clarify their function or to permit their redesign for the production of desirable metabolites (Rios-Estepa and Lange 2007). In contrast to conventional approaches, including breeding and selection, metabolic engineering is defined as the directed modification of metabolism by means of recombinant DNA technology and considering metabolism as a cellular network (Schwender 2008). Hence, any rational approach to the engineering of a biosynthetic pathway in a particular plant species has three primary requirements, viz. knowledge of the synthesis and accumulation patterns of the metabolites of interest, availability of DNA sequences encoding appropriate enzyme isoforms or regulatory factors and effective genetic transformation technologies that permit high-level, targeted transgene expression (Davies 2007). With such enabling knowledge and technologies in place, predictive metabolic engineering strategies may be devised, in which data derived from the application of omic technologies are used to identify key targets, such as flux control points or regulatory proteins (Dixon 2005). However, the detailed HDB data that are essential for targeted metabolic engineering are available in only very few instances. For the vast majority of pathways and species, there is a dearth of information regarding participating genes, key flux points and regulatory factors, as well as on the important influence which cellular and subcellular compartmentalization and metabolite channeling have on pathway functionality. Hence, to date, metabolic engineering of important crop plants has largely relied on a reiterative strategy as a means to identify the appropriate combination of transgenes required for effective pathway modification.

10.4.3.2 Basis of Metabolic Engineering

The capacity of a plant to biosynthesize a valuable metabolite via a specific pathway may be increased either by modifying flux within the endogenous pathway or by conferring novel biosynthetic capacity through the expression of heterologous genes. Various strategies may be used to enhance flux within a native pathway, including increasing the activity of a rate-limiting biosynthetic enzyme, inhibiting the activity of an enzyme that competes with the pathway of interest for substrate or by altering the availability of crucial regulatory factors. The use of heterologous genes is also popular in this regard, as the proteins delivered by such genes may not be subject to the same regulatory control that often confounds attempts to increase enzyme activity by the overexpression of endogenous genes (Trethewey 2004). For reducing production of an undesirable product, down-regulation of the activity of one of the biosynthetic genes has proven to be an effective approach, with RNA interference (RNAi) being the current method-of-choice, due to the better performance of the technology over that achieved with traditional antisense or sense-inhibition constructs. However, regardless of the approach or technology used, the likelihood that a metabolic engineering strategy will meet with success is greatly enhanced if the target species and metabolic pathway have been well characterized.

10.4.3.3 Metabolomics: An Essential Tool

The physiological changes to a plant resulting from transgenesis or fluctuating environmental and developmental cues may be reflected in and, hence, studied at the level of the transcriptome, proteome and metabolome. Although studies at all three levels provide both unique and complementary information, the metabolome is increasingly viewed as an attractive target for investigation of phenotype for a number of reasons. Perturbations wrought by the down-regulation of an endogenous gene, for example, are amplified through the hierarchy of the transcriptome and proteome and hence, are more readily detectable in the metabolome (Kell et al. 2005), even in the face of negligible changes to metabolic fluxes (Raamsdonk et al. 2001). In addition, the frequent lack of absolute correlation amongst changes in the expression of a particular gene and observed levels of the encoded protein, together with the fine regulatory control often exerted over enzyme activity, suggest that the metabolome may be more sensitive to perturbations than either the transcriptome or proteome (Urbanczyk-Wochniak et al. 2003). Hence, the value of metabolomics in the study of metabolic networks is particularly acute for plant species with complex genomes and for which sequence information is still restricted, as is the case for sugarcane.

Regulatory interactions occur within and across all hierarchical levels of biological organization, spanning gene to phenotype, and which may be modified by both internal and external signals. Hence, the perception that metabolic pathways may be effectively studied and engineered in isolation is fundamentally flawed (Trethewey 2004) and reports of both overt and subtle collateral effects of specific genetic manipulations are increasingly common (e.g., Bohmert et al. 2000; Roessner et al. 2001; Regierer et al. 2002; Sweetlove et al. 2003; Geigenberger et al. 2005; Larkin and Harrigan 2007; Coleman et al. 2008; Aluru et al. 2008). Metabolite profiling, for the reasons outlined above, is an appropriate tool to unmask such consequences, thereby providing deeper insights into plant metabolic networks. Such improved insights into the metabolic context in which a particular pathway operates may then be used in the design of a more rational engineering strategy.

A reiterative approach to metabolic engineering that is guided by sequential metabolic profiling has been used to good effect in sugarcane, albeit to a limited extent. In an attempt to reduce carbon allocation to a sink that is in competition with sucrose accumulation in the culm, the activity of UDP-glucose dehydrogenase (UGD) has been down-regulated by means of antisense technology (Bekker 2007) (Table 10-2). This enzyme catalyzes the rate limiting step in the biosynthesis of precursors of both hemicelluloses and pectin. As had been hypothesized, the manipulation resulted in an increase in sucrose accumulation, accompanied by increased sucrose phosphate synthase and UDP-glucose pyrophosphorylase (EC 2.7.7.9) activities, which was highly correlated with the decreased UGD activity. Unexpectedly, however, down-regulation of UGD resulted in an increased pentose:hexose ratio and higher uronic acid and cellulose concentrations. Those results were indicative of a collateral homeostatic effect of UGD down-regulation, in which a cascade of metabolic responses was triggered towards the maintenance of hexose-phosphate equilibrium. The comprehensive metabolite profiling that is reportedly in progress (Bekker 2007) will provide greater clarity of this metabolic network, which will potentially enhance future strategies aimed at engineering sugarcane sucrose accumulation and cell wall composition. Similarly, the expression of two bacterial genes in sugarcane to synthesize p-hydroxybenzoic acid resulted in a diversion of carbon away from the endogenous phenylpropanoid pathway (McQualter et al. 2005). Unanticipated effects of this diversion were a reduction in leaf chlorogenic acid content and an apparent compensatory up-regulation of phenylalanine ammonia-lyase (EC 4.3.1.5), once again highlighting the analytical power offered by a combination of transgenesis and metabolite profiling in the elucidation of metabolic networks.

Demonstration of substantive equivalence of metabolically-engineered plants to the wild type is an essential part of risk assessment that is mandatory prior to commercialization. However, knowledge of the effects of engineering novel pathways on native metabolism is generally gained through the analysis of a small set of metabolites allied to the target pathway; an approach which is inherently biased. Comparative global metabolite profiling, on the other hand, offers an effective and unbiased means to demonstrate substantive equivalence, as was illustrated in the elegant study of Kristensen and colleagues (2005). Those workers used targeted and non-targeted metabolic engineering strategies to examine the effect on the metabolome of the insertion of the high-flux *Sorghum bicolor* dhurrin synthesis pathway into *Arabidopsis thaliana*. The targeted and non-targeted comparison was achieved through the insertion of the entire pathway, as well as varying combinations of the individual biosynthetic components. Lines containing the entire pathway accumulated dhurrin to high levels without significant alteration of either the transcriptome or metabolome, while the incomplete pathways induced metabolic cross-talk, resulting in significant changes.

The appeal of sugarcane as a target crop for metabolic engineering has increased markedly over the past few years, particularly as a potential feedstock for the production of renewable biomaterials and biofuels. The progress made and requirements in this regard have been addressed recently in a comprehensive review by Robert Birch of the University of Queensland (Birch 2007). The next few years will see progress in the genetic engineering of sugarcane and the commercialization of the first transgenic sugarcane genotype is likely to be imminent. In this drive to commercialize, the demonstration of substantive equivalence by metabolomic, and possibly proteomic, approaches may play a vital role in gaining regulatory approval and consumer acceptance of such genetically modified (GM) sugarcane (see Casu et al. in Chapter 9 of this volume). The potential for demonstration that differences in the metabolome amongst sugarcane varieties bred by conventional means are of comparable magnitude to those between a GM line and the wild type, as has been demonstrated for *A. thaliana* ecotypes, *Solanum tuberosum* cultivars and their respective GM lines (Catchpole et al. 2005; Kristensen et al. 2005; Møller and Bak 2005), may sit well with regulatory authorities and the public alike.

10.4.3.4 Synergy between Metabolomics and Transcriptomics

Although metabolome analysis is crucial to rational metabolic engineering, the integration of transcriptome and metabolome data into a single data set is even more powerful, as it can lead to the identification of unknown genes and related regulatory networks within specific metabolic pathways. This approach, however, is restricted to plant species for which a genome sequence is available. In instances where such comprehensive information is available, transcriptome coexpression analysis is a viable approach, whereby publicly available, condition-independent data are compared to condition-specific data generated with a particular metabolic pathway in mind. The application of transcriptome coexpression analysis in conjunction with metabolite profiling in *A. thaliana* has led to the identification of many previously unknown structural genes that encode enzymes of important metabolic pathways (Persson et al. 2005; Wei et al. 2006; Hirai et al. 2007). Such analyses provide a rich, detailed picture of metabolic pathways and introduce a wealth of candidates for genetic and biochemical analyses. As genome sequences and comprehensive transcript expression data for important crop plants continue to enter into the public domain, it is likely that the parallel application of transcriptome coexpression analysis and metabolite profiling will lead to novel strategies to improve yield and genetically tailor the crop to produce high-value alternative products.

10.4.3.5 Predictive Metabolic Engineering: The Case for Non-Model Plant Species

Metabolic control analysis (MCA) is a modeling tool that provides a useful framework for the analysis of complex biochemical reactions, particularly in terms of the contribution of individual enzymes and metabolites to flux control in a steady state (Larkin and Harrigan 2007). Insights gained through MCA into the control structure of pathways have revealed that significant changes in flux are sometimes associated with only modest adjustments to metabolite concentrations (Fell 2005). Hence, metabolite profiling alone may be insufficient to gain a full understanding of the metabolic phenotype, as is prerequisite for predictive metabolic engineering. Flux measurements may provide a useful complementary approach to metabolite profiling in the system-wide characterization of metabolic networks (Ratcliffe and Shachar-Hill 2006; Schwender 2008). For non-model plant species, such as sugarcane, it is unlikely that truly predictive metabolic engineering will be realized in the short- to medium-term. However, a preliminary foundation for a more rational approach to the engineering of sucrose accumulation in sugarcane has already been laid by researchers in South Africa (Rohwer and Botha 2001; Uys et al. 2007). Refinement and elaboration of such modeling strategies, together with the implementation of emerging technologies for flux analysis (Ratcliffe and Shachar-Hill 2006; Schwender 2008), offer the means to identify points of control of metabolism in the sugarcane culm and, hence, the design of rational metabolic engineering events, whether for delivery of a novel product or enhanced sucrose or biomass yield.

10.5 Outlook

The dramatic acceleration of proteomic and metabolomic research over the past few years has been noted by a range of reviewers (Allwood et al. 2008; Guy et al. 2008b; Weckwerth 2008). The exponential nature of progress in the field can be illustrated histogramatically by the number of published journal articles (determined by using relevant Boolean search strings in PubMed) over time (Guy et al. 2008b) or by a simple timeline listing of significant papers (Allwood et al. 2008). Those authors all make the point that exponential increase in omics research outputs is certain to continue. The rising trend towards trans-laboratory efforts in these fields is also likely to be maintained.

Future investigations into protein and metabolite dynamics seem likely to become increasingly purposeful, interrogating specific regulatory processes in depth. Hennig (2007) cites the source-sink relationship as one example of an area that needs close attention if leaf development is to be properly understood. In sugarcane this applies to understanding development in the culm as well as the leaf and further work needs to be done (McCormick 2007, 2008b). A more focused approach to plant development would necessarily involve more refined analyses. While quantification of metabolites is already quite precise, it will be imperative to continue describing in chemical terms the large number of unique metabolic entities that are structural unknowns at present (Stitt and Fernie 2003). To advance understanding of metabolism, these large repertoires of compounds will have to be analyzed at even higher resolution (Fernie 2007). In addition, better identification and quantitation of proteins in complex samples will be essential (Weckwerth 2008), especially for the characterization of signaling pathway components in which proteolytic degradation of negative regulators (Huq 2006) and post-translational modifications such as protein phosphorylation and histone acetylation play important roles. For such purposes, expansion of protein databases will be critical. As studies become more targeted to tissues and cells with specific functions, spatial information about a wider range of proteins and metabolites will become increasingly important.

In the context of more focused research, it makes sense that the number of model species will expand, as predicted by Hennig (2007). New model plants will be required as platforms for the study of developmental processes that do not occur in *Arabidopsis*, for example wood formation, bud dormancy and nodulation. Meanwhile, work on monocotyledonous plants will continue to extend and will complement progress in *Arabidopsis* (Hennig 2007). The most useful models for sugarcane are likely to include those that provide variations in biochemical pathways towards cellulose and other fibrous polymers as well as those representing sucrose metabolism. They might well be parent species rather than a commercial hybrid.

In a more holistic mode, Stitt and Fernie (2003) look forward to seeing high-throughput metabolomic technologies being applied in combination with large sampling strategies to study ecological problems. They argue that this could avoid the limitations of preconception-based, hypothesis-driven research. Such an approach follows, instead, the "discovery science" method of research, which enumerates and describes elements of a system regardless of any notion of how the system functions (Aebersold et al. 2000). Metagenomics is already pioneering aspects of microbial ecology by characterizing the composition of microbial communities and their relationships with the habitats they occupy without the need to isolate and culture community members or understand how they operate. Turnbaugh and Gordon (2008) have suggested that combining metagenomics with metabolomics will shed light on how microbial communities actually function in a specific environment. These approaches could play a role in crop science by effecting better understanding of soil biota and ecological interactions at the soil-root interface.

Commercial applications of metabolite profiling in particular will become more prominent, for example in defining key qualities of crops such as flavor and color, specific properties such as the presence of health-promoting compounds and storage factors such as levels of contamination and spoilage, all of which have implications for management and marketing (Hall 2006; Guy et al. 2008b). However, phenotypes like these cannot usually be explained by single biomarkers, underscoring the importance of further developments in tools for pattern recognition and the analysis of multifactorial states (Weckwerth 2008).

Molecular profiling technologies are now generating large quantities of experimental data, well in advance of systems for their interpretation. Many authors indicate the need for mathematical advances that can meet the challenges of data integration, network construction and modeling. There is a serious lag in the transition "from information to knowledge" (Stitt and Fernie 2003; Hall 2006). Continued collaboration between specialists is a vital route to follow, but there is clearly a need to accelerate opportunities for cross-training in biology, mathematics, statistics and bioinformatics early in academic development.

Acknowledgements

The authors thank Dr. Marna van der Merwe of the Institute of Plant Biotechnology (Stellenbosch University, South Africa) for her critical appraisal of the draft manuscript.

References

Aerbersold R, Mann M (2003) Mass spectrometry-based proteomics. Nature 422: 198–207.

Aerbersold R, Hood LE, Watts JD (2000) Equipping scientists for the new biology. Nat Biotechnol 18: 359.

Aitken K, Jackson P, McIntyre CL (2006) Quantitative trait loci identified for sugar related traits in a sugarcane (*Saccharum* spp.) cultivar x *Saccharum officinarum* population. Theor Appl Genet 112: 1306–1317.

Allwood JW, Ellis DI, Goodacre R (2008) Metabolomic technologies and their application to the study of plants and plant-host interactions. Physiol Planta 132: 117–135.

Aluru M, Xu Y, Guo R, Wang Z, Li S, White W, Wang K, Rodermel S (2008). Generation of transgenic maize with enhanced provitamin A content. J Exp Bot (in press): doi: 10.1093/jxb/ern212.

Asis CA, Shimizu T, Khan MK, Akao S (2003) Organic acid and sugar content in sugarcane stem apoplast solution and their role as carbon source for endophytic diazotrophs. Soil Sci Plant Nutr 49: 915–920.

Baek D, Villen J, Shin C, Camargo FD, Gygi SP, Bartel DP (2008) The impact of microRNAs on protein output. Nature 455: 64–71.

Bekker J (2007) Genetic manipulation of the cell wall composition of sugarcane. PhD Thesis, Univ of Stellenbosch, Stellenbosch, South Africa.

Beranova-Giorgianni S (2003) Proteome analysis by two-dimensional gel electrophoresis and mass spectrometry: strengths and limitations. Trend Anal Chem 22: 273–281.

Bindon KA, Botha FC (2002) Carbon allocation to the insoluble fraction, respiration and triose-phosphate cycling in the sugarcane culm. Physiol Planta 116: 12–19.

Birch RG (2007) Metabolic engineering in sugarcane: Assisting in the transition to a bio-based economy. In: R Verpoorte , AW Alfermann, TSJohnson (eds) Applications of Plant Metabolic Engineering. Springer, Dordrecht, The Netherlands, pp 249–281.

Bohmert K, Balbo I, Kopka J, Mittendorf V, Nawrath C, Poirier Y, Tischendorf G, Trethewey RN, Willmitzer L (2000) Transgenic *Arabidopsis* plants can accumulate polyhydroxybutyrate up to 4% of their fresh weight. Planta 211: 841–845.

Bonnett GD, Hewitt ML, Glassop D (2006) Effects of high temperature on the growth and composition of sugarcane internodes. Aust J Agri Res 57: 1087–1095.

Bosch S (2005) Trehalose and carbon partitioning in sugarcane. PhD Thesis, Univ of Stellenbosch, Stellenbosch, South Africa.

Bosch S, Rohwer JM, Botha FC (2003) The sugarcane metabolome. Proc S Afr Sugar Technol Assoc 77: 129–133.

Bosch S, Grof CPL, Botha FC (2004) Expression of neutral invertase in sugarcane. Plant Sci 166: 1125–1133.

Botha FC, Black KG (2000) Sucrose phosphate synthase and sucrose synthase activity during maturation of internodal tissue in sugarcane. Aust J Plant Physiol 27: 81–85.

Botha FC, Sawyer BJB, Birch RG (2001) Sucrose metabolism in the culm of transgenic sugarcane with reduced soluble acid invertase activity. Proc Int Soc Sugar Cane Technol 24: 588–591.

Bower NI, Casu RE, Maclean DJ, Reverter A, Chapman SC, Manners JM (2005) Transcriptional response of sugarcane roots to methyl jasmonate. Plant Sci 168: 761–772.

Breitling R, Pitt AR, Barrett MP (2006) Precision mapping of the metabolome. Trends Biotechnol 24: 543–548.

Buczynski SR, Thom M, Chourey P, Maretzki A (1993) Tissue distribution and characterisation of sucrose synthase isozymes in sugarcane. J Plant Physiol 142: 641–646.

Burstin J, de Vienne D, Dubreuil P, Damerval C (1994) Molecular markers and protein quantities as genetic descriptors in maize. 1. Diversity among 21 inbred lines. Theor Appl Genet 89: 943–950.

Carpentier SC, Coemans B, Podevin N, Laukens K, Witters E, Matsumura H, Terauchi R, Swennen R, Panis B (2008) Functional genomics in a non-model crop: transcriptomics or proteomics? Physiol Planta 133: 117–130.

Carson DL, Botha FC (2002) Sugarcane ESTs differentially expressed in immature and maturing internodal tissue. Plant Sci 162: 289–300.

Casu RE, Grof CPL, Rae AL, McIntyre CL, Dimmock CM, Manners JM (2003) Identification of a novel transporter homologue strongly expressed in maturing stem vascular tissues of sugarcane by expressed sequence tag and microarray analysis. Plant Mol Biol 52: 371–386.

Casu RE, Manners JM, Bonnett GD, Jackson PA, McIntyre CL, Dunne R, Chapman SC, Rae AL, Grof CPL (2005) Genomics approaches for the identification of genes determining important traits in sugarcane. Field Crops Res 92: 137–147.

Casu RE, Jarmey J, Bonnett GD, Manners JM (2007) Identification of transcripts associated with cell wall metabolism and development in the stem of sugarcane by Affymetrix GeneChip Sugarcane Genome Array expression profiling. Funct Integr Genom 7: 153–167.

Catchpole GS, Beckmann M, Enot DP, Mondhe M, Zywicki B, Taylor J, Hardy N, Smith A, King RD, Kell DB, Fiehn O, Draper J (2005) Hierarchical metabolomics demonstrates substantial compositional similarity between genetically modified and conventional potato crops. Proc Natl Acad Sci USA, 102: 14458–14462.

Chaves DFS, Ferrer PP, de Souza EM, Gruz LM, Monteiro RA, Pedrosa F de O (2007) A two-dimensional protein reference map of *Herbaspirillum seropedicae* proteins. Proteomics 7: 3759–3763.

Cock JH (2003) Sugarcane growth and development. Int Sugar J 105: 540–552.

Coleman HD, Park J-Y, Nair R, Chapple C, Mansfield SD (2008) RNAi-mediated suppression of r-coumaroyl-CoA 3'-hydroxylase in hybrid poplar impacts lignin deposition and soluble secondary metabolism. Proc Natl Acad Sci USA, 105: 4501–4506.

Cook D, Fowler S, Fiehn O, and Thomashow MF (2004) A prominent role for the CBF cold response pathway in configuring the low-temperature metabolome of Arabidopsis. Proc Natl Acad Sci USA, 101: 15243–15248.

Davies KM (2007) Genetic modification of plant metabolism for human health benefits. Mutat Res 622: 122–137.

Debat V, David P (2001) Mapping phenotypes: canalization, plasticity and developmental stability. Trends Ecol Evol 16: 555–561.

de Vienne D, Leonardi A, Damerval C, Zivy M (1999) Genetics of proteome variation for QTL characterization: application to drought-stress responses in maize. J Exp Bot 50: 303–309.

Diaz-Meijia JJ, Perez-Rueda E, Segovia L (2007) A network perspective of metabolism by gene duplication. Genome Biol 8:R26: doi 10.1186/gb-2007-8-2-r26.

Dixon RA (2005) Engineering of plant natural product pathways. Curr Opin Plant Biol 8: 329–336.

Du Y-C, Kawamitsu Y, Nose A, Hiyane S, Murayama S, Wasano K, Uchida Y (1996) Effects of water stress on carbon exchange rate and activities of photosynthetic enzymes in leaves of sugarcane (*Saccharum* sp.). Aust J Plant Physiol 23: 719–726.

Du Y-C, Nose A, Wasano K, Uchida Y (1998) Responses to water stress of enzyme activities and metabolite levels in relation to sucrose and starch synthesis, the Calvin cycle and the C_4 pathway in sugarcane (*Saccharum* sp.) leaves. Aust J Plant Physiol 25: 253–260.

Du Y-C, Nose A, Wasano K (1999) Effects of chilling temperatures on photosynthetic rates, photosynthetic enzyme activities and metabolite levels in leaves of three sugarcane species. Plant Cell Environ 22: 317–324.

Du Y-C, Nose A, Kondo A, Wisano K (2000) Diurnal changes in photosynthesis in sugarcane leaves. II. Enzyme activities and metabolite levels relating to sucrose and starch metabolism. Plant Prod Sci 3: 9–16.

Dubey H, Grover A (2001) Current initiatives in proteomics research: The plant perspective. Curr Sci (India) 80: 262–269.

Dunn MJ, Görg A (2000) Two-dimensional polyacrylamide gel electrophoresis for protein analysis. In: SR Pennington , MJ Dunn (eds) Proteomics: From Protein Sequence to Function. Garland Science Publ, Taylor & Francis Group, London, UK, pp 43–64.

Dunn WB, Ellis DI (2005) Metabolomics: Current analytical platforms and methodologies. Trends Anal Chem. 24: 285–294.

Edwards D, Batley J (2004) Plant bioinformatics: from genome to phenome. Trends Biotechnol 22: 232–237.

Feilner T, Hultschig C, Lee J, Meyer S, Immink RGH, Koenig A, Possling A, Seitz H, Beveridge A, Scheel D, Cahill DJ, Lehrach H, Kreutzberger J, Kersten B (2005) High-throughput identification of potential *Arabisopsis* mitogen-activated protein kinase substrates. Mol Cell Proteom 4: 1558–1568.

Fell DA (2005) Enzymes, metabolites and fluxes. J Exp Bot 56: 267–272.

Fernie AD, Geigenberger P, Stitt M (2005) Flux an important, but neglected, component of functional genomics. Curr Opin Plant Biol 8: 174–182.

Fernie AR (2007) The future of metabolic phytochemistry: Larger numbers of metabolites, higher resolution, greater understanding. Phytochemistry 68: 2861–2880.

Fernie AR, Trethewey RN, Krotsky AJ, Willmitzer L (2004) Metabolite profiling: from diagnostics to systems biology. Nat Rev Mol Cell Biol 5: 1–7.

Fiehn O (2002) Metabolomics: the link between genotypes and phenotypes. Plant Mol Biol 48: 155–171.

Fiehn O, Weckwerth W (2003) Deciphering metabolic networks. Eur J Biochem 270: 579–588.

Gadbury GL, Page GP, Edwards J, Kayo T, Prolla TA, Weindruch R, Permana PA, Mountz JD, Allison DD (2004) Power and sample size estimation in high dimensional biology. Stat Meth Med Res 13: 325–338.

Gallo-Meagher M, Irvine JE (1996) Herbicide resistant transgenic sugarcane plants containing the *bar* gene. Crop Sci 36: 1367–1374.

Geigenberger P, Regierer B, Nunes-Nesi A, Leisse A, Urbanczyk-Wochniak E, Springer F, van Dongen JT, Kossmann J, Fernie AR (2005) Inhibition of *de novo* pyrimidine synthesis in growing potato tubers leads to a compensatory stimulation of the pyrimidine salvage pathway and a subsequent increase in biosynthetic performance. Plant Cell 17: 2077–2088.

Giege P, Heazlewood JL, Roessner-Tunali U, Millar AH, Fernie AR, Leaver CJ, Sweetlove LJ (2003) Enzymes of glycolysis are functionally associated with the mitochondrion in *Arabidopsis* cells. Plant Cell 15: 2140–2151.

Glassbrook N, Ryals J (2001) A systematic approach to biochemical profiling. Curr Opin Plant Biol 4: 186–190.

Glassop D, Roessner U, Bacic A, Bonnett GD (2007) Changes in the sugarcane metabolome with stem development. Are they related to sucrose accumulation? Plant Cell Physiol 48: 573–584.

Glasziou KT (1960) Accumulation and transformation of sugars in sugar cane stalks. Plant Physiol 35: 895–901.

Glasziou KT (1961) Accumulation and transformation of sugars in sugar cane stalks. Origin of glucose and fructose in the inner space. Plant Physiol 36: 175–179.

Glazsiou KT, Gayler KR (1972) Sugar accumulation in sugarcane. Role of cell walls in sucrose transport. Plant Physiol 49: 912–913.

Goldner W, Thom M, Maretzki A (1991) Sucrose metabolism in sugarcane cell suspension cultures. Plant Sci 73: 143–147.

Goulas E, Schubert M, Kieselbach T, Kleczkowski LA, Gardeström P, Schröder W, Hurry V (2006) The chloroplast lumen and stromal proteomes of *Arabidopsis thaliana* show differential sensitivity to short- and long-term exposure to low temperature. Plant J 47: 720–734.

Gray GR, Heath D (2005) A global reorganization of the metabolome in *Arabidopsis* during cold acclimation is revealed by metabolic fingerprinting. Physiol Planta 124: 236–248.

Groenewald J-H, Botha FC (2007a). Down-regulation of pyrophosphate: fructose-6-phosphate 1-phosphotransferase (PFP) activity in sugarcane enhances sucrose accumulation in immature internodes. Transgen Res 17: 85–92.

Groenewald J-H, Botha FC (2007b) Molecular and kinetic characterisation of sugarcane pyrophosphate: fructose-6-phosphate 1-phosphotransferase (PFP) and its possible role in the sucrose accumulation phenotype. Funct Plant Biol 34: 517–525.

Grof CPL, Campbell JA (2001) Sugarcane sucrose metabolism: scope for molecular manipulation. Aust J Plant Physiol 28: 1–12.

Grof CPL, Knight DP, McNeil SD, Lunn JE, Campbell JA (1998) A modified assay method shows leaf sucrose-phosphate synthase activity is correlated with leaf sucrose content across a range of sugarcane varieties. Aust J Plant Physiol 25: 499–502.

Grof CPL, So, CTE, Perroux JM, Bonnett GD, Forrester RI (2006) The five families of sucrose-phosphate synthase genes in *Saccharum* spp. are differentially expressed in leaves and stem. Funct Plant Biol 33: 605–610.

Grof CPL, Albertson PL, Bursle J, Perroux JM, Bonnett GD, Manners JM (2007) Sucrose-phosphate synthase, a biochemical marker of high sucrose accumulation in sugarcane. Crop Sci 47: 1530–1539.

Guy C, Kaplan F, Kopka J, Selbig J, Hincha DK (2008a) Metabolomics of temperature stress. Physiol Planta 132: 220–235.

Guy C, Kopka J, Moritz T (2008b) Editorial: Plant metabolomics coming of age. Physiol Planta 132: 113–116.

Gygi SP, Rochon Y, Franza BR, Aerbersold R (1999) Correlation between protein and mRNA abundance in yeast. Mol Cell Biol 19: 1720–1730.

Hall MC, Dworkin I, Ungerer MC, Purugganan M (2007) Genetics of microenvironmental canalization in Arabidopsis thaliana. Proc Natl Acad Sci USA, 104: 13717–13722.

Hall RD (2006) Plant metabolomics: from holistic hope, to hype, to hot topic. New Phytol 169: 453–468.

Hall R, Beale M, Fiehn O, Hardy N, Sumner L, Bino R (2002) Plant metabolomics: The missing link in functional genomics strategies. Plant Cell 14: 1437–1440.

Halpin C, Barakate A, Askari BM, Abbott JC, Ryan MD (2001) Enabling technologies for manipulating multiple genes on complex pathways. Plant Mol Biol 47: 295–310.

Hanson AD, Rivoal J, Paquet L, Gage DA (1994) Biosynthesis of 3-dimethylsulphoniopropionate in *Wollastonia biflora* (L) DC (Evidence that S-methylmethionine is an intermediate) Plant Physiol 105: 103–110.

Hart SR, Gaskell SJ (2005) LC-tandem MS in proteome characterisation. Trend Anal Chem 24: 566–575.

Hartt E, Kortschack HP, Forbes AJ, Burr GO (1963) Translocation of C in sugarcane. Plant Physiol 38: 305–318.

Hartt E, Kortschack HP, Burr GO (1964) Effects of defoliation, deradication and darkening the blade upon translocation of C in sugarcane. Plant Physiol 39: 15–22.

Hatch MD (2002) C4 photosynthesis: discovery and resolution. Photosynth Res 73: 251–256.

Hatch MD, Glasziou KT (1963) Sugar accumulation cycle in sugarcane. II. Relationship of invertase activity to sugar content and growth rate in storage tissue of plants grown in controlled environments. Plant Physiol 38: 344–348.

Hatch MD, Glasziou KT (1964) Direct evidence for translocation of sucrose in sugarcane leaves and stems. Plant Physiol 39: 180–184.

Hatch MD, Slack CR (1966) Photosynthesis in sugarcane leaves: a new carboxylation reaction and the pathway of sugar formation. Biochem J 101: 103–111.

Hatch MD, Slack CR (1969) Studies on the mechanism of activation and inactivation of pyruvate, phosphate dikinase. Biochem J 112: 549–558.

Hawker JS (1985) Sucrose. In: PM Dey , RA Dixon (eds) Biochemistry and Storage of Carbohydrates in Green Plants. Academic Press, New York, USA, pp 1–51.

Hawker JS, Jenner CM, Niemietz CM (1991) Sugar metabolism and compartmentation. Aust J Plant Physiol 18: 227–237.

Hennig L (2007) Patterns of beauty - omics meets plant development. Trends Plant Sci 12: 287–293.

Hirai MY, Sugiyama K, Sawada Y, Tohge T, Obayashi T, Suzuki A, Araki R, Sakurai N, Suzuki H, Aoki K, Goda H, Nishizawa OI, Shibata D, Saito K (2007) Omics-based identification of *Arabidopsis* Myb transcription factors regulating aliphatic glucosinolate biosynthesis. Proc Natl Acad Sci USA 104: 6478–6483.

Hoepfner SW, Botha FC (2003) Expression of fructokinase isoforms in the sugarcane culm. Plant Physiol Biochem 41: 741–747.

Hoepfner SW, Botha FC (2004) Purification and characterization of fructokinase from the culm of sugarcane. Plant Sci 167: 646–654.

Hood L (1998) Systems biology: new opportunities arising from genomics, proteomics and beyond. Exp Hematol 26: 681.

Huq E (2006) Degradation of negative regulators: a common theme in hormone and light signalling networks? Trends Plant Sci 11: 4–7.

Irvine JE (1975) Relations of photosynthetic rates and leaf and canopy characters to sugarcane yield. Crop Sci 15: 671–676.

Jacobsen KR, Fisher DG, Maretzki A, Moore PH (1992) Developmental changes in the anatomy of the sugarcane stem in relation to phloem unloading and sucrose storage. Bot Acta 105: 70–80.

Jangpromma N, Kitthaisong S, Daduang S, Jaisi P, Thammasirirak S (2007) 18 kDa protein accumulation in sugarcane leaves under drought stress conditions. KMITL Sci Technol J 7(Spl 1ss): 44–54.

Kaplan F, Kopka J, Haskell DW, Zhao W, Schiller KC, Gatzke N, Sung DY, Guy CL (2004) Exploring the temperature-stress metabolome of *Arabidopsis*. Plant Physiol 136: 4159–4168.

Kaplan F, Kopka J, Sung DY, Zhao W, Popp M, Porat R, Guy CL (2007) Transcript and metabolite profiling during cold acclimation of Arabidopsis reveals an intricate relationship of cold-regulated gene expression with modifications in metabolite content. Plant J 50: 967–981.

Kell DB (2004) Metabolomics and systems biology: making sense of the soup. Curr Opin Microbiol 7: 296–307.

Kell DB, Brown M, Davet HM, Dunn WB, Spasic I, Oliver SG (2005) Metabolic footprinting and systems biology: the medium is the message. Nat Rev Microbiol. doi:10.1038/nrmicro1177.

Klamt S, Stelling J (2003) Two approaches for metabolic pathway analysis? Trends Biotechnol 21: 64–69.

Komor E, Thom M, Maretzki A (1981) The mechanism of sugar uptake by sugarcane cell suspension culture. Planta 153: 181–192.

Kortschak HP, Hartt CE, Burr GO (1965) Carbon dioxide fixation in sugarcane leaves. Plant Physiol 40: 209–213.

Krishnan P, Kruger NJ, Ratcliffe RG (2005) Metabolite fingerprinting and profiling in plants using NMR. J Exp Bot 56: 255–265.

Kristensen C, Morant M, Olsen CE, Ekstrøm CT, Galbraith DW, Møller BL, Bak S (2005) Metabolic engineering of dhurrin in transgenic *Arabidopsis* plants with marginal inadvertent effects on the metabolome and transcriptome. Proc Natl Acad Sci USA 102: 1779–1784.

Larher F, Hamelin J, Steward GR (1977) L'acide diméthylsulphonium-3-propanoique de *Spartina anglica*. Phytochemistry 16: 2019–2020.

Larkin P, Harrigan GG (2007) Opportunities and surprises in crops modified by transgenic technology: metabolic engineering of benzylisoquinoline alkaloid, gossypol and lysine biosynthetic pathways. Metabolomics 3: 371–382.

Last R, Willmitzer L (2001) Physiology and metabolism. Curr Opin Plant Biol 4: 179–180.

Lay Jr JO, Liyanage R, Borgmann S, Wilkins CL (2006) Problems with the "omics". Trend Anal Chem 25: 1046–1056.

Lilley KS, Dupree P (2007) Plant organelle proteomics. Curr Opin Plant Biol 10: 594–599.

Lin D, Tabb DL, Yates JR (2003) Large-scale protein identification using mass spectrometry. Biochim Biophys Acta 1646: 1–10.

Lindermayr C, Saalbach G, Durner J (2005) Proteomic identification of S-nitrosylated proteins in *Arabidopsis*. Plant Physiol 137: 921–930.

Lingle SE (1989) Evidence for the uptake of sucrose intact into sugarcane internodes. Plant Physiol 90: 6–8.

Lingle SE (1997) Seasonal internode development and sucrose metabolism in sugarcane. Crop Sci 37: 1222–1227.

Lingle SE (1999) Sugar metabolism during growth and development in sugarcane internodes. Crop Sci 39: 480–486.

Lingle SE, Smith RC (1991) Sucrose metabolism related to growth and ripening in sugarcane internodes. Crop Sci 31: 172–177.

Lingle SE, Tew TL (2008) A comparison of growth and sucrose metabolism in sugarcane germplasm from Louisiana and Hawaii. Crop Sci 48: 1155–1163.

McCormick AJ (2007) Sink regulation of photosynthesis in sugarcane. PhD Thesis, Univ of KwaZulu-Natal, Durban, South Africa.

McCormick AJ, Cramer MD, Watt DA (2008a) Changes in photosynthetic rates and gene expression of leaves during a source-sink perturbation in sugarcane. Ann Bot-London 101: 89–102.

McCormick AJ, Cramer MD, Watt DA (2008b) Differential expression of genes in the leaves of sugarcane in response to sugar accumulation. Trop Plant Biol 1: 142–148.

McQualter RB, Chong BF, Meyer K, van Dyk DE, O'Shea MG, Walton MJ, Viitanen PV, Brumbley SM (2005) Initial evaluation of sugarcane as a production platform for r-hydroxybenzoic acid. Plant Biotechnol J 3: 29–41.

Mehta TS, Zakharkin SO, Gadbury GL, Allison DB (2006) Epistemological issues in omics and high-dimensional biology: give the people what they want. Physiol Genom 28: 24–32.

Ming R, Liu S-C, Moore PH, Irvine J (2001) QTL analysis in a complex autopolyploid: genetic control of sugar content in sugarcane. Genome Res 11: 2075–2084.

Mittler R (2006) Abiotic stress, the field environment and stress combination. Trends Plant Sci 11: 15–19.

Moisyadi S, Harrington HM (1989) Characterization of the heat shock response in cultured sugarcane cells. Plant Physiol 90: 1156–1162.

Moisyadi S, Dharmasiri S, Harrington HM, Lukas TJ (1994) Characterization of a low molecular mass autophosphorylated protein in cultured sugarcane cells and its identification as a nucleoside diphosphate kinase. Plant Physiol 104: 1401–1409.

Møller BL, Bak S (2005) Response to Kutchan: Genetic engineering, natural variation and substantial equivalence. Trends Biotechnol 23: 383.

Moore PH (1995) Temporal and spatial regulation of sucrose accumulation in the sugarcane stem. Aust J Plant Physiol 22: 661–679.

Moore PH (2005) Integration of sucrose accumulation processes across hierarchical scales: towards developing and understanding of the gene-to-crop continuum. Field Crops Res 92: 119–135.

Morgenthal K, Weckwerth W, Steur R (2006) Metabolomic networks in plants: transition from pattern recognition to biological interpretation. Biosystems 83: 108–117.

Morsy MR, Jouve L, Hausman J-F, Hoffmann L, Stewart, JMcD (2007) Alteration of oxidative and carbohydrate metabolism under abiotic stress in two rice (*Oryza sativa* L.) genotypes contrasting in chilling tolerance. J Plant Physiol 164: 157–167.

Mourelatos Z (2008) The seeds of silence. Nature 455: 44–45.

Nielsen J, Oliver S (2005) The next wave in metabolome analysis. Trends Biotechnol 23: 544–5456.

Nikolau BJ, Wurtele ES (eds) (2007) Concepts in Plant Metabolomics. Springer, Dordrecht, The Netherlands.

Nutt KA, O'Shea MG, Allsopp PG (2004) Feeding by sugarcane whitegrubs induces changes in the types and amounts of phenolics in the roots of sugarcane. Environ Exp Bot 51: 155–165.

Oksman-Caldentey K-M, Inzé D (2004) Plant cell factories in the post-genomic era: new ways to produce designer secondary metabolites. Trends Plant Sci 9: 433–440.

Oliver SG, Winson MK, Kell DB, Baganz F (1998) Systematic functional analysis of the yeast genome. Trends Biotechnol 16: 373–378.

Osuna D, Usadel B, Morcuende R, Gibon Y, Bläsing OE, Höhne M, Günter M, Kamlage B, Trethewey R, Scheible WR, Stitt M (2007) Temporal responses of transcripts, enzyme activities and metabolites after adding sucrose to carbon-deprived *Arabidopsis* seedlings. Plant J 49: 463–491.

Papin JA, Stelling J, Price ND, Klamt S, Schuster S, Palsson BØ (2004) Comparison of network-based pathway analysis methods. Trends Biotechnol 22: 400–405.

Papin JA, Price ND, Wiback SJ, Fell DA, Palsson BØ (2003) Metabolic pathways in the post-genome era. Trends Biochem Sci 28: 250–258.

Paquet L, Rathinasabapathi B, Saini H, Zamir L, Gage DA, Huang Z-H, Hanson AD (1994) Accumulation of the compatible solute 3-dimethylsulfoniopropionate in sugarcane and its relatives but not other Graminaceous crops. Aust J Plant Physiol 21: 37–48.

Paul MJ, Primavesi LF, Jhurrea DJ, Zhang Y (2008) Trehalose metabolism and signalling. Annu Rev Plant Biol 59: 417–441.

Peck SC (2005) Update on proteomics in *Arabidopsis*. Where do we go from here? Plant Physiol 138: 591–599.

Perera MADN, Nikolau BJ (2007) Metabolomics of cuticular waxes: a system for metabolomics analysis of a single tissue-type in a multicellular organism. In: BJ Nikolau , ES Wurtele (eds) Concepts in Plant Metabolomics. Springer, Dordrecht, The Netherlands, pp 111–123.

Persson S, Wei H, Milne J, Page GP, Somerville CR (2005) Identification of genes required for cellulose synthesis by regression analysis of public microarray data sets. Proc Natl Acad Sci USA, 102: 8633–3638.

Poolman MG, Assmus HE, Fell DA (2004) Applications of metabolic modelling to plant metabolism. J Exp Bot 55: 1177–1186.

Priesser J, Komor E (1991) Sucrose uptake into vacuoles of sugarcane suspension cells. Planta 186: 109–114.

Quackenbush J (2007) Extracting biology from high-dimensional biological data. J Exp Biol 210: 1507–1517.

Raamsdonk LM, Teusink B, Broadhurst D, Zhang N, Hayes A, Walsh MC, Berden JA, Brindle KM, Kell DB, Rowland JJ, Westerhoff HV, van Dam K, Oliver SG (2001) A functional genomics strategy that uses metabolome data to reveal the phenotype of silent mutations. Nat Biotechnol 19: 45–50.

Rae AL, Perroux JM, Grof CPL (2005) Sucrose partitioning between vascular bundles and storage parenchyma in the sugarcane stem: a potential role for the ShSUT1 sucrose transporter. Planta 220: 817–825.

Ramagopal S (1990) Protein polymorphism in sugarcane revealed by two-dimensional gel analysis. Theor Appl Genet 79: 297–304.

Ratcliffe RG, Shachar-Hill Y (2006) Measuring multiple fluxes through plant metabolic networks. Plant J 45: 490–511.

Regierer B, Fernie AR, Springer F, Perez-Melis A, Leisse A, Koehl K, Willmitzer L, Geigenberger P, Kossmann J (2002) Starch content and yield increase as a result of altering adenylate pools in transgenic plants. Nat Biotechnol 20: 1256–1260.

Reinders A, Sivitz AB, His A, Grof CPL, Perroux JM, Ward JM (2006) Sugarcane ShSUT1: analysis of sucrose transport activity and inhibition by sucralose. Plant Cell Environ 29: 1871–1880.

Rios-Estepa R, Lange BM (2007) Experimental and mathematical approaches to modeling plant metabolic networks. Phytochemistry 68: 2351–2374.

Rizhsky L, Liang H, Shuman J, Shulaev V, Davletova S, Mittler R (2004) When defence pathways collide. The response of *Arabidopsis* to a combination of drought and heat stress. Plant Physiol 134: 1683–1696.

Roach BT (1989) Origin and improvement of the genetic base for sugarcane. Proc Aust Soc Sugar Cane Technol 11: 492–503.

Roessner U, Willmitzer L, Fernie AR (2001) High-resolution metabolic phenotyping of genetically and environmentally diverse potato tuber systems. Identification of phenocopies. Plant Physiol 127: 749–764.

Rohwer JM, Botha FC (2001) Analysis of sucrose accumulation in the sugar cane culm on the basis of *in vitro* kinetic data. Biochem J 358: 437–445.

Rose S, Botha FC (2000) Distribution patterns of neutral invertase and sugar content in sugarcane internodal tissues. Plant Physiol Biochem 38: 819–824.

Rossouw D, Bosch S, Kossmann JM, Botha FC, Groenewald J-H (2007) Down-regulation of neutral invertase activity in sugarcane cell suspension cultures leads to increased sucrose accumulation. Funct Plant Biol 34: 490–498.

Rutherford RS (1998) Prediction of resistance in sugarcane to the stalk borer *Eldana saccharina* Walker (Lepidoptera: Pyralidae) by Near Infrared Spectroscopy on crude budscale extracts: the involvement of chlorogenates and flavonoids. J Chem Ecol 24: 1447–1463.

Sacher JA, Hatch MD, Glasziou KT (1963) Sugar accumulation cycle in sugarcane. III. Physical and metabolic aspects of cycle in immature storage tissues. Plant Physiol 38: 348–354.

Sato S, Soga T, Nishioka T, Tomita M (2004) Simultaneous determination of the main metabolites in rice leaves using capillary electrophoresis mass spectrometry and capillary electrophoresis diode array detection. Plant J 40: 151–163.

Schäfer WE, Rohwer JM, Botha FC (2004a) A kinetic study of sugarcane sucrose synthase. Eur J Biochem 271: 3971–3977.

Schäfer WE, Rohwer JM, Botha FC (2004b) Protein-level expression and localization of sucrose synthase in the sugarcane culm. Physiol Planta 121: 187–195.

Schäfer WE, Rohwer JM, Botha FC (2005) Partial purification and characterization of sucrose synthase in sugarcane. J Plant Physiol 162: 11–20.

Schauer N, Fernie AD (2006) Plant metabolomics: towards biological function and mechanism. Trends Plant Sci 11: 508–516.

Schilling CH, Letscher D, Palsson, BØ (2000) Theory for the systemic definition of metabolic pathways and their use in interpreting metabolic function from a pathway-oriented perspective. J Theor Biol 203: 229–248.

Schuster S, Dandekar T, Fell DA (1999) Detection of elementary flux modes in biochemical networks: a promising tool for pathway analysis and metabolic engineering. Trends Biotechnol 17: 53–60.

Schuster S, Eils R, Prank K (2006) Guest Editorial: 5th Int Conf on Systems Biology, Heidelberg, Oct 9-13, 2004. Biosystems 83: 71–74.

Schwender J (2008) Metabolic flux analysis as a tool in metabolic engineering of plants. Curr Opin Biotechnol 19: 131–137.

Schwender J, Goffmann F, Ohlrogge JB, Shachar-Hill Y (2004) Rubisco without the Calvin cycle improves the carbon efficiency of developing green seeds. Nature 432: 779–782.

Seki M, Umezawa T, Urano K, Shinozaki K (2007) Regulatory metabolic networks in drought stress responses. Curr Opin Plant Biol 10: 296–302.

Selbach M, Schwanhausser B, Thierfelder N, Fang Z, Khanin R, Rajewsky N (2008) Widespread changes in protein synthesis induced by microRNAs. Nature 455: 58–63.

Shulaev V, Cortes D, Miller G, Mittler R (2008) Metabolomics for plant stress response. Physiol Planta 132: 199–208.

Siegal ML, Bergman A (2002) Waddington's canalization revisited: Developmental stability and evolution. Proc Natl Acad Sci USA 99: 10528–10532.

Stitt M, Fernie AR (2003) From measurements of metabolites to metabolomics: An 'on the fly' perspective illustrated by recent studies of carbon-nitrogen interactions. Curr Opin Biotechnol 14: 136–144.

Sugiharto B, Ermawati N, Mori H, Aoki K, Yonekura-Sakakibara K, Yamaya T, Sugiyama T, Sakakibara H (2002) Identification and characterisation of a gene encoding drought-inducible protein localizing in the bundle sheath cell of sugarcane. Plant Cell Physiol 43: 350–354.

Sumner LW, Mendes P, Dixon RA (2003) Plant metabolomics: large-scale phytochemistry in the functional genomics era. Phytochemistry 62: 817–836.

Sweetlove LJ, Last RL, Fernie AR (2003) Predictive metabolic engineering: a goal for systems biology. Plant Physiol 132: 420–425.

Terauchi T, Matsuoka M, Kobayashi M, Nakano H (2000) Activity of sucrose phosphate synthase in relation to sucrose concentration in sugarcane internodes. Jpn J Trop Agri 44: 147–151.

Thurston G, Regan S, Rampitsch C, Xing T (2005) Proteomic and phosphoproteomic approaches to better understand plant-pathogen interactions. Physiol Mol Plant Pathol 66: 3–11.

Trethewey RN (2004) Metabolite profiling as an aid to metabolic engineering in plants. Curr Opin Plant Biol 7: 196–201.

Turnbaugh PJ, Gordon JI (2008) An invitation to the marriage of metagenomics and metabolomics. Cell 134: 708–713.

Turner W, Botha FC (2002) Purification and kinetic properties of UDP-glucose dehydrogenase from sugarcane. Arch Biochem Biophys 407: 209–216.

Urbanczyk-Wochniak E, Luedemann A, Kopka J, Selbig J, Roessner-Tunali U, Willmitzer L, Fernie AR (2003) Parallel analysis of transcript and metabolic profiles: a new approach in systems biology. EMBO Rep 4: 989–993.

Uys L, Botha FC, Hofmeyr J-H, Rohwer JM (2007) Kinetic model of sucrose accumulation in maturing sugarcane culm tissue. Phytochemistry 68: 2375–2392.

Van der Merwe MJ (2005) Influence of hexose-phosphates and carbon cycling in sugarcane. MSc Thesis, Univ of Stellenbosch, Stellenbosch, South Africa.

Villas-Bôas SG, Rasmussen S, Lane GA (2005) Metabolomics or metabolite profiles? Trends Biotechnol 23: 385–386.

Vorster DJ, Botha FC (1998) Partial purification and characterisation of sugarcane neutral invertase. Phytochemistry 49: 651–655.

Vorster DJ, Botha FC (1999) Sugarcane internodal invertases and tissue maturity. J Plant Physiol 155: 470–476.

Vu JCV, Allen LH, Gesch RW (2006) Up-regulation of photosynthesis and sucrose metabolism enzymes in young expanding leaves of sugarcane under elevated growth CO_2. Plant Sci 171: 123–131.

Waddington CH (1977) Tools for Thought. Jonathan Cape, London, UK.

Weckwerth W (2003) Metabolomics in systems biology. Annu Rev Plant Biol 54: 669–689.

Weckwerth W (2008) Integration of metabolomics and proteomics in molecular plant physiology—coping with the complexity by data-dimensionality reduction. Physiol Planta 132: 176–189.

Wei H, Persson S, Mehta T, Srinivasasainagendra V, Chen L, Page GP, Somerville C, Loraine A (2006) Transcriptional coordination of the metabolic network in Arabidopsis. Plant Physiol 142: 762–774.

Welbaum GE, Meinzer FC (1990) Compartmentation of solutes and water in developing sugarcane stalk tissue. Plant Physiol 93: 1147–1153.

Whittaker A, Botha FC (1997) Carbon partitioning during sucrose accumulation in sugarcane internodal tissue. Plant Physiol 115: 1651–1659.

Whittaker A, Botha FC (1999) Pyrophosphate: D-fructose-6-phosphate 1-phosphotransferase activity patterns in relation to sucrose storage across sugarcane varieties. Physiol Planta 107: 379–386.

Wilkins MR, Appel RD, Van Eyk JE., Chung MCM, Görg A, Hecker M, Huber L.A, Langen H, Link AJ, Paik Y-K, Patterson SD, Pennington SR, Rabilloud T, Simpson RJ, Weiss W, Dunn MJ (2006) Guidelines for the next 10 years of proteomics. Proteomics 6: 4–8.

Williams DM, Cole PA (2001) Kinase chips hit the proteomics era. Trends Biochem Sci 26: 271–273.

Wolters DA, Washburn MP, Yates JR (2001) An automated multidimensional protein identification technology for shotgun proteomics. Anal Chem 73: 5683–5690.

Xing T, Ouellet T, Miki BL (2002) Towards genomic and proteomic studies of protein phosphorylation in plant-pathogen interactions. Trends Plant Sci 7: 224–230.

Zhu YJ, Komor E, Moore PH (1997) Sucrose accumulation in the sugarcane stem is regulated by the difference between the activities of soluble acid invertase and sucrose phosphate synthase. Plant Physiol 115: 609–616.

Role of Bioinformatics as a Tool for Sugarcane Research

Rosanne E. Casu

ABSTRACT

Technological advances have resulted in the generation of vast amounts of molecular data that require organization and integration before they can efficiently aid further research. This chapter examines the bioinformatic tools that are currently available to aid molecular biology research for genetic improvement in sugarcane. It concentrates on those that specifically organize sugarcane sequences, expression data and genetic information but examines databases and tools created to either organize or analyze data from other genetically relevant species where a particular data form does not yet exist for sugarcane, e.g. a sequenced genome. It will provide a possible template for genetic and genomic data organization for sugarcane that will aid integration of current data with that generated in the future to aid application of this knowledge for varietal improvement.

Keywords: sugarcane, bioinformatics, database, genome, comparative genomics, transcriptome, integration

11.1 Introduction

Technology advances have resulted in an explosion of biological data that requires organization and analysis before it can facilitate further research and deliver benefits. Bioinformatics seeks to classify this data and present it

CSIRO Plant Industry, Queensland Bioscience Precinct, 306 Carmody Road, St. Lucia, QLD, 4067, Australia and CRC Sugar Industry Innovation through Biotechnology, Level 5, John Hines Building, The University of Queensland, St Lucia, QLD, 4072, Australia; e-mail: *Rosanne.Casu@csiro.au* for ease of use or to develop tools that allow additional exploitation of the data by discovering links that were not previously apparent. For many species, both plant and animal, extraordinary resources in the form of a genome, transcriptome, proteome and metabolome are often available. In addition, there may be DNA marker-rich genetic maps and populations of individuals that have been scored for a variety of traits of either scientific or economic interest. Genetically altered stocks displaying known phenotypes may also have been generated. This array of information had previously only been generated for model organisms, e.g. *Escherichia coli*, *Saccharomyces cerevisiae*, *Drosophila melanogaster*, *Caenorhabditis elegans*, *Mus musculus*, *Arabidopsis thaliana* and *Oryza sativa* (Issel-Tarver et al. 2002; Chen et al. 2005; Ouyang et al. 2007; Bult et al. 2008; Swarbreck et al. 2008; Keseler et al. 2009; Tweedie et al. 2009), but is now starting to be created for non-model species including several economically important crop species, e.g. wheat, barley, sorghum and sugarcane.

Sugarcane is an interspecific hybrid tropical C4 grass, which can accumulate up to approximately 50% sucrose dry weight in its stems (Bull and Glasziou 1963). It is the main crop for sugar production worldwide and, more recently, has become a focus for the production of ethanol, especially in Brazil (Goldemberg 2007).

At present, the major bioinformatic resource available for sugarcane consists of over 250,000 expressed sequence tags (ESTs) generated by four different research groups (Carson and Botha 2000, 2002; Vettore et al. 2001, 2003; Casu et al. 2003, 2004; Ma et al. 2004; Bower et al. 2005), detailed further by Casu et al (2009). In addition, there are approximately 9,500 genome survey sequences (GSSs) and several sugarcane bacterial artificial chromosome (BAC) sequences (*http://www.ncbi.nlm.nih.gov/*). Other genetic data are also available in the form of genetic maps (da Silva et al. 1995; Grivet et al. 1996; Guimarães et al. 1999; Hoarau et al. 2001; Aitken et al. 2005, 2007, 2008; Reffay et al. 2005; Garcia et al. 2006; Raboin et al. 2006), but no genome sequence exists for sugarcane at present.

However, there is a wealth of diverse yet organized molecular data available for other closely related species such as rice and sorghum (reviewed below), which could be mined for information that could be exploited for use in sugarcane. These databases could also serve as templates for identifying sugarcane molecular data that would be valuable to generate. They could also provide data structures for the organization of sugarcane molecular data that will be generated in the future.

Without this planned approach to the generation of future data and its integration with existing data into databases and analysis tools, the application of systems biology to increase our understanding of the biology of sugarcane will be difficult to attain. Consequently, our ability to make significant gains in the genetic improvement of sugarcane for traits of interest such as sucrose accumulation, water use efficiency, nitrogen use efficiency and pest and disease resistance will be affected.

In this chapter, the current status of bioinformatics and bioinformatic tools developed for sugarcane is reviewed. In addition, databases and tools developed for other species that can either be used directly to inform sugarcane research or whose structure could be adapted to organize either current or future sugarcane data are surveyed (Table 11-1).

11.2 Gene and Genome Databases

The sequencing of the nuclear genome of sugarcane remains a significant challenge due to its complex genetic structure and phenomenal size which is estimated at 10,000 megabases for a modern variety (D'Hont and Glaszmann 2001). Modern sugarcane varieties are derived from two progenitor species, *Saccharum officinarum* L., which has $2n = 80$ chromosomes and *Saccharum spontaneum* L. for which $2n$ varies from 40 to 128 chromosomes (Sreenivasan et al. 1987). Commercial varieties are interspecific hybrids, which are both polyploid and aneuploid in nature, with chromosome numbers varying between 100 and 130 (D'Hont et al. 1996). These genetic anomalies present a challenge for the sequencing of the sugarcane genome, however, a base genome may well be within reach, given recent advances in sequencing techniques and algorithms for sequence clustering and assembly (Pettersson et al. 2009).

However, one genome is available for sugarcane—the chloroplast genome. This sequence, 141,182 bp in length, was published in 2004 by both Asano et al. (2004 - NC_006084 [AP006714]) and Calsa Júnior et al. (2004 - NC_005878 [AE009947]). Both versions can be interrogated at NCBI but the best comparative access is available at the Chloroplast Genome database *http://chloroplast.cbio.psu.edu/*.

An organism search of the NCBI nucleic acid databases on 31st August, 2009 using the term "Saccharum*[Organism]", which retrieves all sequences belonging to the *Saccharum* genus, yields 1,472 nucleotide sequences (including 482 mRNA sequences), 256,895 expressed sequence tags (ESTs) and 10,699 genome survey sequences (GSSs). At present, the latter group of sequences has not been organized but it would be entirely possible to cluster the available sequences for further analysis. However, unless there a particular research need, it is unlikely to be done until at least a draft sugarcane genome sequence is being prepared.

Since sequencing of the sugarcane genome is swiftly gaining feasibility, it is imperative that the sugarcane research community evaluate data structures and databases being used by other genome sequencing initiatives in order to determine whether they are suitable for sugarcane, given its unusual genetic structure. In addition, considerable thought will need to be

Table 11-1 Relevant URLs for publicly available databases and bioinformatics tools.

Type	Name	URL	Reference
Portals	Gramene	http://www.gramene.org/	Liang et al. 2008
	JCVI	http://www.jcvi.org/	–
Species genome databases	Rice Genome Annotation Project	http://rice.plantbiology.msu.edu/	Yuan et al. 2005; Ouyang et al. 2007
	BGI-RIS	http://rice.genomics.org.cn	Zhao et al. 2004
	Sorghum bicolor genome sequence at Phytozome	http://www.phytozome.net/sorghum	–
	Sorghum bicolor genome sequence at JGI	http://genome.jgi-psf.org/Sorbi1/Sorbi1.info.html	–
	Maize Genome Sequencing Project	http://www.maizesequence.org/	–
	MaizeGDB	http://www.maizegdb.org/	Lawrence et al. 2004
	Brachypodium distachyon Information Resource	http://www.brachypodium.org	Garvin et al. 2008
Gene transcript databases	DFCI Sugarcane Gene Index	http://compbio.dfci.harvard.edu/tgi/cgi-bin/tgi/gimain.pl?gudb=s_officinarum	–
	TIGR Plant Transcript Assemblies	http://plantta.tigr.org/	Childs et al. 2007
	UniGene	http://www.ncbi.nlm.nih.gov/sites/entrez?db=unigene	Wheeler et al. 2003
	Plant Genome Database: PlantGDB	http://www.plantgdb.org/	Dong et al. 2005, 2004
Gene Expression databases	Gene Expression Omnibus (GEO)	http://www.ncbi.nlm.nih.gov/geo/	Edgar et al. 2002; Barrett et al. 2007, 2008
	ArrayExpress	http://www.ebi.ac.uk/microarray-as/ae/	Parkinson et al. 2008

Category	Name	URL	Reference
	PLEXdb	http://www.plexdb.org	Wise et al. 2007
	Babelomics	http://www.babelomics.org/	Al-Shahrour et al. 2006
	GOrilla	http://cbl-gorilla.cs.technion.ac.il/	Eden et al. 2009
	EasyGO	http://bioinformatics.cau.edu.cn/easygo/main.html	Zhou and Su 2007
	Arabidopsis Co-Expression Tool (ACT)	http://www.arabidopsis.leeds.ac.uk/act/	Manfield et al. 2006
	Genevestigator	https://www.genevestigator.ethz.ch/	Hruz et al. 2008
	CSB.DB Co-Response Databases (CoRDBs@CSB.DB)	http://csbdb.mpimp-golm.mpg.de/csbdb/dbcor/cor.html	Steinhauser et al. 2004
Molecular marker and genetic map databases	GrainGenes	http://wheat.pw.usda.gov/GG2/index.shtml	–
	Gramene	http://www.gramene.org/	Liang et al. 2008
	TropGENE DB	http://tropgenedb.cirad.fr/index.html	Ruiz et al. 2004
Protein or metabolome databases	Plant Proteome Annotation program (PPAP)	http://au.expasy.org/sprot/ppap/	Aubourg et al. 2005; Schneider et al. 2005; Schneider et al. 2004
	Multinational Arabidopsis Steering Committee	http://www.masc-proteomics.org/index.html	Weckwerth et al. 2008
	GMD@CSB.DB-The Golm Metabolome Database	http://csbdb.mpimp-golm.mpg.de/csbdb/gmd/home/gmd_sm.html	Steinhauser et al. 2004
Integrated databases and tools	The Arabidopsis Information Resource	http://www.arabidopsis.org/index.jsp	Swarbreck et al. 2008
	Gramene	http://www.gramene.org/	Liang et al. 2008
	FLAGdb++	http://urgv.evry.inra.fr/projects/FLAGdb++/HTML/index.shtml	Samson et al. 2004
	CSB.DB (CSB.DB@MPIMP)-A Comprehensive Systems-Biology Database	http://csbdb.mpimp-golm.mpg.de/	Steinhauser et al. 2004

given to basic nomenclature, the use of various ontologies and to the need to cross-reference large amounts of associated data that currently resides in numerous other databases, both public and private.

11.3 Comparative Genome Databases

Due to the genetic anomalies discussed above, which will greatly complicate the genome sequencing of sugarcane, comparative genome analysis will be important to assist in gene alignment, particularly for the alignment of sequences derived from homoeologous chromosomes. Several plant genome databases are either currently available or under development (Cannon et al. 2006; Swarbreck et al. 2008) but the most useful for comparative genome analysis in sugarcane are those devoted to rice (both *Oryza sativa* ssp. *japonica* and *indica*) (Zhao et al. 2004; Ouyang et al. 2007), *Sorghum bicolor* (Paterson et al. 2009) and *Brachypodium distachyon* (Garvin et al. 2008).

Comparative databases are most useful when presented as part of an "umbrella portal". The most comprehensive portal currently available for comparative grass genomics is Gramene (Liang et al. 2008). This is an open source data resource for comparative genome analysis that has the added benefit of curation, vastly increasing the utility of the data available. The resources are drawn from public projects covering a variety of disciplines including genomic and EST sequencing, analysis of protein structure and function, genetic and physical mapping, gene and quantitative trait loci (QTL) localizations as well as descriptions of phenotypic characteristics and mutations. Assembled nuclear genomes are currently available for six grass species as well as two species of *Arabidopsis*, grapevine and *Populus*. A particularly comprehensive database of genetic markers is presented, which can be searched by species, simple sequence repeat (SSR) primer pair and type. It is also tied to a series of comparative genome maps from a variety of grass species, together with a wealth of trait information. At present, there is very little sugarcane data presented in the Gramene database since it has been focused on the other crop grasses, however, it is likely that this will change as soon as sufficient publicly available trait and genetic information can be organized.

Another major portal is located at the J. Craig Venter Institute (JCVI—formerly The Institute for Genomic Research, TIGR). Plant genomics resources here are now restricted to various genome database projects, the Plant Transcript Assemblies (reviewed in more detail below) and plant genomics BLAST servers interrogating the Plant Transcript Assemblies and sequences from *Medicago truncatula*, maize, wheat, *Arabidopsis thaliana* and Plant Repeats.

The rice genome is the most advanced and best annotated monocot genome, with rice being considered the "model grass". Draft genomes were initially published in 2002 for both *Oryza sativa* L. ssp. *japonica* ("the Rice Genome") (Goff et al. 2002) and *O. sativa* L. ssp. *indica* (Yu et al. 2002), and sequence information for both genomes has undergone considerable revision since then. The genome of rice is highly relevant to sugarcane even though it is taxonomically a member of the BEP clade rather than the PACCAD clade of the Poaceae (*http://www.ncbi.nlm.nih.gov/Taxonomy/*). The sequencing of the rice genome was done through a consortium approach with the Rice Genome Annotation project (*http://rice.plantbiology.msu.edu/*) being originally hosted at The Institute for Genomic Research. It is currently hosted by Michigan State University with the current release being Pseudomolecule Assembly Release 6.1, officially released on 3 June 2009 (Ouyang et al. 2007). The database is divided into seven sections. The Pseudomolecules includes the current pseudomolecules, the Plant GO Slims assignment of rice proteins, paralogous families, alternatively spliced genes and putative SSRs in the pseudomolecules. A highly customizable Genome Browser has numerous tracks that can be activated. Most useful for comparative genomics are the *Sorghum bicolor* Gene Models and various transcript assemblies, the most relevant being Other Poaceae Transcript Assemblies, which includes sugarcane transcripts. The *in silico* mapping section presents the rice genetic markers where 13,895 marker sequences were used to integrate genetic and physical map positions for the rice genetic markers by aligning them to Pseudomolecules Release 5 of the rice genome. It also focuses on *in silico* mapping of 58,023 flanking sequence tags (FSTs), generated for rice insertional mutational populations. The rice gene expression database consists of an anatomy viewer, which presents gene expression evidence for rice loci based on EST data; a tissue-specific expression tool that uses EST frequency data identify a set of genes that are highly expressed in certain tissue, with the output being a list of rice gene models, oligonucleotides mapped to the gene models, putative functional annotation for gene models, and frequency of the matched ESTs; an expressed genes tool to be used for the identification of gene expression evidence for a rice locus from full length cDNA, EST, MPSS, SAGE and proteomic projects; and, finally, a rice multi-platform microarray search tool, which interrogates five different oligonucleotide microarray platforms. Comprehensive search functions include sequence similarity (BLAST), putative function, locus, domain, motif and Gene Ontology (GO) term retrieval. The final sections consist of a community annotation facility, which serves community-curated rice gene families and also a sequence download facility. Some of this information is also presented at Gramene (see above) as part of their comparative grass genome database. The genome of *O. sativa* L. ssp. *indica* (Yu et al. 2002) has also undergone systematic revision and annotation, which is presented at BGI-RIS—*http://rice.genomics.org.cn*—(Zhao et al. 2004). The genomes of both *indica* and *japonica* subspecies have been annotated for gene content, repetitive elements and SNPs, and are presented in a genome browser in a similar fashion to that of the Rice Genome Annotation Project. Taken together, these databases represent the most complete annotation of any grass genome.

Sorghum bicolor is a member of the Andropogoneae tribe within the Panicoideae (part of the PACCAD clade), together with *Miscanthus*, the *Saccharum* species and others (Pan et al. 2000). It is the most genetically similar crop plant to sugarcane and is tractable genetically due to having a diploid genome rather than an aneuploid polyploid genome like sugarcane. Its genome is estimated at ~736 Mb and is considered to be relatively small, particularly in comparison to sugarcane (Paterson 2006). The *S. bicolor* genome was recently sequenced (Paterson et al. 2009) and the first chromosome-based assembly and annotation of it is currently available at Phytozome (*http://www.phytozome.org*) and at the Joint Genome Institute (JGI; *http://genome.jgi-psf.org/Sorbi1/Sorbi1.info.html*), having been released on 26th March 2008. This was particularly important to the sugarcane research community, since this is the second monocot and first C4 plant genome to be sequenced. The Phytozome site currently offers a sequence similarity search facility, a sequence download service and a comprehensive genome browser. For comparative genomics, particularly in relation to sugarcane, the most useful feature is the presentation of the sorghum-rice syntenic segments. This gave the first direct comparison between the two genomes since those offered by Gramene and the rice genome sequencing project until recently only present the shorter ASBs. Additionally, the Poaceae TIGR TAs are presented, similar to that available for the rice genome and also the rice peptide blastx. The latter two tracks are useful in determining transcriptionally active parts of the sorghum genome and, by implication, those that may well be transcriptionally active in sugarcane, even if no sugarcane transcript has been localized to these parts of the genome. The JGI also offers a comprehensive suite of tools for genome analysis, including a genome browser, sequence similarity searching, Gene Ontology assignments, mapping to the KEGG pathways and annotations to the EuKaryotic Orthologous Groups (KOG).

The maize (*Zea mays*) genome sequence has been completed (December 2008). This is the third monocot and second C4 plant genome and will consolidate the excellent information already available now from sorghum since it is the next most related crop to sugarcane after sorghum. The current release is considered to mark the completion of the main sequencing effort with the focus changing now to gap filling and annotation. It is expected that fully-annotated phylogenetic trees describing evolutionary relationships between maize, rice, and sorghum gene orthologs will be released in collaboration with Gramene in the next three months. The main site for the Maize Genome Sequencing Project (*http://www.maizesequence.org/*) provides the main portal to search for and retrieve data. It contains a genome browser, a sequence similarity and name search facility and also contains a section for sequence retrieval. As the annotation of this genome improves, it is likely to become almost as valuable to the sugarcane research community as the sorghum genome. An additional resource for maize sequence is MaizeGDB (*http://www.maizegdb.org/*). This is a community database for a variety of biological information including genetic data (maps, loci, QTL, genetic stocks, cytogenetics and variations for alleles and polymorphisms), genomics data (molecular markers and probes, and sequences) and functional characterization (gene products, images, metabolic pathways and mutant phenotypes). It also serves some information about the Maize Genome Sequencing Project but that data is best viewed at the Project's own web site (Lawrence et al. 2004).

Sequencing of the *B. distachyon* (L.) Beauv genome has also commenced. This species is also considered an excellent model for functional genomics research in temperate grasses, cereals and other plants, e.g. switchgrass since it possesses a small genome (approximately 300 Mb) and is available as diploid, tetraploid and hexaploid accessions. Other physical assets include a small size, self-fertility, a short life cycle unlike some of crop grasses and simple growth requirements for experimental use. It is taxonomically in the BEP clade like rice and, when sequenced, will provide additional comparative data to rice, sorghum and maize. Sequencing of the *Brachypodium* genome has now reached 8x coverage and is presented with a Genome Browser and sequence similarity search facility at *http://www.brachypodium.org* as well as access to genetic stocks (Garvin et al. 2008).

11.4 Gene Expression Databases

11.4.1 Sequence Cluster Databases

Over 250,000 ESTs have been generated from an assortment of sugarcane varieties and tissues (see elsewhere in this volume, Carson and Botha 2000, 2002; Vettore et al. 2001, 2003; Casu et al. 2003, 2004; Ma et al. 2004; Bower et al. 2005). Private databases organizing EST data were developed by both the SUCEST (Telles et al. 2001) and Australian projects (Casu et al. 2003) to organize project data but neither of these databases were released publicly. By 2004, all of the projects had lodged all of their sugarcane ESTs at GenBank, which presented an opportunity for systematic organization of all of this data into a transcriptome. Four publicly available sequence cluster databases have been developed, offering different types and levels of organization.

The Gene Indices were developed originally at The Institute for Genomic Research (now the J. Craig Venter Institute—JCVI) but are now maintained at the Computational Biology and Functional Genomics Laboratory at the Dana-Farber Cancer Institute and Harvard School of Public Health (DFCI) (*http://compbio.dfci.harvard.edu/tgi/*). The Sugarcane Gene Index is in its second major release (version 2.2), which was re-clustered on 29th July 2008. Input sequences consisted of 255,635 ESTs and 499 mRNAs (all derived from GenBank). The index presents these sequences organized into 40,016 Theoretical Contigs (TCs), 76,529 singleton ESTs and 43 singleton mRNAs, giving a total of 116,588 unique sequences. This Gene Index will be re-clustered at least annually or if there is an increase in new sequences by 10% or more than 25,000 new sequences are deposited. Numerous analysis tools and methods for extracting data are available including sequence similarity searching using the BLAST algorithm, sequence reports that can be generated using TC identifiers, GenBank accession numbers or keywords, a listing of all annotations for both TCs and ESTs, the ability to search the EST libraries by keywords or tissue origin and the ability to download ESTs and TCs originating from one library. Functional annotation and analysis is also available with the prediction of alternative splice variants, digital northerns by comparing EST expression as calculated by EST number between different libraries, classification of the TCs by Gene Ontology vocabularies, association of TCs with various metabolic and signaling pathways, and, finally, predicted discriminatory 70-mer oligonucleotides.

Since the devolvement of the Gene Indices to the DFCI, the JCVI now offers the Plant Transcript Assemblies (TAs). This is located under "Plant Genomics—Resources" *http://www.jcvi.org/cms/research/groups/plant-genomics/resources/ and also at http://plantta.jcvi.org/* (Childs et al. 2007). Like the Sugarcane Gene Index, the input sequences used to build the plant TAs are expressed transcripts collected from dbEST (ESTs) and the NCBI GenBank nucleotide database (full length and partial cDNAs). Analysis capability is currently limited to TA name search and to sequence similarity searching using the BLAST algorithm. In the case of sugarcane, currently there is an artificial separation between two sets of sugarcane sequences in NCBI each having different taxon IDs. Since the input sequences are clustered within each taxon ID by the JVCI, this has led to the production of two transcript assemblies rather than just the one representing commercial sugarcane. The Sugarcane Community is currently having discussions to resolve artificial distinctions and to remove ambiguity in naming which will, undoubtedly, lead to the production of a single appropriate sugarcane TA which will be more accurate and, therefore, more useful.

Sugarcane sequence clusters have also been produced by the UniGene project at NCBI (*http://www.ncbi.nlm.nih.gov/sites/entrez?db = unigene*) and by PlantGDB (*http://www.plantgdb.org/*). Due to the method of clustering and the fact that only sequences ascribed to one of the two NCBI taxon IDs has been used, UniGene build #13 (the most recent) contains only 15,594 cluster sets. Unfortunately, this is well short of the number of clusters contained within both the Sugarcane Gene Index and the Sugarcane TAs, therefore, limiting its utility. The sugarcane sequence clusters produced by PlantGDB also only contain sequences ascribed to one of the NCBI taxon IDs, again limiting their utility.

11.4.2 *Transcript Expression Analysis*

Transcript expression databases have been devised to hold and curate high through-put experimental data. The Gene Expression Omnibus (GEO—*http:/ /www.ncbi.nlm.nih.gov/projects/geo/*) is a public repository for a variety of macroarray, microarray (both single and dual channel), serial analysis of gene expression (SAGE), mass spectrometry peptide profiling and quantitative sequence data. ArrayExpress (*http://www.ebi.ac.uk/microarray-as/ae/*) also organizes high-throughput data (Parkinson et al. 2008). Unlike the sequence databases maintained by NCBI, EMBL and DDBJ, these databases are not synchronized and do contain quite different experiments within them. Sugarcane high-throughput profiling experiments have so far only been lodged with GEO. At present, 17 experiments have been lodged at GEO, but only one curated dataset is present due to an acknowledged backlog. The raw data of each of the lodged experiments is available for download but GEO is not, and nor was it designed to be, an efficient data interrogation tool. In the future, as more sugarcane high-throughput experiments are lodged, particularly when more are lodged that use the same platform, this will become an extremely useful repository for data mining and reanalysis of existing data, and therefore, adding more value to the original experiments. On-line resources have already been developed to analyze plant transcript expression, e.g. PLEXdb—*http://www.plexdb.org/* (Wise et al. 2007), however, these do not yet contain sugarcane gene expression profiling data even if the array platform data is already present.

Transcript expression experiments are also more valuable if the products represented are annotated in an organized, consistent manner. The Gene Ontology (GO) project developed three hierarchically structured controlled vocabularies (ontologies) that describe gene products in the context of their associated biological processes, cellular components and molecular functions, all in a species-independent manner (Ashburner et al. 2000; The Gene Ontology Consortium 2008). This project originally commenced as a collaboration between FlyBase, the Saccharomyces Genome Database and the Mouse Genome database, but has expanded to now include 16 consortium members, including The *Arabidopsis* Information Resource (TAIR) and Gramene. The GO terms quickly replaced the individual researcher-derived product classifications for genes and array elements and there are a variety of bioinformatic tools available to enrich GO term-annotated gene lists, particularly for model organisms, e.g. Babelomics— *http://www.babelomics.org/* (Al-Shahrour et al. 2006) and GOrilla—

http://cbl-gorilla.cs.technion.ac.il/ (Eden et al. 2009). The only publicly available GO Term enrichment tool that specifically targets agronomic organisms including sugarcane is EasyGO—*http://bioinformatics.cau.edu.cn/easygo/* (Zhou and Su 2007). This tool can identify enriched GO terms for gene models, gene loci, protein coding genes, Ensembl, RefSeq and UniProt gene products, Gene Index entries and microarray oligonucleotides or probe sets for up to 17 organisms, also including the Sugarcane Gene Index and the Sugar Cane Affymetrix Genome Array probe sets. It is easy to use and quite interactive. The recent addition of GO term annotation of the Sugarcane Gene Index (29 November 2008) will make this tool extremely useful in the future.

The gold standard for transcript expression experiment interrogation is to identify genuine co-expression of microarray elements or probe sets in order to identify new genes in pathways. The *Arabidopsis* Co-Expression Tool (ACT—*http://www.arabidopsis.leeds.ac.uk/act/*) is an excellent example of a publicly available co-expression analysis tool (Manfield et al. 2006). However, this tool would only be useful in providing leads for sugarcane if the sugarcane probe sets could be accurately mapped through to those from *Arabidopsis*, and if the biology is also relevant across both species. Other tools include Genevestigator which presents datasets from six organisms including rice and barley (*https://www.genevestigator.ethz.ch/*) (Hruz et al. 2008) and CSB.DB who maintain Co-Response Databases of various model organisms, i.e. *Escherichia coli, Saccharomyces cerevisiae* and *Arabidopsis thaliana* at *http://csbdb.mpimp-golm.mpg.de/csbdb/dbcor/cor.html* (Steinhauser et al. 2004).

11.5 Molecular Marker and Genetic Map Databases

The key to efficient use of genetic maps and resources is the ability to examine these maps interactively and to speedily identify markers for further development either for mapping in other populations or as a diagnostic tool. Several excellent examples already exist for other crop species, e.g. wheat and rice. GrainGenes (*http://wheat.pw.usda.gov/GG2/index.shtml*) consists of a set of resources serving the Triticeae and Oat research communities. Their "short list" consists of 18 wheat maps, 19 barley maps, five rye maps and seven oat maps, all of which are searchable using various criteria. Similarly, Gramene hosts map data for numerous species with a focus on rice (Liang et al. 2008). Both of these databases use CMap, a web-based tool from GMOD (*http://gmod.org/wiki/CMap*), which allows comparison of genetic and physical maps. In sugarcane, the number of available maps is much smaller. Two *Saccharum spontaneum* maps generated in 1993 and 1995, respectively, are presented at GrainGenes, however, neither of these maps is in current use. The only publicly available mapping resource for sugarcane so far has been developed by CIRAD as part of TropGENE-DB (*http://tropgenedb.cirad.fr/*) (Ruiz et al. 2004). This database was initially developed with modules for sugarcane, cocoa and banana to serve as a management tool for genetic and genomic information on the tropical crops studied at CIRAD. This database now contains nine modules—banana, cocoa, coconut, coffee, cotton, oil palm, rice, rubber tree and sugarcane. The sugarcane module presents information on six genetic maps, which all have at least one parent in common (Grivet et al. 1996; Asnaghi et al. 2000; Hoarau et al. 2001; Rossi et al. 2003; Raboin et al. 2006). There is a search capability for molecular markers, QTL and clones as well as associated phenotypic data, particularly pest and disease ratings. No other genomic data is available yet, however, this database is certainly a good model to build on further capability .

Another possible approach for adding value to marker data and genetic maps is to integrate these data into existing databases that organize trait data for plant breeding programs. One example in sugarcane is SPIDNet, the plant breeding database that underpins the BSES-CSIRO Joint Venture Program for variety improvement in Australia. The capability of this database is currently being increased by including marker data in order to assist in the introduction of marker-assisted selection (P. Lethbridge, pers. comm.), however, it is not publicly available at the present time.

11.6 Protein and Metabolome Databases

Proteomics and metabolomics, the analysis of the total protein and metabolite complement, respectively, of an organism, constitute the current final links in the chain connecting phenotype to genotype. Even in organisms as thoroughly researched as *Arabidopsis*, there is considerable doubt about the exact number of protein-coding genes due to the discovery of errors made by automated gene finding algorithms when annotating genomes for possible genes. This has been partially addressed by the alignment of full-length cDNAs (Haas et al. 2002) but is more comprehensively remedied by manual annotation of the proteome. For plants, an excellent example is the Plant Proteome Annotation Program (PPAP), hosted by EXPASY at *http://au.expasy.org/sprot/ppap/* as part of UniPROT, which is systematically annotating plant-specific proteins and protein families, with a major emphasis on *Arabidopsis* and rice (Schneider et al. 2004, 2005; Aubourg et al. 2005). This will more carefully identify genuine protein-coding genes and facilitate further research in this area. This resource will become even more valuable as other relevant genomes become available. For *Arabidopsis*, proteomics data organization is already quite advanced under the auspices of the proteomics subcommittee of the Multinational Arabidopsis Steering Committee (*http://www.masc-proteomics.org/index.html*; Weckwerth et al. 2008). Numerous proteomics-related databases have been gathered together at this portal and this serves as an example for possible data organization when proteomics research in sugarcane becomes more advanced (Watt et al. 2009).

Metabolomics studies are also in their infancy in sugarcane (Bosch et al. 2003; Glassop et al. 2007) but careful thought needs to be given to the organization and presentation of data to be generated in the future. One example of a possible database to interact with is the Golm Metabolome Database (*http://csbdb.mpimp-golm.mpg.de/csbdb/gmd/gmd.html*). This database provides access to mass spectra libraries, metabolite profiling experiments and other pertinent data (Kopka et al. 2005). Another possibility is KNApSAcK (*http://kanaya.naist.jp/KNApSAcK/*) a tool for the analysis of metabolites which already contains information pertaining to 17 *Saccharum* genus metabolites. This database emphasizes the biological origins of the compounds contained within it and the data can be extracted in a variety of ways. In addition, it provides a tool for the analysis of mass spectrum data.

11.7 Integration of Different Data

As alluded to above, systems biology requires integrated access to the full spectrum of "omics" data that has been well organized and profiled. These tools are already advanced for organisms such as *Escherichia coli* and other bacteria (Steinhauser et al. 2004; Keseler et al. 2009) as well as for yeast and mice (Issel-Tarver et al. 2002; Begley et al. 2007; Smith et al. 2007; Bult et al. 2008). Examples in the plant area are currently restricted to the model plants, especially *A. thaliana*. The best example for *A. thaliana* is The *Arabidopsis* Information Resource (TAIR), hosted at *http://www.arabidopsis.org/* (Swarbreck et al. 2008). This resource collects and organizes a wealth of genetic and genomic data and integrates this with information on seed stocks, markers, publication and information on the *Arabidopsis* research community. It is updated regularly, curates community submission and also provides connections to other *Arabidopsis* resources, e.g. the *Arabidopsis* Biological Resource Center (ABRC) at the Ohio State University. This type of data organization extracts maximum benefit from all of the research performed on an organism and will allow for new insights to be more easily gained than if the resources were distributed and not related. Others plant resources integrating various data including genetic and genomics data include Gramene (Liang et al. 2008), FLAGdb++ (*http://urgv.evry.inra.fr/projects/FLAGdb++/HTML/index.shtml;* Samson et al. 2004) and CSB.DB (Steinhauser et al. 2004).

11.8 Conclusion

In conclusion, there is a wealth of molecular data available for sugarcane, especially transcript clusters and gene expression profiles. The transcript clusters are easy to access and analyze, and data underpinning gene expression profiles can also be easily downloaded for reanalysis, if required. A number of genetic maps have also been generated from mapping populations relevant to various breeding programs, however, not all of these have been integrated yet. The sequencing of the *S. bicolor* genome has provided an excellent resource for use in comparative genomics but it cannot replace a sugarcane genome. The sugarcane research community now has an opportunity to integrate the existing valuable genetic and transcriptomic data with a genome and other molecular data in order to provide a true "systems biology" portal to aid future advances in the genetic improvement of sugarcane for important traits. The foundation for this has already been laid through experience with other plants which will facilitate planning for the effort in sugarcane.

Acknowledgements

The author acknowledges Dr. John M. Manners, Dr. Graham D. Bonnett and Dr. Louise Thatcher for critical reading of this manuscript.

References

Aitken KS, Jackson PA, McIntyre CL (2005) A combination of AFLP and SSR markers provides extensive map coverage and identification of homo(eo)logous linkage groups in a sugarcane cultivar. Theor Appl Genet 110: 789–801.

Aitken KS, Jackson PA, McIntyre CL (2007) Construction of a genetic linkage map for *Saccharum officinarum* incorporating both simplex and duplex markers to increase genome coverage. Genome 50: 742–756.

Aitken KS, Hermann S, Karno K, Bonnett GD, McIntyre LC, Jackson PA (2008) Genetic control of yield related stalk traits in sugarcane. Theor Appl Genet 117: 1191–1203.

Al-Shahrour F, Minguez P, Tarraga J, Montaner D, Alloza E, Vaquerizas JM, Conde L, Blaschke C, Vera J, Dopazo J (2006) BABELOMICS: a systems biology perspective in the functional annotation of genome-scale experiments. Nucl Acids Res 34: W472–476.

Asano T, Tsudzuki T, Takahashi S, Shimada H, Kadowaki K-I (2004) Complete nucleotide sequence of the sugarcane (*Saccharum officinarum*) chloroplast genome: a comparative analysis of four monocot chloroplast genomes. DNA Res 11: 93–99.

Ashburner M, Ball CA, Blake JA, Botstein D, Butler H, Cherry JM, Davis AP, Dolinski K, Dwight SS, Eppig JT, Harris MA, Hill DP, Issel-Tarver L, Kasarskis A, Lewis S, Matese JC, Richardson JE, Ringwald M, Rubin GM, Sherlock G (2000) Gene Ontology: tool for the unification of biology. Nat Genet 25: 25–29.

Asnaghi C, Paulet F, Kaye C, Grivet L, Deu M, Glaszmann JC, D'Hont A (2000) Application of synteny across Poaceae to determine the map location of a sugarcane rust resistance gene. Theor Appl Genet 101: 962–969.

Aubourg S, Brunaud V, Bruyere C, Cock M, Cooke R, Cottet A, Couloux A, Dehais P, Deleage G, Duclert A, Echeverria M, Eschbach A, Falconet D, Filippi G, Gaspin C, Geourjon C, Grienenberger J-M, Houlne G, Jamet E, Lechauve F, Leleu O, Leroy P, Mache R, Meyer C, Nedjari H, Negrutiu I, Orsini V, Peyretaillade E, Pommier C, Raes J, Risler J-L, Riviere S, Rombauts S, Rouze P, Schneider M, Schwob P, Small I,

Soumayet-Kampetenga G, Stankovski D, Toffano C, Tognolli M, Caboche M, Lecharny A (2005) GeneFarm, structural and functional annotation of *Arabidopsis* gene and protein families by a network of experts. Nucl Acids Res 33: D641–646.

Barrett T, Troup DB, Wilhite SE, Ledoux P, Rudnev D, Evangelista C, Kim IF, Soboleva A, Tomashevsky M, Edgar R (2007) NCBI GEO: mining tens of millions of expression profiles—database and tools update. Nucl Acids Res 35: D760–765.

Barrett T, Troup DB, Wilhite SE, Ledoux P, Rudnev D, Evangelista C, Kim IF, Soboleva A, Tomashevsky M, Marshall KA, Phillippy KH, Sherman PM, Muertter RN, Edgar R (2008) NCBI GEO: archive for high-throughput functional genomic data. Nucl Acids Res 37: D885–D890.

Begley DA, Krupke DM, Vincent MJ, Sundberg JP, Bult CJ, Eppig JT (2007) Mouse Tumor Biology Database (MTB): status update and future directions. Nucl Acids Res 35: D638–642.

Bosch S, Rohwer JM, Botha FC (2003) The sugarcane metabolome. Proc Ann Congr—S Afr Sugar Technol Assoc, pp 129–133.

Bower NI, Casu RE, Maclean DJ, Reverter A, Chapman SC, Manners JM (2005) Transcriptional response of sugarcane roots to methyl jasmonate. Plant Sci 168: 761–772.

Bull TA, Glasziou KT (1963) The evolutionary significance of sugar accumulation in *Saccharum*. Aust J Biol Sci 16: 737–742.

Bult CJ, Eppig JT, Kadin JA, Richardson JE, Blake JA, the Mouse Genome Database Group (2008) The Mouse Genome Database (MGD): mouse biology and model systems. Nucl Acids Res 36: D724–728.

Calsa Júnior T, Carraro DM, Benatti MR, Barbosa AC, Kitajima JP, Carrer H (2004) Structural features and transcript-editing analysis of sugarcane (*Saccharum officinarum* L.) chloroplast genome. Curr Genet 46: 366–373.

Cannon SB, Sterck L, Rombauts S, Sato S, Cheung F, Gouzy J, Wang X, Mudge J, Vasdewani J, Schiex T, Spannagl M, Monaghan E, Nicholson C, Humphray SJ, Schoof H, Mayer KFX, Rogers J, Quétier F, Oldroyd GE, Debellé F, Cook DR, Retzel EF, Roe BA, Town CD, Tabata S, Van de Peer Y, Young ND (2006) Legume genome evolution viewed through the *Medicago truncatula* and *Lotus japonicus* genomes. Proc Natl Acad Sci USA, 103: 14959–14964.

Carson D, Botha F (2000) Preliminary analysis of expressed sequence tags for sugarcane. Crop Sci 40: 1769–1779.

Carson D, Botha F (2002) Genes expressed in sugarcane maturing internodal tissue. Plant Cell Rep 20: 1075–1081.

Casu RE, Grof CPL, Rae AL, McIntyre CL, Dimmock CM, Manners JM (2003) Identification of differentially expressed transcripts from maturing stem of sugarcane by *in silico* analysis of stem expressed sequence tags and gene expression profiling. Plant Mol Biol 54: 503–517.

Casu RE, Dimmock CM, Chapman SC, Grof CPL, McIntyre CL, Bonnett GD, Manners JM (2004) Identification of a novel sugar transporter homologue strongly expressed in maturing stem vascular tissues of sugarcane by expressed sequence tag and microarray analysis. Plant Mol Biol 52: 371–386.

Casu RE, Hotta CT, Souza GM (2010) Functional genomics: Transcriptomics of sugarcane —current status and future prospects. In: RJ Henry , C Kole (eds) Genetics, Genomics and Breeding of Sugarcane. Science Publ, Enfield, New Hampshire, USA, pp 167–192.

Chen N, Harris TW, Antoshechkin I, Bastiani C, Bieri T, Blasiar D, Bradnam K, Canaran P, Chan J, Chen C-K, Chen WJ, Cunningham F, Davis P, Kenny E, Kishore R, Lawson D, Lee R, Muller H-M, Nakamura C, Pai S, Ozersky P, Petcherski A, Rogers A, Sabo A, Schwarz EM, Van Auken K, Wang Q, Durbin R, Spieth J, Sternberg PW, Stein LD (2005) WormBase: a comprehensive data resource for *Caenorhabditis* biology and genomics. Nucl Acids Res 33: D383–389.

Childs KL, Hamilton JP, Zhu W, Ly E, Cheung F, Wu H, Rabinowicz PD, Town CD, Buell CR, Chan AP (2007) The TIGR Plant Transcript Assemblies database. Nucl Acids Res 35: D846–851.

D'Hont A, Glaszmann JC (2001) Sugarcane genome analysis with molecular markers: a first decade of research. In: DM Hogarth (ed) Proc 24th Congr Int Society of Sugar Cane Technol, Aust Soc Sugar Cane Technol, Brisbane, Australia, pp 556–559.

D'Hont A, Grivet L, Feldmann P, Rao S, Berding N, Glaszmann JC (1996) Characterisation of the double genome structure of modern sugarcane cultivars (*Saccharum* spp.) by molecular cytogenetics. Mol Gen Genet 250: 405–413.

Dong Q, Schlueter SD, Brendel V (2004) PlantGDB, plant genome database and analysis tools. Nucl Acids Res 32: D354–359.

Dong Q, Lawrence CJ, Schlueter SD, Wilkerson MD, Kurtz S, Lushbough C, Brendel V (2005). Comparative plant genomics resources at PlantGDB. Plant Physiol 139: 610–618.

Eden E, Navon R, Steinfeld I, Lipson D, Yakhini Z (2009) GOrilla: a tool for discovery and visualization of enriched GO terms in ranked gene lists. BMC Bioinformat 10: 48.

Edgar R, Domrachev M, Lash AE (2002) Gene Expression Omnibus: NCBI gene expression and hybridization array data repository. Nucl Acids Res 30: 207–210.

Garcia AAF, Kido EA, Meza AN, Souza HMB, Pinto LR, Pastina MM, Leite CS, Silva J, Ulian EC, Figueira A, Souza AP (2006) Development of an integrated genetic map of a sugarcane (*Saccharum* spp.) commercial cross, based on a maximum-likelihood approach for estimation of linkage and linkage phases. Theor Appl Genet 112: 298–314.

Garvin DF, Gu Y-Q, Hasterok R, Hazen SP, Jenkins G, Mockler TC, Mur LAJ, Vogel JP (2008) Development of genetic and genomic research resources for *Brachypodium distachyon*, a new model system for grass crop research. Crop Sci 48: S-69–84.

Glassop D, Roessner U, Bacic A, Bonnett GD (2007) Changes in the sugarcane metabolome with stem development. Are they related to sucrose accumulation? Plant Cell Physiol 48: 573–584.

Goff SA, Ricke D, Lan T-H, Presting G, Wang R, Dunn M, Glazebrook J, Sessions A, Oeller P, Varma H, Hadley D, Hutchison D, Martin C, Katagiri F, Lange BM, Moughamer T, Xia Y, Budworth P, Zhong J, Miguel T, Paszkowski U, Zhang S, Colbert M, Sun W-l, Chen L, Cooper B, Park S, Wood TC, Mao L, Quail P, Wing R, Dean R, Yu Y, Zharkikh A, Shen R, Sahasrabudhe S, Thomas A, Cannings R, Gutin A, Pruss D, Reid J, Tavtigian S, Mitchell J, Eldredge G, Scholl T, Miller RM, Bhatnagar S, Adey N, Rubano T, Tusneem N, Robinson R, Feldhaus J, Macalma T, Oliphant A, Briggs S (2002) A draft sequence of the rice genome (*Oryza sativa* L. ssp. *japonica*). Science 296: 92–100.

Goldemberg J (2007) Ethanol for a sustainable energy future. Science 315: 808–810.

Grivet L, D'Hont A, Roques D, Feldmann P, Lanaud C, Glaszmann JC (1996) RFLP mapping in cultivated sugarcane (*Saccharum* spp.): genome organization in a highly polyploid and aneuploid interspecific hybrid. Genetics 142: 987–1000.

Guimarães C, Honeycutt R, Sills G, Sobral B (1999) Genetic maps of *Saccharum officinarum* L. and *Saccharum robustum* Brandes & Jew. ex grassl. Genet Mol Biol 22: 125–132.

Haas B, Volfovsky N, Town C, Troukhan M, Alexandrov N, Feldmann K, Flavell R, White O, Salzberg S (2002) Full-length messenger RNA sequences greatly improve genome annotation. Genome Biol 3: research0029.1-research0029.12.

Hoarau J-Y, Offmann B, D'Hont A, Risterucci A-M, Roques D, Glaszmann J-C, Grivet L (2001) Genetic dissection of a modern sugarcane cultivar (*Saccharum* spp.). I. Genome mapping with AFLP markers. Theor Appl Genet 103: 84–97.

Hruz T, Laule O, Szabo G, Wessendorp F, Bleuler S, Oertle L, Widmayer P, Gruissem W, Zimmermann P (2008) Genevestigator V3: A reference expression database for the meta-analysis of transcriptomes. Adv Bioinformat 2008: Article ID 420747, doi:10.1155/2008/420747.

Issel-Tarver L, Christie KR, Dolinski K, Andrada R, Balakrishnan R, Ball CA, Binkley G, Dong S, Dwight SS, Fisk DG, Harris M, Schroeder M, Sethuraman A, Tse K, Weng S, Botstein D, Michael Cherry J, Christine G, Gerald RF (2002) *Saccharomyces* genome database. Meth Enzymol 350: 329–346.

Keseler IM, Bonavides-Martinez C, Collado-Vides J, Gama-Castro S, Gunsalus RP, Johnson DA, Krummenacker M, Nolan LM, Paley S, Paulsen IT, Peralta-Gil M, Santos-Zavaleta A, Shearer AG, Karp PD (2009) EcoCyc: A comprehensive view of *Escherichia coli* biology. Nucl Acids Res. 37: D464–470.

Kopka J, Schauer N, Krueger S, Birkemeyer C, Usadel B, Bergmuller E, Dormann P, Weckwerth W, Gibon Y, Stitt M, Willmitzer L, Fernie AR, Steinhauser D (2005) GMD@CSB.DB: the Golm Metabolome Database. Bioinformatics 21: 1635–1638.

Lawrence CJ, Dong Q, Polacco ML, Seigfried TE, Brendel V (2004) MaizeGDB, the community database for maize genetics and genomics. Nucl Acids Res 32: D393–397.

Liang C, Jaiswal P, Hebbard C, Avraham S, Buckler ES, Casstevens T, Hurwitz B, McCouch S, Ni J, Pujar A, Ravenscroft D, Ren L, Spooner W, Tecle I, Thomason J, Tung C-w, Wei X, Yap I, Youens-Clark K, Ware D, Stein L (2008) Gramene: a growing plant comparative genomics resource. Nucl Acids Res. 36: D947–953.

Ma HM, Schulze S, Lee S, Yang M, Mirkov E, Irvine J, Moore P, Paterson A (2004) An EST survey of the sugarcane transcriptome. Theor Appl Genet 108: 851–863.

Manfield IW, Jen C-H, Pinney JW, Michalopoulos I, Bradford JR, Gilmartin PM, Westhead DR (2006) Arabidopsis Co-expression Tool (ACT): web server tools for microarray-based gene expression analysis. Nucl Acids Res 34: W504–509.

Ouyang S, Zhu W, Hamilton J, Lin H, Campbell M, Childs K, Thibaud-Nissen F, Malek RL, Lee Y, Zheng L, Orvis J, Haas B, Wortman J, Buell CR (2007) The TIGR Rice Genome Annotation Resource: improvements and new features. Nucl Acids Res 35: D883–887.

Pan YB, Burner DM, Legendre BL (2000) An assessment of the phylogenetic relationship among sugarcane and related taxa based on the nucleotide sequence of 5S rRNA intergenic spacers. Genetica 108: 285–295.

Parkinson H, Kapushesky M, Kolesnikov N, Rustici G, Shojatalab M, Abeygunawardena N, Berube H, Dylag M, Emam I, Farne A, Holloway E, Lukk M, Malone J, Mani R, Pilicheva E, Rayner TF, Rezwan F, Sharma A, Williams E, Bradley XZ, Adamusiak T, Brandizi M, Burdett T, Coulson R, Krestyaninova M, Kurnosov P, Maguire E, Neogi SG, Rocca-Serra P, Sansone S-A, Sklyar N, Zhao M, Sarkans U, Brazma A (2008) ArrayExpress update—from an archive of functional genomics experiments to the atlas of gene expression. Nucl Acids Res 37: D868–D872.

Paterson AH (2006) Leafing through the genomes of our major crop plants: strategies for capturing unique information. Nat Rev Genet 7: 174–184.

Paterson AH, Bowers JE, Bruggmann R, Dubchak I, Grimwood J, Gundlach H, Haberer G, Hellsten U, Mitros T, Poliakov A, Schmutz J, Spannagl M, Tang H, Wang X, Wicker T, Bharti AK, Chapman J, Feltus FA, Gowik U, Grigoriev IV, Lyons E, Maher CA, Martis M, Narechania A, Otillar RP, Penning BW, Salamov AA, Wang Y, Zhang L, Carpita NC, Freeling M, Gingle AR, Hash CT, Keller B, Klein P, Kresovich S, McCann MC, Ming R, Peterson DG, Mehboob ur R, Ware D, Westhoff P, Mayer KFX, Messing J, Rokhsar DS (2009) The *Sorghum bicolor* genome and the diversification of grasses. Nature 457: 551–556.

Pettersson E, Lundeberg J, Ahmadian A (2009) Generations of sequencing technologies. Genomics 93: 105–111.

Raboin L-M, Oliveira KM, Lecunff L, Telismart H, Roques D, Butterfield M, Hoarau J-Y, D'Hont A (2006) Genetic mapping in sugarcane, a high polyploid, using bi-parental progeny: identification of a gene controlling stalk colour and a new rust resistance gene. Theor Appl Genet 112: 1382–1391.

Reffay N, Jackson PA, Aitken KS, Hoarau J-Y, D'Hont A, Besse P, McIntyre CL (2005) Characterisation of genome regions incorporated from an important wild relative into Australian sugarcane. Mol Breed 15: 367–381.

Rossi M, Araujo PG, Paulet F, Garsmeur O, Dias VM, Chen H, Van Sluys M-A, D'Hont A (2003) Genomic distribution and characterization of EST-derived resistance gene analogs (RGAs) in sugarcane. Mol Genet Genom 269: 406–419.

Ruiz M, Rouard M, Raboin LM, Lartaud M, Lagoda P, Courtois B (2004) TropGENE-DB, a multi-tropical crop information system. Nucl Acids Res 32: D364–367.

Samson F, Brunaud V, Duchene S, De Oliveira Y, Caboche M, Lecharny A, Aubourg S (2004) FLAGdb++: a database for the functional analysis of the *Arabidopsis* genome. Nucl Acids Res 32: D347–350.

Schneider M, Tognolli M, Bairoch A (2004). The Swiss-Prot protein knowledgebase and ExPASy: providing the plant community with high quality proteomic data and tools. Plant Physiol Biochem 42: 1013–1021.

Schneider M, Bairoch A, Wu CH, Apweiler R (2005) Plant protein annotation in the UniProt knowledgebase. Plant Physiol 138: 59–66.

da Silva J, Honeycutt RJ, Burnquist W, Al-Janabi SM, Sorrells ME, Tanksley SD, Sobral BWS (1995) *Saccharum spontaneum* L. 'SES 208' genetic linkage map combining RFLP- and PCR-based markers. Mol Breed 1: 165–179.

Smith CM, Finger JH, Hayamizu TF, McCright IJ, Eppig JT, Kadin JA, Richardson JE, Ringwald M (2007) The mouse Gene Expression Database (GXD): 2007 update. Nucl Acids Res 35: D618–623.

Sreenivasan T, Ahloowalia B, Heinz D (1987) Cytogenetics. In: D Heinz (ed) Sugarcane Improvement through Breeding. Elsevier Press, Amsterdam, The Netherlands, pp 211–253.

Steinhauser D, Usadel B, Luedemann A, Thimm O, Kopka J (2004) CSB.DB: a comprehensive systems-biology database. Bioinformatics 20: 3647–3651.

Swarbreck D, Wilks C, Lamesch P, Berardini TZ, Garcia-Hernandez M, Foerster H, Li D, Meyer T, Muller R, Ploetz L, Radenbaugh A, Singh S, Swing V, Tissier C, Zhang P, Huala E (2008) The *Arabidopsis* Information Resource (TAIR): gene structure and function annotation. Nucl Acids Res 36: D1009–1014.

Telles GP, Braga MDV, Dias Z, Lin T-L, Quitzau JAA, da Silva FR, Meidanis J (2001) Bioinformatics of the sugarcane EST project. Genet. Mol. Biol. 24: 9–15.

The Gene Ontology Consortium (2008) The Gene Ontology project in 2008. Nucl Acids Res 36: D440–444.

Tweedie S, Ashburner M, Falls K, Leyland P, McQuilton P, Marygold S, Millburn G, Osumi-Sutherland D, Schroeder A, Seal R, Zhang H, The FlyBase Consortium (2009) FlyBase: enhancing *Drosophila* Gene Ontology annotations. Nucl Acids Res 37: D555–559.

Vettore AL, da Silva FR, Kemper EL, Arruda P (2001) The libraries that made SUCEST. Genet Mol Biol 24: 1–7.

Vettore AL, da Silva FR, Kemper EL, Souza GM, da Silva AM, Ferro MIT, Henrique-Silva F, Giglioti EA, Lemos MVF, Coutinho LL, Nobrega MP, Carrer H, Franca SC, Bacci M, Jr., Goldman MHS, Gomes SL, Nunes LR, Camargo LEA, Siqueira WJ, Van Sluys M-A, Thiemann OH, Kuramae EE, Santelli RV, Marino CL, Targon MLPN, Ferro JA, Silveira HCS, Marini DC, Lemos EGM, Monteiro-Vitorello CB, Tambor JHM, Carraro DM, Roberto PG, Martins VG, Goldman GH, de Oliveira RC, Truffi D, Colombo CA, Rossi M, de Araujo PG, Sculaccio SA, Angella A, Lima MMA, de Rosa VE, Jr, Siviero F, Coscrato VE, Machado MA, Grivet L, Di Mauro SMZ, Nobrega FG, Menck CFM, Braga MDV, Telles GP, Cara FAA, Pedrosa G, Meidanis J, Arruda

P (2003) Analysis and functional annotation of an expressed sequence tag collection for tropical crop sugarcane. Genome Res 13: 2725–2735.

Watt D, Butterfield M, Huckett B (2010) Proteomics and metabolomics. In: RJ Henry , C Kole (eds) Genetics, Genomics and Breeding of Sugarcane. Science Publ, Enfield, New Hampshire, USA, pp 193–228.

Weckwerth W, Baginsky S, Wijk K, Heazlewood JL, Millar H (2008) The multinational *Arabidopsis* steering subcommittee for proteomics assembles the largest proteome database resource for plant systems biology. J Proteom Res 7: 4209–4210.

Wheeler DL, Church DM, Federhen S, Lash AE, Madden TL, Pontius JU, Schuler GD, Schriml LM, Sequeira E, Tatusova TA, Wagner L (2003) Database resources of the National Center for Biotechnology. Nucl Acids Res 31: 28–33.

Wise R, Caldo R, Hong L, Shen L, Cannon E, JA D (2007) BarleyBase/PLEXdb: a unified expression profiling database for plants and plant pathogens. In: D Edwards (ed) Methods in Molecular Biology, vol 406. Humana Press, Totowa, NJ, USA, pp 347–363.

Yu J, Hu S, Wang J, Wong GK, Li S, Liu B, Deng Y, Dai L, Zhou Y, Zhang X, Cao M, Liu J, Sun J, Tang J, Chen Y, Huang X, Lin W, Ye C, Tong W, Cong L, Geng J, Han Y, Li L, Li W, Hu G, Huang X, Li W, Li J, Liu Z, Li L, Liu J, Qi Q, Liu J, Li L, Li T, Wang X, Lu H, Wu T, Zhu M, Ni P, Han H, Dong W, Ren X, Feng X, Cui P, Li X, Wang H, Xu X, Zhai W, Xu Z, Zhang J, He S, Zhang J, Xu J, Zhang K, Zheng X, Dong J, Zeng W, Tao L, Ye J, Tan J, Ren X, Chen X, He J, Liu D, Tian W, Tian C, Xia H, Bao Q, Li G, Gao H, Cao T, Wang J, Zhao W, Li P, Chen W, Wang X, Zhang Y, Hu J, Wang J, Liu S, Yang J, Zhang G, Xiong Y, Li Z, Mao L, Zhou C, Zhu Z, Chen R, Hao B, Zheng W, Chen S, Guo W, Li G, Liu S, Tao M, Wang J, Zhu L, Yuan L, Yang H (2002) A draft sequence of the rice genome (*Oryza sativa* L. ssp. *indica*). Science 296: 79–92.

Yuan Q, Ouyang S, Wang A, Zhu W, Maiti R, Lin H, Hamilton J, Haas B, Sultana R, Cheung F, Wortman J, Buell CR (2005) The Institute for Genomic Research Osa1 rice genome annotation database. Plant Physiol 138: 18–26.

Zhao W, Wang J, He X, Huang X, Jiao Y, Dai M, Wei S, Fu J, Chen Y, Ren X, Zhang Y, Ni P, Zhang J, Li S, Wang J, Wong GK-S, Zhao H, Yu J, Yang H, Wang J (2004) BGI-RIS: an integrated information resource and comparative analysis workbench for rice genomics. Nucl Acids Res 32: D377–382.

Zhou X, Su Z (2007) EasyGO: Gene Ontology-based annotation and functional enrichment analysis tool for agronomical species. BMC Genom 8: 246.

Future Prospects

Frederik C Botha

ABSTRACT

Over the past two decades many sugarcane R&D programs have diverted significant portions of their investment towards biotechnology-based activities. Many exciting developments occurred as a result and these led to a better understanding of the sugarcane genome, gene mapping, development of molecular markers and genetic manipulation. This expansion in knowledge and the development of technologies should greatly assist in realizing the enormous potential of sugarcane as a biomass feedstock for energy production and bioprocessing in a biorefinery. There is much unexploited genetic potential to further enhance the biomass production capacity of sugarcane. High selection pressure for biomass probably will have a negative impact on the sugar content but not on total sugar production per unit land area. Progress on genetic improvement of sugarcane will be highly dependent on gaining a better understanding of feedback inhibition, probably through sugar signaling, and understanding relationships between gene expression and metabolic flux. The complexity of the metabolic system, with all the potential permutations and resource limitations, will necessitate that metabolic modeling becomes an integral part of the future R&D investment.

Keywords: biomass, energy production, genetic potential, gene expression, metabolic flux

BSES Ltd, PO Box 86, Indooroopilly, QLD 4068, Australia; e-mail: *fbotha@bses.org.au*

12.1 Introduction

The content of this book reflects a remarkable progress, since the pioneering publications by (Bower and Birch 1992; Rathus and Birch 1992; Wu et al. 1992), in genetic analysis, gene expression and genetic manipulation of sugarcane.

For the industry that for the past century has utilized the crop for primarily sucrose production it is increasingly evident that little or any progress is made in increasing the sucrose content of sugarcane despite this being the main focus for research and development (R&D). Since the early 1990s many programs have diverted significant portions of their R&D investment towards biotechnology-based activities. Although there is great appreciation for the exciting developments that occurred in gaining a better understanding of the sugarcane genome, gene mapping, development of molecular markers and genetic manipulation, funders are increasingly pessimistic on the potential of this technology to impact on profitability of a sugarcane-based production system. There is a growing concern among the funders of sugarcane R&D whether this is a worthwhile investment. It is very important that researchers and research managers take note, and realize, that unless one of these approaches leads to some significant improvement soon further investment will be jeopardized.

Today there is an increasing realization that sugarcane can be used for many other applications including as a biomass feedstock for energy production and for bioprocessing in a biorefinery. In a bioenergy business the drive for germplasm improvement and production will focus on maximizing energy yield (Joule) per unit area in the shortest possible cropping cycle. For a successful biorefinery application, however, not only the production cost but also the quality of the feedstock is important. There are several reasons why sugarcane could be a preferred feedstock for a future biomass driven economy. These include its exceptional biomass production capacity, which could be further enhanced, a rapid volume of information on its genome structure and development of molecular markers to facilitate introgression and broadening of the germplasm base for breeding programs and well established genetic modification (GM) technology to introduce novel traits.

The purpose of this chapter is not to define specific future R&D priorities for sugarcane research, but rather to sketch the potential future use for this crop and some of the current limitations in realizing its full potential. It is against this background that researchers and the industry will have to formulate their future research priorities, and this should be based on very specific business objectives. The biggest danger for an industry that is so entrenched in dealing with primarily one product that it becomes complacent in defining its own future strategic objectives and setting research priorities.

In most sugar industries the research agenda is defined by the researchers, and not as it should be by the business. There will be so many new future opportunities that, without clearly defined strategic objectives, research activities could become too diluted and hence result in failure.

Although there is clearly an exciting potential future for sugarcane production and utilization, there should be no doubt that the realization of the full potential of sugarcane will require a significant further investment in R&D and very specifically on integration of genomic, gene expression, compartmentation and metabolic flux information.

12.2 A Biomass Driven Future Economy

The current world economy is completely dominated by technologies, which rely on fossil energy and fossil carbon, and there are growing concerns about this dependence on fossil fuels. These concerns broadly relates to the following issues. It is a non-renewable energy and carbon source, there is a huge negative impact of fossil fuel exploitation on the environment and the volatility in the price and availability of fossil fuel creates economic and social uncertainty and instability. The high demand for fossil carbon will eventually deplete the existing stocks, with consequences not only in the area of energy but also in the wider chemical industry (Benning and Pichersky 2008). It is therefore no surprise that there are unprecedented investments from governments worldwide to reduce the fossil fuel dependence (Anon 2006; Herrera 2006; Schubert 2006; Service 2007). The impact of this is evident from the fact that energy technology start-ups doubled between 2005 and 2006, and this number was 10 times higher than in 1999 (Anon 2006). It is, however, important to remember that it is not only that energy is derived from fossil fuel, but most of the basic precursors for the synthesis of drugs, plastics and other industrially important products comes from the petrochemical industry and is derived from fossil carbon. Although there are evidently many other potential sources of renewable energy, biomass presents the only alternative carbon source.

Biomass was the primary feedstock for many manufacturing processes at the beginning of the 20th century. However, in the past little note was taken on the environmental footprint that was left as a consequence of the large exploitation of fossil fuel. Consequently, with the advent of cheap fossil fuels biomass was readily displaced as an important energy and carbon source in the developed world. The only biomass derived products, which remained important, were those where production cost was similar between fossil fuel or biomass derived precursors, or those which are extremely difficult to manufacture, e.g. compounds with chirally pure precursors (Benning and Pichersky 2008). With the dramatic escalation in the cost of fossil fuel the competitiveness of biomass is rapidly increasing.

However, other factors such as carbon trading, carbon taxes, consumer preferences and trade agreements, will undoubtedly accelerate the development of a future economy that utilizes much more biomass than what was the case in the previous century. This will create an expanding array of exciting new business and employment opportunities for farmers and other businesses and will generate many new opportunities and create expansion opportunities for the use of existing agricultural crops.

12.3 Is There a Role for Sugarcane in the Future Biomass Economy

Obviously there is much debate on which crops will be best suited for feedstock in a biomass economy. It is therefore not surprising that numerous crops are being evaluated for commercial energy farming. These species include woody crops and grasses/herbaceous plants (all perennial crops), starch and sugar crops and oilseeds. There will not be a simple answer, especially if the focus is on delivery of renewable carbon for product synthesis. In general, the characteristics of the ideal energy crop will include, high yield (maximum production of dry matter per hectare), low energy input to produce, low production cost, correct composition with the least contaminants and the lowest nutrient requirement (McKendry 2002). A factor that increasingly carries a premium is the environmental footprint of crop production with water usage and soil degradation being two of the major considerations.

Due to its exceptional ability to produce biomass, sugarcane, and some other C4 plants, will always be important in a biomass dependent economy. Generally C4 species such as sugarcane will outperform C3 species in the ability to accumulate biomass. Cellulosic crops like sugarcane can produce on average 22 $gm^{-2}d^{-1}$ over an annual growth cycle and that by far outweighs that of C3 plants (13 $gm^{-2}d^{-1}$) (Monteith 1978). At an oil price of US$50 $barrel^{-1}$ the energy cost for petroleum is approximately $9 GJ^{-1}, and for gasoline US$14 GJ^{-1} assuming a price of US$1.67 $gallon^{-1}$. At the same time the price per GJ for energy derived from biomass feedstock will vary between US$3 and US$13.8 depending on the particular crop. These numbers suggest that when crude oil prices reach US$100 $barrel^{-1}$, as we have recently experienced, even the least efficient biomass producers presents a solid business case!

Although the view exists that 29% more energy is consumed in the production of maize ethanol (that what is returned in the final ethanol) (Pimentel and Patzek 2005), this opinion is not widely accepted. The approach followed by these authors appears to be flawed in many aspects. Most studies show that there is a net gain of at least 50% (1.5:1) with a maize-based production system and as much as a 10:1 gain in a sugarcane system.

12.3.1 *Exceptional Biomass Accumulation and Water Use Efficiency*

It is important to note that the numbers that are used in these calculations reflect the average biomass gain over the total growth cycle. For sugarcane, the maximum above ground biomass accumulation rate can be up to 55 $gm^{-2}d^{-1}$. When these maximum production numbers are used, the already impressive and highly competitive production cost of sugarcane as a renewable energy (Lynd et al. 2008), becomes even better.

A sugarcane crop of 37 ton ha^{-1} will contain approximately 600 GJ of energy. Taking into account the current production costs in Australia, would equate to a cost of approximately US$1.60 GJ^{-1}. Taking only the energy in the sugar component in an average crop into consideration the production cost will be US$3.86 GJ^{-1} (Botha 2009). This implies that if we only consider the energy in the water soluble fraction the production costs for energy is at least three times lower than that for energy from crude oil.

In addition to its ability to rapidly produce biomass, sugarcane varieties bred for rainfed conditions have a water use efficiency (Kg water Kg^{-1} dry mass) of 240 and this is significantly better than that for other crops (Thompson 1986; Rao and Cramer 1994). Despite this outstanding characteristic of sugarcane there are generally wasteful applications of water in sugarcane production systems and this is one of the critical issues to be improved in future production.

12.3.2 *The Old Concept of Energy Cane is Now More Relevant than Ever*

The concept "energy cane" has evolved to distinguish between different sugarcane management systems (Alexander 1985). During the past decade, this term has been increasingly used to describe a variety of aspects, which in addition to a changed management system could also imply high biomass varieties.

The "energy cane" production system ensures that all aerial biomass is regarded as valuable. Approximately 25% of the standing total aerial biomass is present in the tops (young internodes and leaves) of sugarcane with the rest being the culm tissue (Rípoli et al. 2000; Singels et al. 2005). Significant differences in total biomass yield exists between a sucrose and "energy cane" production system (Alexander 1985). Utilizing the same varieties, yields almost doubled in an "energy cane" approach. It is important to note that this increased biomass realization is not purely due to more lignocelluloses but also significantly more total fermentable solids. Gains are therefore possible in both biofuel production and in the cogeneration of electricity.

12.3.2.1 Better Utilization of the Crop

Undoubtedly much more can be achieved through genetic gain in sugarcane. However, it is equally important to point out that much gain can be made through a change in mindset from sugarcane as a sucrose producer to one of sugarcane as an energy crop and full realization of the available energy. For the purpose of discussion ethanol production is used as an example.

Although the production cost of ethanol from sugarcane, especially in warmer climates, is much more efficient than the lowest for any biofuel (Anon 2004), much more is possible even without further improvement of the crop. Currently all bioethanol from sugarcane is entirely derived from the water-soluble fraction and therefore the lignocellulosic fraction that contains most of the energy is excluded. Although current technology does not allow the effective conversion of this fraction to biofuel, it is an international research focus.

The ethanol yield from sugarcane in Brazil showed a dramatic increase from 600 liters ha^{-1} year^{-1} in 1975 to more than 1,400 litres ha^{-1} year^{-1} in 2006 (Milton 2007). Several factors probably contributed to this increase including less exhaustion of molasses leaving a better quality fermentation feedstock and superior varieties. It is important to note that the current practice in sugarcane bioethanol production is still dominated by the utilization of molasses. If all the soluble sugars in cane juice are converted to ethanol (no sucrose production) yield in excess of 5,000 litres ha^{-1} year^{-1} are possible.

Theoretically 450 litres of ethanol can be produced from one ton dry lignocelluloses material such as baggase. The complexity of the chemical composition of baggase currently makes this currently unachievable (Knauf and Moniruzzman 2004). With current technology probably no more than 50% of this potential can be realized. However, it is not unrealistic to assume that new mill technologies, such as boiler and steam utilization efficiency, will result in an improvement of baggase utilization to the extent that 40% of the baggase becomes surplus and available for ethanol production. This would result in significant enhanced ethanol yields per area of land.

12.3.3 Much More is Possible

Considering only the aboveground portion of the sugarcane plant, the radiation use efficiency (RUE) of sugarcane is 1.75g MJ^{-1} (Muchow et al. 1994). Owing to the fact that probably more than 20% of the total biomass is below ground this implies a RUE of > 2.0 g MJ^{-1}. Total yield varies dramatically between genotypes and environmental conditions. Reported estimated record yields vary between 80 and 85 tons dry mass ha^{-1}year^{-1} (Moore et al. 1998). Given the radiation use efficiency, this implies a realization of only about 40% of the theoretical potential. This implies that there are major biomass yield-limiting constraints.

Biomass accumulation is a function of the total photosynthetic active radiation (PAR), the radiation use efficiency (RUE), leaf area and carbon partitioning. For most plant species the primary photosynthetic product, which is produced in the leaf (source) and then translocated to other parts of the plant (sink), is sucrose. In C4 plants like sugarcane, the process initially involves the production of organic acids in the source however sucrose is still the final translocated product from the source to sink. It is well-documented that metabolism in both the source and sink is important for biomass production. However, most studies to date suggest that sink strength and the ability of the sink to draw sucrose toward it, plays a major role in the ability of the plant to accumulate biomass. In sugarcane the major sinks would be the leafroll, young leaves, internodes and the root system. Partitioning of biomass between these different sinks in sugarcane is not constant and can vary depending on the age, genotype and environmental conditions (Singels et al. 2005). The metabolism of sucrose in the sink tissue is not constant and allocation of the reduced carbon in the sink to different cellular constituents varies (Botha et al. 1996; Bindon and Botha 2002; Singels et al. 2005). The interactions between photosynthesis, carbon partitioning and sucrose accumulation is still poorly understood (Smith 2008) but undoubtedly the accumulation of sucrose especially early in the cropping cycle is a major contributor to realize more of the theoretical maximum. A much better understanding of carbohydrate storage and turnover in relation to carbon assimilation and plant growth is required, both for improvement of starch and sugar crops and for attempts to increase biomass production in second-generation biofuel crops (Smith 2008).

12.3.4 *Potential for Expansion into other Geographical Areas*

A large variation in dry matter accumulation and partitioning exists between different genotypes and will be strongly influenced by environmental conditions (Singels et al. 2005; Smit and Singels 2006). Growth patterns in areas with cold winter temperatures will be very different from that of the tropical regions. For a sucrose-based production system, cooler winter temperatures are a positive event as the carbon allocation to the culm is favored at the expense of trash and tops (Singels et al. 2005), which causes a natural ripening of the cane. However, this would be a negative event for maximum biomass production.

However, it is well known that large variation exists within the sugarcane genetic complex for environmental factors such as low temperature and dry conditions. With an emphasis on these parameters significant gains can probably be made to expand the production areas for sugarcane, especially if the emphasis moves to a biomass and total fermentables rather than a sucrose focus.

12.3.5 New Uses

Although sugarcane is primarily associated with sucrose production, it is good source of soluble carbohydrate. In fact, parts of the crop currently regarded as a nuisance factor in the extraction and processing of sucrose, such as the tops and leaves have similar levels of total fermentables than the mature culm. It is well known that these sugars and other organic components in cane juice and molasses can be utilized for the production of ethanol, synthetic rubber, and other high value products such as fructose, dextran, sucrose esters, organic acids, mannitol, sorbitol and biosolvents. The non-water soluble component (baggase), which is largely the lignocellulosic component, can be utilized to generate electricity, to manufacture paper and building materials. A combination of molasses and baggase is used for animal feed. Probably the best-known use of baggase is the production of furfural. For many years, furfural served primarily as a catalyst in the manufacturing of a large number of products but recently has become a registered nematocide (Cropguard®), which also shows suppression of soilborne fungi such as *Pythium, Fusarium, Phytophthora* and *Rhizoctonia (http://www.cropguard.co.za/)*.

The use of sugarcane in folk medicine has been well reported with at least 20 different uses for sugarcane, which ranges from basic issues such as dressing for wounds, antiseptics, the treatment of inflammation and cancer (Watt and Breyer-Brandwijk 1962). As expected, for a plant material with a wide spectrum of uses, chemical analyses confirm the chemical complexity of sugarcane leaves and culm material (Duke 2003). Modern metabolic profiling studies also showed that at least two hundred chemicals are present in the sugarcane culm (Bosch 2005; Van der Merwe 2005).

In addition to the already vast array of available metabolic precursors in the sugarcane plant, recent work has shown that new characteristics can successfully be introduced to sugarcane. Two different approaches were followed to genetically modify sugarcane to produce sugars other than sucrose in the vacuole (Nell 2007; Wu and Birch 2007). Interestingly, in both approaches the newly introduced activity does not only lead to the production of a new product but also an increase in sucrose content. For example, the production of small quantities of isomaltulose through the expression of sucrose isomerase leads to a doubling in sucrose content in glasshouse grown plants (Wu and Birch 2007).

In a very different approach, three bacterial enzymes were introduced into sugarcane to allow the conversion of acetyl-coenzyme A (acetyl-CoA) to the polyester polyhydroxybuterate (PHB) (McQualter et al. 2005; Petrasovits et al. 2007). The results showed that significant levels of PHB can be produced in sugarcane, especially when the new biosynthetic pathway was targeted to the plastid. However, a complex spatial and temporal pattern of accumulation of the polymer was also observed (Purnell et al. 2007). Accumulation of PHB in culm tissue was at least 16 times lower than in leaf tissue. Although other factors as discussed by these authors could be important, a likely reason could be that the substrate availability is limited in the sugarcane culm.

12.4 Some R&D Challenges

12.4.1 Multipurpose High Biomass Varieties

With the growing realization that biofuels will play a significant role in the future and that total biomass per area of land is the critical driver, many groups have already started aggressive searches for high biomass canes. Great progress has been reported in selecting high biomass clones. As can be expected, selection for such clones coincide with lower sucrose concentrations in the tissue (Terajima et al. 2005; Autrey and Chang 2006). These superior high biomass varieties have been designated "monster cane" (Terajima et al. 2005). As with "energy cane" production systems (Alexander 1985) high biomass cane varieties also deliver similar amounts of sucrose but vastly improved amounts of lignocellulose and total soluble sugars per hectare. Even considered within the current sugar-ethanol production system, these canes can deliver significantly more biomass and similar amounts of sucrose per hectare. Also an increase in mill efficiency as far as baggase utilization is concerned could produce additional biomass that can be used.

Recently released results showed that in an introgression program where commercial Australian varieties were crossed with wild relatives originating from China that total biomass could be increased by more than 40% while total fermentable sugars remained similar and fiber content was only slightly increased (Anon 2008). Often the mistake is made to assume that high fiber canes are necessarily synonymous with low sucrose and high biomass. Commercial type sucrose varieties have high biomass yield (fresh weight), and in these varieties there is no negative correlation between fiber content and sucrose or brix. It is only when wild material is included, that there is a negative correlation (due to the 35% fiber, 2% brix type-genotypes).

However, these wild canes often do not produce high biomass but rather would add useful characteristics such as tolerance to environmental stress and good ratoonability. It is also likely that breeding programs could make significant gains in selection for characteristics such as fiber quality and other desired chemical constituents.

12.4.2 A Reliable Transformation System

Understandably there is a keen interest in using the excellent biomass production platform in sugarcane to add or remove single genes or simple metabolic pathways. This has been the focal point of many research programs of sugar industries worldwide (see Chapter 9). Many examples of the successful addition of new genetic material to sugarcane exist. Broadly, these include: enhanced resistance to pest and diseases (Arencibia et al. 1997; Nutt et al. 1999; McAllister et al. 2004), herbicide resistance (GalloMeagher and Irvine 1996; Enriquez-Obregon et al. 1998; Falco et al. 2000; Leibbrandt and Snyman 2003), viral resistance (Joyce et al. 1998; Ingelbrecht et al. 1999; Gilbert et al. 2005), human collagen (Santa-Ana 2007), redirection of carbon in metabolism (Botha 2008) and alternative high value non-protein product.

However, most of these activities have focused on germplasm that can easily be transformed and not necessarily on elite varieties. Most groups have had limited success in routine genetic transformation of a wide range of varieties. In addition, there has been limited success in expressing new traits in sugarcane under field conditions. Both these issues need to be resolved before the potential of this technology can significantly contribute to sugarcane improvement.

12.4.3 Unraveling Feedback Inhibition and Mechanisms

Several levels of feedback inhibition occurs which negatively impacts on biomass production. Although there may be several other factors involved, it would be safe to assume that at least a significant portion of this yield limitation in sugarcane is associated with the accumulation of sucrose in the culm. In sugarcane, like many other crops which have been studied, sucrose accumulation results in feedback inhibition of photosynthesis in the source tissue (McCormick et al. 2006, 2008). This is achieved through a complex change in the expression patterns of a number of enzymes in the sink leaf.

In the young and active growing sugarcane culm, carbon partitioning is favored towards the soluble non-sucrose and structural components. During this period of growth the RUE of sugarcane is substantially higher than during the later stages of growth where sucrose accumulates to high levels (Singels et al. 2005). This aspect forms the basis for energy cane production where sucrose levels in the tissue is kept relatively low in comparison to a sucrose production system.

The fact that RUE is negatively influenced by the accumulation of sucrose implies that an energy efficient sugarcane system will always depend on at least one of the following: delaying or prevention of sucrose accumulation through specific agronomic practices (Alexander 1985), or overriding the control mechanisms through genetic improvement. Recently it was shown that sucrose levels can be significantly increased by expression of sucrose isomerase in sugarcane (Wu and Birch 2007). Not only was the total sugar concentration increased but photosynthesis was also stimulated. One would therefore assume that these plants should also exhibit a total increase in biomass.

12.4.4 Understanding Relationships between Gene Expression and Metabolic Flux

Continued progress in crop production is dependent on a good knowledge and understanding of the genome and its expression. In comparison to other crops, little effort was put into the characterization of the sugarcane transcriptome until pioneering genome expression studies led to the first expressed sequence tags (ESTs) to the international database (Carson et al. 2002; Casu et al. 2005). At present, sugarcane genome projects have led to more than 300,000 sugarcane ESTs that are now available to researchers (see Chapter 11). This wealth of information allows the identification of genetic shortfalls and potential targets for genetic improvement. This information, for example, facilitates the discovery of unidentified metabolic pathways in sugarcane (Bosch 2005), characterization and improvement of nutrient uptake systems (Smith 1997), responses of the crop to stress (Watt 2003) and tissue specific expression of specific alleles (Grof et al. 2006).

There is a rapidly increasing volume of literature that suggests that the relationship between gene expression, transcript levels, enzyme activity and metabolic flux is endlessly more complex than initially anticipated. Data covering aspects of nitrogen metabolism, plant responses to elevated CO_2 and altered photosynthesis reveal that in some cases a tight coordination between enzyme and transcript levels but in many instances major changes in enzyme activity occur completely independently from changes in transcript levels (Scheible et al. 2000; Matt et al. 2001; Matt et al. 2002; Benning and Stitt 2004; Fernie et al. 2005). Gene expression profiling in the sugarcane culm also indicates a weak correlation with the drastic redirection of carbon flux towards sucrose accumulation. There are many challenges in gaining a better understanding of sucrose accumulation in the internodes of the sugarcane culm (Rae et al. 2005; Uys et al. 2007; Botha 2008).

12.4.5 Modeling of Metabolism

The failure to alter metabolism through targeting some of the enzyme reactions, which were deemed crucial showed that the early assumptions that were used to try and alter sugarcane metabolism in the culm were incorrect. The unexpected enhancement of sucrose through the introduction of an alternative sugar biosynthetic activity clearly demonstrates that we have no understanding of the metabolic control of sucrose accumulation in the culm (Wu and Birch 2007). Work done has already indicated that gene expression profiles are inconsistent with a tissue only involved in sucrose accumulation. In fact, these studies indicated that probably a small portion of total gene expression is directed at sucrose metabolism.

The complexity of the system and all the potential permutations makes it virtually impossible to identify the reactions that could be good targets for genetic manipulation in an attempt to alter metabolism (Rohwer and Botha 2001; Uys et al. 2007). Instead of a cumbersome gene-by-gene manipulation strategy, kinetic modeling offers an attractive alternative.

12.4.6 Substrate Availability

One of the least studied areas and an aspect that is critically important for future exploitation of the crop is compartmentation of metabolism and substrate availability. Labelling studies indicated that the sugar pools are highly compartmentalized. Exploitation of these sugars therefore will require targeting of enzymes to specific compartments. In addition, producing a fructan in the vacuole of sugarcane did not result in any glucose accumulation indicating that the reducing sugar is not only rapidly removed from the vacuole but also reintroduced into metabolism through phosphorylation (Nell 2007). In addition, labelling studies also showed that more than 80% of the incoming carbon is channelled towards sucrose and hence leaves very little for utilization in other biosynthetic processes (Bindon and Botha 2002).

There is very limited information to suggest that sucrose, once stored in the vacuole can be remobilized in sugarcane. A significant reason why transgenic sugarcane do not accumulate high levels of PBH in the culm (Petrasovits et al. 2007) could be a direct result of the absence of adequate substrate.

12.5 Conclusions

The next few decades will see a massive reintroduction of biomass as a primary feedstock for energy, biofuels and electricity production as well as for the bioprocessing and biorefinery industries. Sugarcane has all the characteristics to be one of the pillars on which such a futuristic biofuels industry could be built. Its extremely efficient biosynthetic capacity, water use efficiency and untapped genetic potential make it the prime candidate as a feedstock for new enterprises in the "biomass" era that has started.

Traditional sugarcane production systems focused on optimal sucrose concentrations in the millable culm. This system is not necessarily optimally designed and developed for biomass and biofuel production. The biofuel production capacity from sugarcane as a feedstock under favorable conditions is much greater than what is currently realized within the sucrose production system.

To realize the opportunities that the future holds require a mindset change not only from researchers but also the producers and processors of sugarcane. Whether it would be possible for current role players to make the mindset transition that would be required to realize this potential needs to be seen. One may expect that the realization of this potential will be driven by new stakeholders entering the production system as energy rather than sucrose producers.

Apart from realizing its unexploited genetic potential as an energy crop, the future will undoubtedly see major improvements in sugarcane through introgression of new germplasm and genetic manipulation. Although the technology is well-advanced to improve the agronomic performance of the crop as well as broadening the useful metabolic precursors as end-products in sugarcane through genetic manipulation there are many serious shortfalls and gaps in our knowledge that needs to be resolved. The rapidly growing knowledge base on the expressed component of the sugarcane genome could greatly contribute to further crop improvement provided that we succeed in a full integration of metabolic flux and enzyme levels with transcript levels.

Reference

Alexander AG (1985) The Energy Cane Alternative. Elsevier, New York, USA.

Anonymous (2004) Biofuels for Transport: An international Perspective: *http:// www.iea.org/textbase/nppdf/free/2004/biofuels2004.pdf.*

Anonymous (2006) Investing in Clean Energy. The Economist, November 18, 2006.

Anonymous (2008) Queensland scientists breeding for Hi-Energy Canes: *http:// www.crcsugar.com/News/tabid/56/xmmid/407/xmid/181/xmview/2/Default.aspx.*

Arencibia A, Vazques RI, Prieto D, Tellez P, Carmona ER, Coego A, Hernandez L, De la Riva GA, Selman-Housein G (1997) Transgenic sugarcane resistant to stem borer attack. Mol Breed 3: 247–255.

Autrey JC, Chang KTKF (2006) The multifunctional role of cane sugar industry in Mauritius: *http://www.issct.org/cpabst06.htm.*

Benning C, Stitt M (2004) Physiology and metabolism: Reacting to the full complexity of metabolic pathways in a postgenomic era. Curr Opin Plant Biol 7: 231–234.

Benning C, Pichersky E (2008) Harnessing plant biomass for biofuels and biomaterials. Plant J 54: 533–535.

Bindon KA, Botha FC (2002) Carbon allocation to the insoluble fraction, respiration and triose-phosphate cycling in the sugarcane culm. Physiol Planta 116: 12–19.

Bosch S (2005) Trehalose and carbon partitioning in sugarcane. PhD Thesis Univ of Stellenbosch, South Africa, pp 1–195.

Botha FC (2008) Precision breeding to improve the usefulnesss of sugarcane. Sugar Cane Int 26: 11–14.

Botha, FC (2009) Energy yield and cost in a sugarcane biomass system. Proc Aust Soc Sugar Cane Technol 31: 1–10.

Botha FC, Whittaker A, Vorster DJ, Black KG (1996) Sucrose accumulation rate, carbon partitioning and expression of key enzyme activities in sugarcane stem tissue. In: JR Wilson, DM Hogarth, JA Campbell, AL Garside (eds) Sugarcane: research towards efficient and sustainable production. CSIRO division of Tropical Crops and Pastures, Brisbane, Australia, pp 98–101.

Bower R, Birch RG (1992) Transgenic sugarcane plants via microprojectile bombardment. Plant J 2: 409–416.

Carson DL, Huckett BI, Botha FC (2002) Sugarcane ESTs differentially expressed in immature and maturing internodal tissue. Plant Sci 162: 289–300.

Casu RE, Manners JM, Bonnett GD, Jackson PA, McIntyre CL, Dunne R, Chapman SC, Rae AL, Grof CPL (2005) Genomics approaches for the identification of genes determining important traits in sugarcane. Field Crops Res 92: 137–147.

Duke JA (2003) *Saccharum officinarum* L. Handbook of energy crops: http://www.hort.purdue.edu/newcrop/duke_energy/Saccharum_officinarum.html.

Enriquez-Obregon GA, Vazquez-Padron RI, Prieto-Samsonov DL, De la Riva GA, Selman-Housein G (1998) Herbicide-resistant sugarcane (*Saccharum officinarum* L.) plants by *Agrobacterium*-mediated transformation. Planta 206: 20–27.

Falco MC, Neto AT, Ulian EC (2000) Transformation and expression of a gene for herbicide resistance in a Brazilian sugarcane. Plant Cell Rep 19: 1188–1194.

Fernie AR, Geigenberger P, Stitt M (2005) Flux an important, but neglected, component of functional genomics. Curr Opin Plant Biol 8: 174–182.

GalloMeagher M, Irvine JE (1996) Herbicide resistant transgenic sugarcane plants containing the bar gene. Crop Sci 36: 1367–1374.

Gilbert RA, Gallo-Meagher M, Comstock JC, Miller JD, Jain M, Abouzid A (2005) Agronomic evaluation of sugarcane lines transformed for resistance to Sugarcane mosaic virus strain E. Crop Sci 45: 2060–2067.

Grof CPL, So CTE, Perroux JM, Bonnett GD, Forrester RI (2006) The five families of sucrose-phosphate synthase genes in *Saccharum* spp. are differentially expressed in leaves and stem. Funct Plant Biol 33: 605–610.

Herrera S (2006) Bonkers about Biofuels. 24th edn. pp 755–760.

Ingelbrecht IL, Irvine JE, Mirkov TE (1999) Posttranscriptional gene silencing in transgenic sugarcane. Dissection of homology-dependent virus resistance in a monocot that has a complex polyploid genome. Plant Physiol 119: 1187–1197.

Joyce PA, McQualter RB, Handley JA, Dale JL, Smith GR (1998) Transgenic sugarcane resistant to sugarcane mosaic virus. Proc Aust Soc Sugar Cane Technol 20: 204–210.

Knauf M, Moniruzzman M (2004) Lignocellulosic biomass processing: A perspective. Int Sugar J 106: 147–150.

Leibbrandt NB, Snyman SJ (2003) Stability of gene expression and agronomic performance of a transgenic herbicide-resistant sugarcane line in South Africa. Crop Sci43: 671–677.

Lynd L, Laser MS, Bransby D, Dale BE, Davison B, Hamilton R, Himmel M, Keller M, McMillan JD, Sheehan J, Wyman CE (2008) How biotech can transform biofuels. Nat Biotechnol 26: 169–172.

Matt P, Geiger M, Walch-Lui P, Engels C, Krapp A, Stitt M (2001) The immediate cause of the diurnal changes of nitrogen metabolism in leaves of nitrate-replete tobacco: a major imbalance between the rate of nitrate reduction and the rates of nitrate uptake and ammonium metabolism during the first part of the light period. Plant Cell Environ 24: 177–190.

Matt P, Krapp A, Haake V, Mock HP, Stitt M (2002) Decreased Rubisco activity leads to dramatic changes of nitrate metabolism, amino acid metabolism and the levels of phenylpropanoids and nicotine in tobacco antisense RBCS transformants. Plant J 30: 663–667.

McAllister CD, Bischoff KP, Gravois KA, Schexnayder HP (2004) Transgenic Bt-corn affects sugarcane borer in Louisiana. SW Entomol 29: 263–269.

McCormick AJ, Cramer MD, Watt DA (2006) Sink strength regulates photosynthesis in sugarcane. New Phytol 171: 759–770.

McCormick AJ, Cramer MD, Watt DA (2008) Regulation of photosynthesis by sugars in sugarcane leaves. J Plant Physiol 165: 1817–1829.

McKendry P (2002) Energy production from biomass (part 1): overview of biomass. Bioresour Technol 83: 37–46.

McQualter RB, Chong BF, Meyer K, Van Dyk DE, O'Shea MG, Walton NJ, Vitanen PV, Brumbley SM (2005) Initial evaluation of sugarcane as a production platform for p-hydroxybenzoic acid. Plant Biotechnol J 3: 29–41.

Milton M (2007) Ethanol from Brazil and the USA: *http://www.energybulletin.net/21064.html.*

Monteith JL (1978) Reassessment of the maximum growth rates for C3 and C4 crops. Exp Agric 14: 1–5.

Moore PH, Botha FC, Furbank RT, Grof CPL (1998) Potential for overcoming physio-biochemical limits to sucrose accumulation. In: BA Keating , JR Wilson (eds) Intensive Sugarcane Production: Meeting the Challenges Beyond 2000. CAB Int, New York, USA, pp 141–155.

Muchow RCSMF, Wood AW, Keating BA (1994) Radiation interception and biomass accumulation in a sugarcane crop grown under irrigated tropical conditions. Aust J Agri Res 45: 37–49.

Nell JS (2007) Genetic Manipulation of Sucrose-storing Tissue to Produce Alternative Products. PhD Thesis. Univ of Stellenbosch. South Africa, pp 1–97.

Nutt KA, Allsopp PG, Mcghie TK, et al. (1999) Transgenic sugarcane with increased resistance to canegrubs. pp 171–176.

Petrasovits L, Purnell MP, Nielsen LK, Brumbley SM (2007) Production of polyhydroxybutyrater in sugarcane. Plant Biotechnol J 5: 162–172.

Pimentel D, Patzek TW (2005) Ethanol production using corn, switchgrass, and wood; Biodiesel production using soybean and sunflower. Nat Resour Res 14: 65–76.

Purnell MP, Petrasovits L A, Nielsen LK, Brumbley SM (2007) Spatio-temporal characterization of polyhydroxybuterate accummulation in sugarcane. Plant Biotechnol J 5: 173–184.

Rae AL, Grof CPL, Casu RE, Bonnett GD (2005) Sucrose accumulation in the sugarcane stem: pathways and control points for transport and compartmentation. Field Crops Res 92: 159–168.

Rao IM, Cramer GR (1994) Plant nutrition and crop improvement in adverse soil conditions. *In:* MJ Chrispeels, DE Sadava (eds) Plants, genes and crop biotechnology, Jones and Bartlett Int., London pp 270–303.

Rathus C, Birch RG (1992) Stable transformation of callus from elecroporated sugarcane protoplasts. Plant Sci 82: 81–89.

Rípoli TCC, Molina WF, Rípoli MLC (2000) Energy potential of sugar cane biomass in brazil. Sci Agri 57: 677–681.

Rohwer JM, Botha FC (2001) Analysis of sucrose accumulation in the sugar cane culm on the basis of in vitro kinetic data. Biochem J 358: 437–445.

Santa-Ana R (2007). Texas A&M, ECOR Corporation, sign deal to produce health-related proteins: *http://agnews.tamu.edu/dailynews/stories/SOIL/Apr2103a.htm.*

Scheible WR, Krapp A, Stitt M (2000) Reciprocal diurnal changes of phosphoenolpyruvate carboxylase expression and cytosolic pyruvate kinase, citrate synthase and NADP-isocitrate dehydrogenase expression regulate organic acid metabolism during nitrate assimilation in tobacco leaves. Plant Cell Environ 23: 1155–1167.

Schubert C (2006) Can biofuels finally take center stage? Nat Biotechnol 24: 777–784.

Service RF (2007) Cellulosic ethanol. Biofuel researchers prepare to reap a new harvest. Science 16: 1488–1491.

Singels A, Donaldson RA, Smit MA (2005) Improving biomass production and partitioning in sugarcane: theory and practice. Field Crops Res 92: 291–303.

Smith A (2008) Prospects for increasing starch and sucrose yields for bioethanol production. Plant J 54: 546–558.

Smit MA, Singels A (2006) The response of sugarcane canopy development to water stress. Crops Research 98: 91–97.

Terajima Y, Matsuoka M, Ujihara K, Irei S, Fukuhara S, Sakaigaichi T, Ohara S, Hayano T, Sugimoto A (2005) The simultaneous production of sugar and biomass ethanol using high-biomass sugarcane derived from inter-specific and inter-generic cross in Japan: *http://unit.aist.go.jp/internat/biomassws/02workshop/reports/20051213PP01-02p.pdf.*

Thompson GD (1986) Water use by sugarcane. SA Sugar J 60: 593–600.

Uys L, Botha FC, Hofmeyr JH, Rohwer JM (2007) Kinetic model of sucrose accumulation in maturing sugarcane culm tissue. Phytochemistry 68: 2375–2392.

Van der Merwe MJ (2005) Influence of hexose-phosphates and carbon cycling on sucrose accumulation in sugarcane Spp. MSC Thesis. Univ of Stellenbosch, South Africa, pp 1–89.

Watt DA (2003) Aluminium-responsive genes in sugarcane: identification and analysis of expression under oxidative stress. J Exp Bot 54: 1163–74.

Watt JM, Breyer-Brandwijk MG (1962) The medicinal and poisonous plants of southern and eastern Africa E.&S. Livingstone, Ltd., London.

Wu K, Burnquist W, Sorrells ME, Tew TL, Moore PH, Tanksley SD (1992) The detection and estimation of linkage in polyploids using single-dose restriction fragments. Theor Appl Genet 83: 294–300.

Wu LG, Birch RG (2007) Doubled sugar content in sugarcane plants modified to produce a sucrose isomer. Plant Biotechnol J 5: 109–117.

Subject Index

Color Plate Section

Chapter 4

Box 4-1 Comparison of (a) linkage analysis and (b) association analysis (Cardon and Bell 2001).

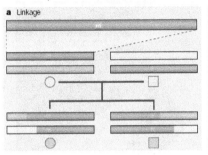

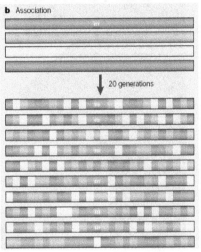

Linkage analysis and association analysis have the same ultimate goal: to detect regions of the genome associated with a trait of interest. They are also based on the same fundamental principle of the co-inheritance of adjacent DNA variants. However, the two approaches have different implementations and consequently different strengths and weaknesses.

Linkage analysis focuses on detecting recent mutations by tracing their descent through families. There have been only a few generations of recombination since the mutation event (indicated by m), so there are large blocks of the genome, which will be co-inherited. This is advantageous in that it increases the power to detect the region. Conversely, it means that the location of the mutation cannot be determined more precisely than 10-20 cM.

Association analysis focuses on much older mutations, which have spread throughout a population. The increased number of generations of recombination means that much smaller regions are co-inherited with the original mutation. Thus a causal gene can be more precisely located by examining the trait-genotype relationship in the broader population. However, much larger sample sizes may be required to detect association since there is greater variation in the population.

Chapter 6

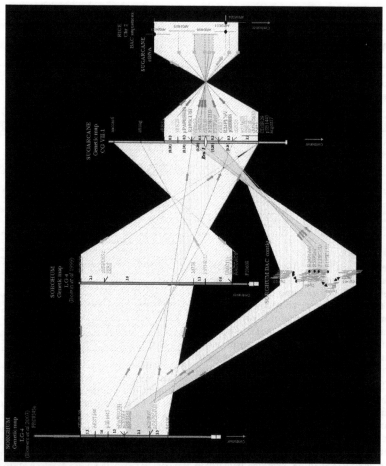

Figure 6-1 Saturation of the sugarcane target genetic region using various genomic resources.

Step 1: Three markers originally surrounding *Bru1* in sugarcane (orange arrows) were mapped on the sorghum genetic map. Double green arrows link common markers between the two sorghum genetics maps (Boivin et al. 1999 and Bowers et al. 2003). The green arrows point to the 4 new sorghum markers that could be mapped on sugarcane.

Step 2: Four sorghum probes (indicated in italic in the Bowers et al. 2003 sorghum genetic map) were used to construct a sorghum BAC contig. Four BAC-ends markers derived from this contig could be mapped in the sugarcane target region (pointed by green arrows).

Step 3: Two probes were used to identify the corresponding rice orthologous physical map. Orange arrows point to the rice BACs identified. Seven sugarcane cDNA with homology to the rice orthologous sequence were mapped on the sugarcane genetic map (red arrows). Markers used to build the sugarcane physical map are indicated in bold. Figure from Le Cunff et al. (2008).

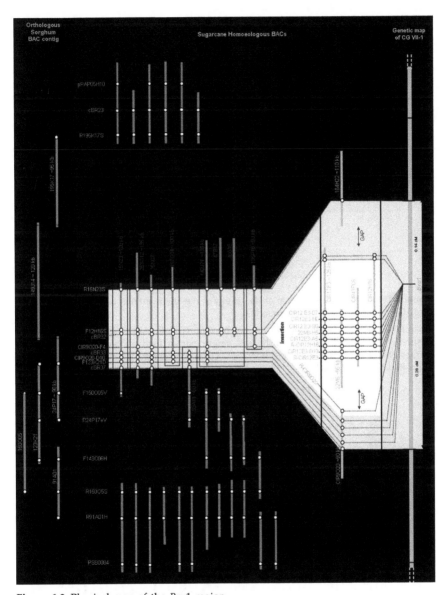

Figure 6-2 Physical map of the *Bru*1 region.

BAC clones are represented by vertical lines: orange for the target haplotype, brown for the hom(e)ologous haplotypes and green for sorghum. Dotted-lines and white rounds indicate the localization of probes used on the sorghum or/and sugarcane physical map or/and on the genetic map of the *Bru*1 region in sugarcane cv R570. Boxes assemble BAC clones for the same haplotype. Probes in green represent those from the sorghum genetic or physical map, those in red are from the sugarcane cDNA library, and those in orange are BAC-ends or subclones of sugarcane BACs. Figure from Le Cunff et al. (2008).

Chapter 7

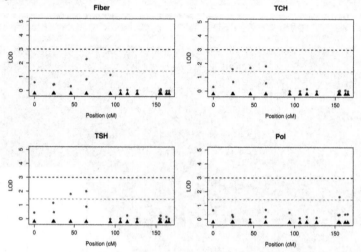

Figure 7-1 Graphic representation of the results of the Single marker analysis for sugarcane data (Garcia et al. 2009, unpub results).

Fiber: Fiber percent; TCH: tons of cane per hectare; TSH: tons of sugar per hectare; Pol: g of sucrose per kg per 100 g of fresh cane; single marker method: blue points; molecular markers: black triangle; dashed red line: LOD value considering 1% of significance level as a threshold; dashed black line: LOD = 3.

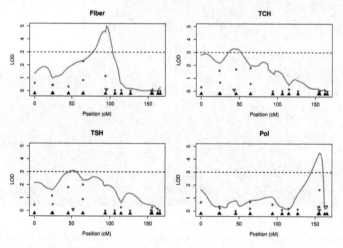

Figure 7-2 Graphic representation comparing the results of Single marker analysis and Interval Mapping for QTL detection in sugarcane (Garcia et al. 2009, unpublished results).

Fiber: Fiber percent; TCH: tonnes of cane per hectare; TSH: tonnes of sugar per hectare; Pol: g of sucrose per kg per 100 g of fresh cane; single marker method: blue points; interval mapping: continued red line; molecular markers: black triangle; green triangles indicate the position of QTL mapped based on IM; and dashed black line: LOD = 3 as the threshold.

Milton Keynes UK
Ingram Content Group UK Ltd.
UKHW020023071024
449327UK00032B/2908